Numerical Methods in Computational Mechanics

Numerical Methods in Computational Mechanics

Jamshid Ghaboussi
Xiping Steven Wu

CRC Press
Taylor & Francis Group
Boca Raton London New York

CRC Press is an imprint of the
Taylor & Francis Group, an **informa** business
A SPON PRESS BOOK

CRC Press
Taylor & Francis Group
6000 Broken Sound Parkway NW, Suite 300
Boca Raton, FL 33487-2742

First issued in paperback 2018

CRC Press is an imprint of Taylor & Francis Group, an Informa business

ISBN: 978-1-4987-4675-5 (hbk)
ISBN: 978-0-367-02802-2 (pbk)

Library of Congress Cataloging-in-Publication Data

Names: Ghaboussi, J., author. | Wu, Xiping, 1962- author.
Title: Numerical methods in computational mechanics / by Jamshid Ghaboussi and Xiping Wu.
Description: Boca Raton, FL : Taylor & Francis Group, LLC, CRC Press is an imprint of Taylor & Francis Group, an Informa Business, [2017] | Includes bibliographical references and index.
Identifiers: LCCN 2016024432| ISBN 9781498746755 (hardback : acid-free paper) | ISBN 9781498746786 (ebook)
Subjects: LCSH: Numerical analysis. | Mechanics, Applied--Mathematics. | Engineering mathematics.
Classification: LCC TA335 .G43 2017 | DDC 620.1001/518--dc23
LC record available at https://lccn.loc.gov/2016024432

Visit the Taylor & Francis Web site at
http://www.taylorandfrancis.com

and the CRC Press Web site at
http://www.crcpress.com

Contents

List of figures

Preface

Computational mechanics is a well-established field. It is widely used in engineering practice and in engineering education and research. Most applications of computational mechanics use linear and nonlinear finite element methods to solve problems in civil engineering, mechanical engineering, aerospace engineering, agricultural engineering, and many other fields of engineering and science. In most cases, commercial software or specialized software developed for specific applications is used. The situation is now very different from early days of development of computational mechanics in 1970s and 1980s when many researchers and industrial users required developing their own computer software. Nowadays, most users of computational mechanics use existing commercial software, and very few have to develop their own specialized software. Understanding the underlying methodology and theory is probably not as important for users of the commercial computational mechanics software as it is for the developer of the software. Nevertheless, understanding the underlying methodology and theory makes them more informed and effective users of the existing computational mechanics software.

Computational mechanics can be looked at three levels. The underlying theoretical mechanics describes the linear and nonlinear behavior of the material at multiscales, from the system to material behavior to, in some cases, microstructure. To arrive at the computational level, we need to move from continuum to discretized systems. This is accomplished by finite element, finite difference, and meshless methods, and other methods such as particle system, discrete element, and discrete finite element methods. The next level requires dealing with numerical methods and actual computational implementation. This book addresses this last level of computational mechanics.

There are three main tasks in performing numerical methods in computational solid mechanics: solution of system of equations; computation of eigenvalues and eigenvectors of discrete systems; and numerical integration of dynamic equations of motion. It is important to point out that these tasks are present for linear as well as nonlinear systems. In addition, there are other subtasks specifically for nonlinear problems.

In the numerical method, we also deal with the theoretical background of these methods and their computational implementation. The main difference between the theory and computation is that in the theory we deal with the mathematics that is exact, whereas in computation there are finite precision and a round-off error. Practical implementation of computational mechanics must satisfy several conditions in order to serve a useful purpose. It must be consistent with the actual problem being solved. It must be reasonably accurate to be acceptable. Computational results are inherently approximations of the exact solution of the problem. The level of approximation must be known within a reasonable level of reliability, and it must be with an acceptable level of accuracy. Finally, it must be stable. Stability is an important concept in computational mechanics and is discussed in detail in this book.

When a computational method is developed and implemented and is consistent and stable with an acceptable level of accuracy, it must also be efficient to be useful. Numerical methods can be implemented in many different ways. Efficiency is an important feature. There are problems that are computationally extremely time consuming. For example, in the discrete element method or discrete finite element method, practical solutions can only be obtained in most computers for a reasonably small number of particles. Some practical problems with a large number of particles may take several months to simulate. Although theoretically it is possible to solve these problems, practically they are too inefficient to be useful.

This book is intended to serve as a reference book for a graduate course. It is assumed that the reader has a general familiarity with computational mechanics and the finite element method. We start with a review of matrix analysis and a brief review of the finite element method in Chapters 1 and 2. This is followed by three chapters covering solutions of systems of equations that are encountered in almost every computational mechanics problem: Chapter 3 covers direct solution methods; Chapter 4 covers iterative solution methods; and, Chapter 5 covers conjugate gradient methods. These methods are mainly for the solution of a linear system of equations. Next, Chapter 6 briefly covers nonlinear solution methods. Nonlinear problems in computational mechanics are covered extensively in an accompanying book (Ghaboussi, Pecknold, and Wu, *Nonlinear Computational Solid Mechanics*, CRC Press). Eigenvalue solution methods are covered in Chapter 7 and the direct solution of dynamic equations of motion are covered in Chapters 8 and 9.

We believe that this book will not only enable the readers to become effective users of the existing software in computational mechanics, it will also be very helpful in developing new software, when needed.

Authors

Jamshid Ghaboussi is emeritus professor of civil and environmental engineering at University of Illinois, Urbana-Champaign. He earned his doctoral degree at University of California, Berkeley. He has more than 40 years of teaching and research experience in computational mechanics and soft computing with applications in structural engineering, geo-mechanics, and bio-medical engineering. Dr. Ghaboussi has published extensively in these areas and is the inventor of five patents, mainly in the application of soft computing and computational mechanics. He is a coauthor of the book *Nonlinear Computational Solid Mechanics* (CRC Press) and the author of the book *Soft Computing in Engineering* (CRC Press). In recent years, he has been conducting research on complex systems and has coauthored a book *Understanding Systems: A Grand Challenge for 21st Century Engineering* (World Scientific Publishing).

Xiping Wu earned his PhD degree at University of Illinois, Urbana-Champaign, in 1991, in structural engineering. His professional career includes academic teaching and research, applied research for frontier development of oil and gas resources, and project delivery and management. He is a principal engineer in civil and marine engineering with Shell International Exploration & Production Inc. Dr. Wu has more than 25 years of experience in onshore and offshore structural engineering, arctic engineering, deepwater floating systems, plant design, LNG development, and capital project management. He has published more than 40 research papers in computational mechanics, numerical methods, arctic engineering, structural integrity and reliability, and applications of soft computing to civil engineering.

Chapter 1

Review of matrix analysis

1.1 VECTORS AND MATRICES

In engineering mechanics, we have worked with scalars, vectors, matrices, and spaces such as work, forces, stiffness matrices, and the three-dimensional Euclidean space. Then, what is a vector space? In addition, what is required to define a vector space? A vector space is a mathematical concept. Without use of rigorous mathematical definitions, we can state that a vector space is a nonempty set V of elements or vectors **a**, **b**, **c**, …, in which two algebraic operations called *vector addition* and *vector multiplication by scalars* are defined. Clearly, if the scalars for vector multiplication are real numbers, a real vector space is obtained; if the scalars are complex numbers, we have a complex vector space. In computational structural mechanics, we usually deal with real vector spaces. In some cases, we need to work with complex vector spaces. In general, a vector space includes sets of vectors, matrices, functions, and operators. The concept of vector space is fundamental in functional analysis, which has direct implications in finite element analysis.

The most elementary vector spaces in computational mechanics are the Euclidean vector spaces $\mathbf{R}^n$, n = 1, 2, 3, …, where n is the dimensionality of the vector space. For example, $\mathbf{R}^3$ is the Euclidean three-dimensional vector space. Any nonzero vector in $\mathbf{R}^3$ can be represented by directed line elements. We can easily illustrate the operations of vector addition and vector multiplication by scalars in $\mathbf{R}^3$. The vector addition and scalar multiplication are carried out on the components of the vectors. Geometrically, the vector addition operation follows the so-called parallelogram rule.

Before we discuss vectors and matrices further, we introduce for convenience the general notational convention used in this book for representing vectors and matrices. We denote a *vector space* as V or $\mathbf{R}^n$ for the Euclidean vector space in uppercase boldfaced letters. A *vector* is denoted by a lowercase boldfaced letter and a *matrix* by an uppercase boldfaced letter, such as a load vector **p** and a stiffness matrix **K**. The *length* of a vector **x** is usually called the *Euclidean norm* of the vector, which is denoted by $|\mathbf{x}|$. Of course, there are other norms that will be discussed in detail in Section 1.8. Since a *scalar* is a quantity described by a single number, it is represented by a normal nonboldfaced letter, either in lowercase or in uppercase letters or Greek letters, such as a_{ij} and α, β, and γ. To illustrate this convention, for example, a vector **a** in the n-dimensional Euclidean space $\mathbf{R}^n$ is represented as $\mathbf{a} \in \mathbf{R}^n$, and a matrix of size n × n as $\mathbf{A} \in \mathbf{R}^{n\times n}$. An n × n matrix can be considered as consisting of n column vectors or n row vectors. If we list the components of the vector **a** and matrix **A** in the n-dimensional Euclidean space, they are expressed as follows:

$$\mathbf{a} = \begin{Bmatrix} a_1 \\ \vdots \\ a_n \end{Bmatrix}; \; \mathbf{A} = \begin{bmatrix} a_{11} & \cdots & a_{1n} \\ \vdots & \cdots & \vdots \\ a_{n1} & \cdots & a_{nn} \end{bmatrix} \tag{1.1}$$

Obviously, the components of a vector and each element within a matrix are scalar quantities.

There are some unique vectors and matrices. A *unit vector* is a vector of unit length, and a *null* or zero vector **0** is a vector with all zero components. A unit matrix (or *identity matrix*) **I** is a square matrix with only unit diagonal terms and all zero off-diagonal terms. We can easily observe that a unit matrix consists of n unit column vectors or n unit row vectors.

1.1.1 Linear combination and linear independence

We now discuss some fundamental concepts related to vectors within a vector space. Suppose that $\mathbf{a}_1, \mathbf{a}_2, \ldots, \mathbf{a}_k$ are vectors in V; the *linear combination* of these k vectors is represented as follows:

$$\sum_{j=1}^{k} \alpha_j \, \mathbf{a}_j = \alpha_1 \, \mathbf{a}_1 + \alpha_2 \, \mathbf{a}_2 + \cdots + \alpha_k \, \mathbf{a}_k \tag{1.2}$$

where α_j, j = 1, 2, ..., k, are scalar constants. We state that a set of k vectors is *linearly dependent* if there exists a set of numbers or scalars $\{\alpha_1, \ldots, \alpha_k\}$ not all zeroes, such that the following relation is satisfied:

$$\sum_{j=1}^{k} \alpha_j \, \mathbf{a}_j = 0; \text{ for } \alpha_j \neq 0, \; 1 \leq j \leq k \tag{1.3}$$

As a consequence, if a set of k vectors is linearly dependent, any one vector can be written as a linear combination of other (k − 1) vectors within the set:

$$\mathbf{a}_m = \sum_{j=1,\, j\neq m}^{k} \frac{\alpha_j}{\alpha_m} \mathbf{a}_j \tag{1.4}$$

It follows that a set of k vectors is *linearly independent* if the condition for linear dependence is violated, that is, the following relation is satisfied:

$$\sum_{j=1}^{k} \alpha_j \, \mathbf{a}_j = 0; \text{ for } \alpha_j = 0, \; 1 \leq j \leq k \tag{1.5}$$

This means that none of the vectors can be represented as a linear combination of the rest of the vectors within a *linearly independent* vector set. For example, in a three-dimensional Euclidean space with a Cartesian coordinate system defined, we can observe that any three vectors on a plane are linearly dependent, and any three vectors along the three mutually orthogonal coordinate axes are linearly independent. If we have four vectors in this space, then they should be linearly dependent.

We define the *maximal linearly independent set* of a vector space as the largest number of linearly independent vectors possible in a vector space. In other words, this maximal set is not a proper subset of any other linearly independent set. Therefore, the *dimension of a vector space*, dim(V), is the number of vectors in a maximal set. Obviously, any maximal linearly independent set is a *basis* for the vector space V. In a three-dimensional Euclidean space, the unit vectors, **i**, **j**, and **k**, the fundamental orthogonal triad, form the basis of the vector space $\mathbf{R}^3$. If a set of vectors $\{e_1, \ldots, e_k\}$ is a basis, then any vector $a \in \mathbf{R}^n$ can be expressed as a linear combination of the basis vectors in the vector space:

$$a = \sum_{j=1}^{k} \alpha_j \, e_j \tag{1.6}$$

In this equation, α_j are components of the vector **a** along the basis vectors, and e_j are basis vectors.

In mathematics, we can work with infinite-dimensional vector spaces, but in computational mechanics, the vector spaces are of finite dimensions. For visualization purpose, we often represent vectors in the three-dimensional Euclidean space.

1.1.2 Basis vectors

Since any maximal linearly independent set of vectors constitutes a basis for a vector space, there are many ways to generate a set of basis vectors. For example, in the three-dimensional Euclidean space, a set of *general basis vectors* can be of the following form:

$$e_1 = \begin{Bmatrix} 1 \\ 0 \\ 1 \end{Bmatrix}; \; e_2 = \begin{Bmatrix} 1 \\ 1 \\ 1 \end{Bmatrix}; \; e_3 = \begin{Bmatrix} 0 \\ 0 \\ 1 \end{Bmatrix} \tag{1.7}$$

Obviously, a set of vectors that are orthogonal to each other should naturally be a set of basis vectors. Two vectors **a** and **b** are orthogonal to each other when the projection of one vector onto the direction of the other vector is zero:

$$a \cdot b = \sum_{j=1}^{n} a_j \, b_j = 0 \tag{1.8}$$

We can simply generate a set of *orthogonal basis vectors* as follows:

$$e_1 = \begin{Bmatrix} 1 \\ -1 \\ 0 \end{Bmatrix}; \; e_2 = \begin{Bmatrix} 1 \\ 1 \\ 0 \end{Bmatrix}; \; e_3 = \begin{Bmatrix} 0 \\ 0 \\ 1 \end{Bmatrix} \tag{1.9}$$

Furthermore, in practice, we usually use *orthonormal basis vectors* in carrying out numerical computations. By definition, an orthonormal basis vector has a unit norm, or the length of the vector is 1, and each vector is orthogonal to other vectors as well. We can easily arrive at the following set of orthonormal basis vectors along each Cartesian coordinate axis in a three-dimensional Euclidean space:

$$e_1 = \begin{Bmatrix} 1 \\ 0 \\ 0 \end{Bmatrix}; \; e_2 = \begin{Bmatrix} 0 \\ 1 \\ 0 \end{Bmatrix}; \; e_3 = \begin{Bmatrix} 0 \\ 0 \\ 1 \end{Bmatrix} \tag{1.10}$$

These three set of basis vectors are schematically illustrated in Figure 1.1.

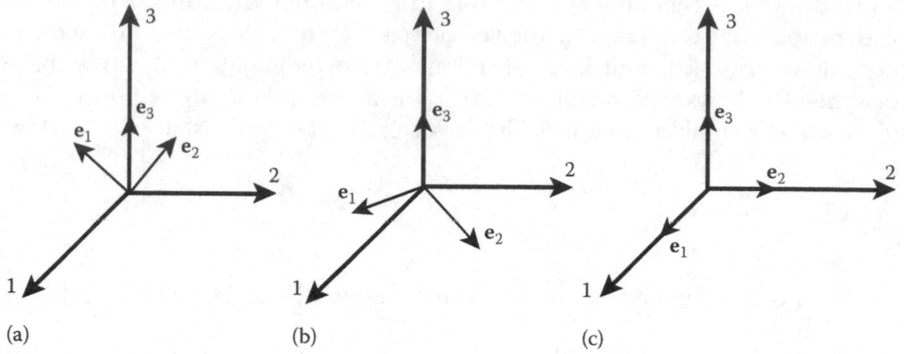

Figure 1.1 Different basis vectors in a three-dimensional Euclidean space: (a) general basis vectors; (b) orthogonal basis vectors; and (c) orthonormal basis vectors.

1.2 SUBSPACES: SUM, UNION, INTERSECTION, DIRECT SUM, AND DIRECT COMPLEMENT OF SUBSPACES

We state that S is a subspace of vector space V if $\dim(S) \leq \dim(V)$. For example, a line is a one-dimensional subspace, and a plane is a two-dimensional subspace in the three-dimensional Euclidean space V or $\mathbf{R}^3$. For vectors belonging to a subspace, any combination of them also belongs to the same subspace:

$$\text{if } \mathbf{a} \in S \text{ and } \mathbf{b} \in S, \text{ then } \alpha\mathbf{a} + \beta\mathbf{b} \in S \tag{1.11}$$

1.2.1 Sum of subspaces

If S_1 and S_2 are subspaces of V, then $S_1 + S_2$ is also a subspace of V. For example, considering two straight lines in a plane, the vector summation of these two lines lies in the plane and the plane is a subspace of V:

$$S_1 + S_2 = \{\mathbf{a} + \mathbf{b} : \mathbf{a} \in S, \ \mathbf{b} \in S\} \tag{1.12}$$

1.2.2 Union of subspaces

If S_1 and S_2 are subspaces of V, then the union of them, $S_1 \cup S_2$, is not a subspace of V. This is obvious in considering two intersecting lines in a plane. A vector within the union of these two lines can be on either line only:

$$S_1 \cup S_2 = \{\mathbf{a} : \mathbf{a} \in S_1 \text{ or } \mathbf{a} \in S_2\} \tag{1.13}$$

1.2.3 Intersection of subspaces

If S_1 and S_2 are subspaces of V, then the intersection of them, $S_1 \cap S_2$, is also a subspace of V. For example, the intersection of two planes is a line that is a subspace of V:

$$S_1 \cap S_2 = \{\mathbf{a} : \mathbf{a} \in S_1 \text{ and } \mathbf{a} \in S_2\} \tag{1.14}$$

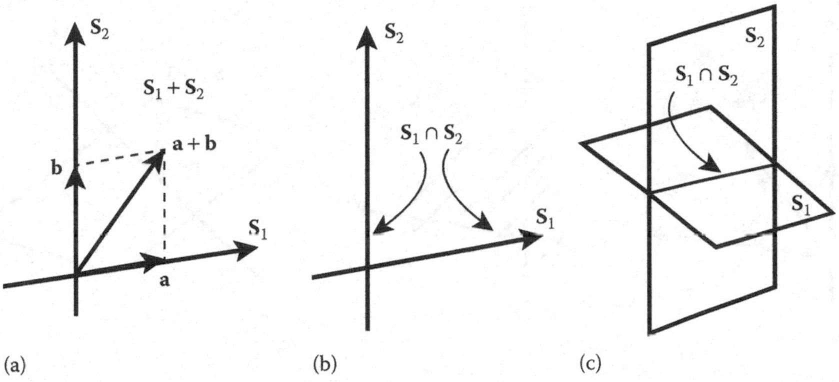

Figure 1.2 (a) Sum, (b) union, and (c) intersection of subspaces in a three-dimensional space.

The concepts of sum, union, and intersection of subspaces are schematically illustrated in the three-dimensional space in Figure 1.2.

1.2.4 Decomposition theorem

If S_1 and S_2 are subspaces of V and $a \in S_1 + S_2$, then the decomposition $a = a_1 + a_2$, such that $a_1 \in S_1$ and $a_2 \in S_2$, is unique if and only if $S_1 \cap S_2 = 0$. For example, we can decompose a vector in a plane uniquely into the summation of two vectors along the coordinate axes in the plane. In statics, a force in the three-dimensional space can be decomposed uniquely into its components along the Cartesian coordinate axes.

1.2.5 Direct sum of subspaces

We can have the direct sum of subspaces if the intersection between them is a null set. If S_1 and S_2 are subspaces of V and if $S_1 \cap S_2 = 0$, then the direct sum of S_1 and S_2 is $S_1 + S_2 = S_1 \oplus S_2$. Obviously, the dimensions of subspaces all add up with a direct sum. We thus have $\dim(S_1 \oplus S_2) = \dim(S_1) + \dim(S_2)$. For example, if S_1 is a two-dimensional subspace, a plane, and S_2 a line in the vector space, then the direct sum of S_1 and S_2 forms a three-dimensional subspace in the vector space. The direct sum of two one-dimensional subspaces forms a plane.

1.2.6 Direct complement

If S_1 and S_2 are subspaces of the vector space V and $S_1 \oplus S_2 = V$, then S_1 is the direct complement of S_2. For example, in a Euclidean space, a line that is perpendicular to a plane is a direct complement of the plane.

The operations of decomposition, direct sum, and direct complement of subspaces are illustrated in Figure 1.3.

1.3 VECTOR OPERATIONS

So far, we have discussed the fundamental concepts of vectors and subspaces in a vector space. Now we introduce some of the basic vector operations that are commonly used in computational mechanics.

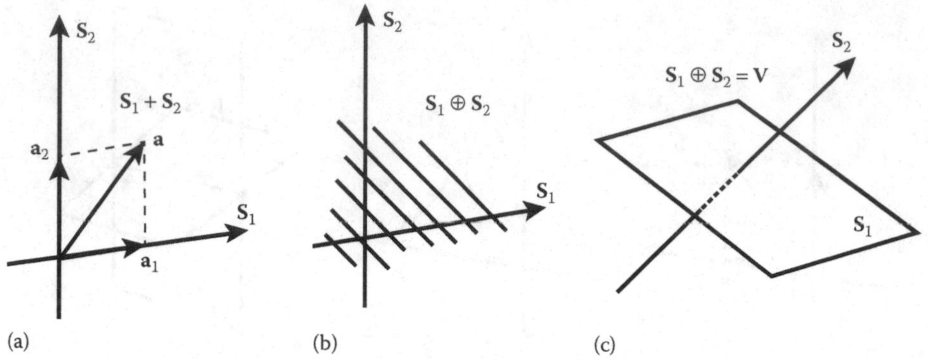

Figure 1.3 (a) Decomposition, (b) direct sum, and (c) direct complement of subspaces in a three-dimensional space.

1.3.1 Inner products

There are many ways to multiply a vector by another vector. First of all, the *inner product* or *dot product* of two vectors is a scalar quantity defined as follows:

$$\mathbf{a}^T \mathbf{b} = \mathbf{a} \cdot \mathbf{b} = \sum_{j=1}^{n} a_j \, b_j \tag{1.15}$$

The superscript T represents the transpose of a vector. With the above definition, the length or Euclidean norm of a vector **a** is as follows:

$$|\mathbf{a}| = (\mathbf{a} \cdot \mathbf{a})^{1/2} \tag{1.16}$$

Through geometric interpretation, we have $\mathbf{a} \cdot \mathbf{b} = |\mathbf{a}||\mathbf{b}|\cos(\mathbf{a}, \mathbf{b})$. If nonzero vectors **a** and **b** are orthogonal to each other, or cos(a, b) = 0, then the inner product is equal to zero. That is, if $\mathbf{a} \perp \mathbf{b}$, then $\mathbf{a} \cdot \mathbf{b} = 0$. This can easily be seen in Figure 1.4a. We can thus state that two nonzero vectors are orthogonal if and only if their inner product is zero. We can also verify the following useful relationships:

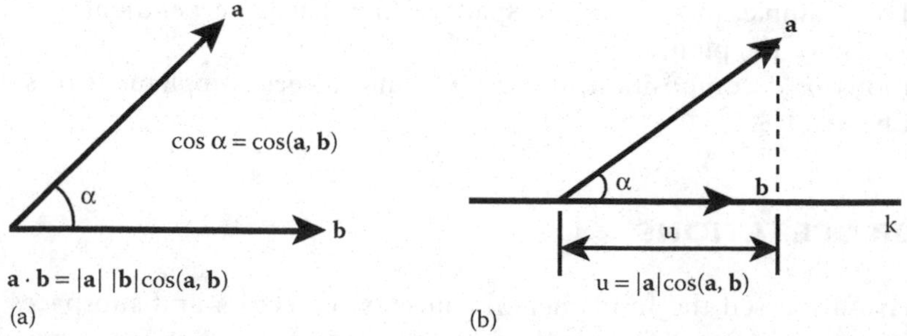

Figure 1.4 (a) Inner product of vectors **a** and **b** and (b) the projection of vector **a** in the direction of vector **b**.

$$\begin{cases} \mathbf{a} \cdot \mathbf{b} \le |\mathbf{a}|\ |\mathbf{b}|;\ \text{Schwartz Inequality} \\ |\mathbf{a}| + |\mathbf{b}| \ge |\mathbf{a} + \mathbf{b}|;\ \text{Triangular Inequality} \\ |\mathbf{a} + \mathbf{b}|^2 + |\mathbf{a} - \mathbf{b}|^2 = 2\left(|\mathbf{a}|^2 + |\mathbf{b}|^2\right);\ \text{Parallelogram Equality} \end{cases} \tag{1.17}$$

With the definition of the inner product between two vectors, we can define the *projection* or component of vector **a** in the direction of vector **b**, u, as illustrated in Figure 1.4b:

$$u = |\mathbf{a}| \cos(\mathbf{a}, \mathbf{b}) = \frac{\mathbf{a} \cdot \mathbf{b}}{|\mathbf{b}|} \tag{1.18}$$

We note that u is a scalar quantity representing the length of the orthogonal projection of **a** on a straight line in the direction of **b**. As an example, the work done by a force **F** in the displacement **d** is defined as the inner product of **F** and **d**.

1.3.2 Orthogonal complement

Assume that S is a subspace of V; then $S^\perp$ is the orthogonal complement of S such that

$$S^\perp = \{\mathbf{a} : \mathbf{a} \cdot \mathbf{b} = 0,\ \mathbf{b} \in S\} \tag{1.19}$$

Obviously, $S^\perp$ is a subspace of V in which all vectors are orthogonal to S. In addition, a vector space is the direct sum of any subspace and its orthogonal complement, $V = S \oplus S^\perp$.

1.3.3 Orthogonal decomposition

The decomposition of a vector into orthogonal complement subspaces, **a** = **b** + **c**, is unique for $\mathbf{a} \in V$, $\mathbf{b} \in S$, and $\mathbf{c} \in S^\perp$. For example, in the three-dimensional Euclidean space, a vector has a unique orthogonal decomposition along any Cartesian coordinate axis and the plane normal to that axis.

1.3.4 Vector products

We have seen that the *inner product* of two n-dimensional vectors results in a scalar quantity. The *outer product* of two vectors is a matrix. If vectors $\mathbf{a} \in \mathbf{R}^n$ and $\mathbf{b} \in \mathbf{R}^m$, then the *outer product* of these two vectors is a matrix:

$$\mathbf{a}\ \mathbf{b}^T = \mathbf{A} = \begin{Bmatrix} a_1 \\ \vdots \\ a_n \end{Bmatrix} \langle b_1\ \cdots\ b_m \rangle = \begin{bmatrix} a_1b_1 & \cdots & a_1b_m \\ \vdots & & \vdots \\ a_nb_1 & \cdots & a_nb_m \end{bmatrix} \tag{1.20}$$

The *vector product* or *cross product* of two vectors is a vector whose direction is perpendicular to the plane formed by the two vectors, as shown in Figure 1.5a. The direction of the resulting vector can be determined by the "right-hand rule." The length of the resulting vector is equal to the area of the parallelogram formed with **a** and **b** as adjacent sides. In the Euclidean space, if we describe two vectors **a** and **b** in terms of their components along

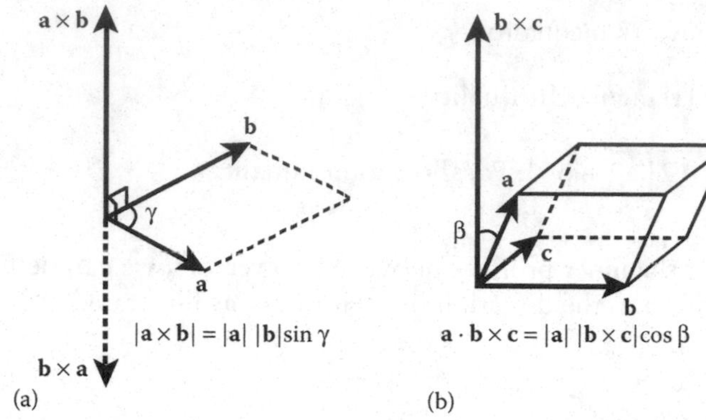

Figure 1.5 Determination of (a) vector product and (b) scalar triple product.

the Cartesian coordinate axes, $\mathbf{a} = a_1\mathbf{i} + a_2\mathbf{j} + a_3\mathbf{k}$ and $\mathbf{b} = b_1\mathbf{i} + b_2\mathbf{j} + b_3\mathbf{k}$, then we have the vector product of **a** and **b** as follows:

$$\mathbf{a} \times \mathbf{b} = \mathbf{c} = \begin{vmatrix} \mathbf{i} & \mathbf{j} & \mathbf{k} \\ a_1 & a_2 & a_3 \\ b_1 & b_2 & b_3 \end{vmatrix} \tag{1.21}$$

The length of the vector **c** as the area of the parallelogram with **a** and **b** as adjacent sides can be determined from $|\mathbf{c}| = |\mathbf{a}||\mathbf{b}|\sin(\mathbf{a}, \mathbf{b})$. If vectors **a** and **b** have the same or opposite direction, then the vector product is the zero vector. This suggests that two vectors are linearly dependent if their vector product is the zero vector. Furthermore, it can be verified that the cross multiplication of vectors is anticommutative and nonassociative:

$$\begin{cases} \mathbf{a} \times \mathbf{b} = -(\mathbf{b} \times \mathbf{a}); \text{ Anticommutative} \\ (\mathbf{a} \times \mathbf{b}) \times \mathbf{c} \neq \mathbf{a} \times (\mathbf{b} \times \mathbf{c}); \text{ Nonassociative} \end{cases} \tag{1.22}$$

1.3.5 Scalar triple product

In a three-dimensional Euclidean space, the scalar triple product of three vectors **a**, **b**, and **c** can be computed by the following relation, where we use determinants that will be discussed in Section 1.4.6:

$$\begin{aligned} \mathbf{a} \cdot (\mathbf{b} \times \mathbf{c}) &= (a_1\mathbf{i} + a_2\mathbf{j} + a_3\mathbf{k}) \begin{vmatrix} \mathbf{i} & \mathbf{j} & \mathbf{k} \\ b_1 & b_2 & b_3 \\ c_1 & c_2 & c_3 \end{vmatrix} \\ &= \begin{vmatrix} a_1 & a_2 & a_3 \\ b_1 & b_2 & b_3 \\ c_1 & c_2 & c_3 \end{vmatrix} \end{aligned} \tag{1.23}$$

It can be easily observed from Figure 1.5b that the absolute value of the scalar triple product is equal to the volume of the parallelepiped with **a**, **b**, and **c** as adjacent edges. Obviously, if the triple product of three vectors is zero, which means that these vectors are on the same plane, then these three vectors must be linearly dependent. There are other repeated vector products that can be expressed in terms of the basic types of vector product described in this section. For example, we can verify the following formula:

$$\mathbf{b}\times(\mathbf{c}\times\mathbf{d})=(\mathbf{b}\cdot\mathbf{d})\,\mathbf{c}-(\mathbf{b}\cdot\mathbf{c})\,\mathbf{d} \tag{1.24}$$

which implies the following Lagrangian identity:

$$(\mathbf{a}\times\mathbf{b})\cdot(\mathbf{c}\times\mathbf{d})=(\mathbf{a}\cdot\mathbf{c})\,(\mathbf{b}\cdot\mathbf{d})-(\mathbf{a}\cdot\mathbf{d})\,(\mathbf{b}\cdot\mathbf{c}) \tag{1.25}$$

1.4 MATRICES

A matrix is a rectangular array of numbers that can be considered as a mapping that transforms a vector **b** into vector **c**. A linear transformation from a vector $\mathbf{b}\in\mathbf{R}^m$ into a vector $\mathbf{c}\in\mathbf{R}^n$ is expressed as $\mathbf{c}=\mathbf{A}\mathbf{b}$, where $\mathbf{A}\in\mathbf{R}^{nxm}$:

$$\begin{Bmatrix} c_1 \\ \vdots \\ c_n \end{Bmatrix}=\begin{bmatrix} a_{11} & \cdots & a_{1m} \\ \vdots & & \vdots \\ a_{n1} & \cdots & a_{nm} \end{bmatrix}\begin{Bmatrix} b_1 \\ \vdots \\ b_m \end{Bmatrix} \tag{1.26}$$

In the matrix **A**, the numbers a_{11}, …, a_{nm} are the *elements* of the matrix. The horizontal arrays of numbers are *row* vectors, and the vertical arrays are *column* vectors. A matrix with n rows and m columns is called an n × m matrix. A matrix can be expressed in terms of its general element enclosed in brackets such as $\mathbf{A}=\left[a_{ij}\right]$. A submatrix is the remaining part of the matrix after eliminating some rows and columns from the original matrix.

A general matrix is *rectangular* where $n\neq m$. It can be proved that all real n × m matrices form a real vector space of dimension n × m, or $\mathbf{R}^{nxm}$. In computational mechanics, we often deal with *square* matrices in which the number of rows is equal to the number of columns, that is, n = m. The number of rows of a square matrix is called its *order*. Another concept is the *rank* of a matrix, which is defined as the maximum number of linearly independent row vectors of a matrix. For example, in linear finite element analysis, the equilibrium equation of a structure can be expressed as $\mathbf{p}=\mathbf{K}\mathbf{u}$, where **p** is the load vector, **u** is the displacement vector, and **K** is the global stiffness matrix of the structural system. The stiffness matrix is always a square matrix.

There are many types of square matrices depending on the characteristics of its elements. Let us consider a square matrix $\mathbf{A}=\left[a_{ij}\right]$. If the elements $a_{ij}=a_{ji}$, then matrix **A** is a *symmetric* matrix. However, if the elements $a_{ij}=-a_{ji}$, we have a *skew-symmetric* matrix. A *diagonal* matrix consists of nonzero diagonal elements and zero off-diagonal elements, $a_{ij}=0$ for $i\neq j$. As has been discussed earlier, an identity matrix **I** is a diagonal matrix with all unit diagonal elements.

In numerical analysis, we often encounter various types of *triangular* matrices. A triangular matrix is a square matrix with either zero lower off-diagonal terms or zero upper off-diagonal terms. If the elements below the diagonal are zero, then we have an *upper triangular matrix* that has the following form:

$$\begin{cases} a_{ij}\neq 0 \text{ for } j\geq i \\ a_{ij}=0 \text{ for } j<i \end{cases}\quad\begin{bmatrix} a_{11} & \cdots & \cdots & a_{1n} \\ 0 & a_{22} & \cdots & a_{2n} \\ \vdots & \vdots & \ddots & \vdots \\ 0 & 0 & \cdots & a_{nn} \end{bmatrix} \tag{1.27}$$

and a *lower triangular matrix* has all zero elements above the diagonal:

$$\begin{cases} a_{ij} \neq 0 \text{ for } j \leq i \\ a_{ij} = 0 \text{ for } j > i \end{cases} \quad \begin{bmatrix} a_{11} & 0 & \cdots & 0 \\ a_{21} & a_{22} & \cdots & 0 \\ \vdots & \vdots & \ddots & \vdots \\ a_{n1} & a_{n2} & \cdots & a_{nn} \end{bmatrix} \tag{1.28}$$

A *strictly upper triangular matrix* is an upper triangular matrix with zero diagonal terms:

$$\begin{cases} a_{ij} \neq 0 \text{ for } j > i \\ a_{ij} = 0 \text{ for } j \leq i \end{cases} \quad \begin{bmatrix} 0 & a_{12} & \cdots & \cdots & a_{1n} \\ 0 & 0 & a_{23} & \cdots & a_{2n} \\ \vdots & \vdots & \ddots & \ddots & \vdots \\ 0 & 0 & \cdots & 0 & a_{n-1,n} \\ 0 & 0 & \cdots & 0 & 0 \end{bmatrix} \tag{1.29}$$

and a *strictly lower triangular matrix* is a lower triangular matrix with zero diagonal terms:

$$\begin{cases} a_{ij} \neq 0 \text{ for } j < i \\ a_{ij} = 0 \text{ for } j \geq i \end{cases} \quad \begin{bmatrix} 0 & 0 & \cdots & \cdots & 0 \\ a_{21} & 0 & \cdots & \cdots & 0 \\ \vdots & a_{32} & \ddots & \vdots & \vdots \\ \vdots & \vdots & \ddots & 0 & 0 \\ a_{n1} & a_{n2} & \cdots & a_{n,n-1} & 0 \end{bmatrix} \tag{1.30}$$

There are also some other types of useful matrices such as the *Hessenberg* matrices. An *upper* Hessenberg matrix is an expanded upper triangular matrix with one layer of nonzero elements on the lower side of the diagonal terms:

$$\begin{cases} a_{ij} \neq 0 \text{ for } j \geq i-1 \\ a_{ij} = 0 \text{ for } j < i-1 \end{cases} \quad \begin{bmatrix} a_{11} & \cdots & \cdots & \cdots & a_{1n} \\ a_{21} & a_{22} & \cdots & \cdots & a_{2n} \\ 0 & a_{32} & \ddots & \vdots & \vdots \\ \vdots & \ddots & \ddots & \ddots & \vdots \\ 0 & \cdots & 0 & a_{n,n-1} & a_{nn} \end{bmatrix} \tag{1.31}$$

Similarly, a *lower* Hessenberg matrix is a lower triangular matrix with one layer of nonzero elements on the upper side of the diagonal terms:

$$\begin{cases} a_{ij} \neq 0 \text{ for } j \leq i+1 \\ a_{ij} = 0 \text{ for } j > i+1 \end{cases} \quad \begin{bmatrix} a_{11} & a_{12} & 0 & \cdots & 0 \\ \vdots & a_{22} & a_{23} & \ddots & \vdots \\ \vdots & \vdots & \ddots & \ddots & 0 \\ \vdots & \vdots & \vdots & \ddots & a_{n-1,n} \\ a_{n1} & a_{n2} & \cdots & \cdots & a_{nn} \end{bmatrix} \tag{1.32}$$

A *tridiagonal* matrix is a square matrix with one nonzero element on either side of the nonzero diagonal terms:

$$\begin{cases} a_{ij} \neq 0 \text{ for } i-1 \leq j \leq i+1 \\ a_{ij} = 0 \text{ otherwise} \end{cases} \quad \begin{bmatrix} a_{11} & a_{12} & & & \\ a_{12} & a_{22} & a_{23} & & \\ & a_{32} & \ddots & \ddots & \\ & & \ddots & \ddots & a_{n-1,n} \\ & & & a_{n,n-1} & a_{nn} \end{bmatrix} \tag{1.33}$$

This is also a *banded* matrix, which means that nonzero elements are within a narrow band around the diagonal. For a banded matrix **A**, the zero elements within the matrix are defined as $a_{ij} = 0$ for $j > i + m_A$, where m_A is called the *half-bandwidth* of **A**. The *bandwidth* of **A** is obviously $2m_A + 1$. A tridiagonal matrix has a bandwidth of 3. Structural stiffness matrices obtained in finite element analysis are usually banded symmetric matrices. The bandedness of a stiffness matrix can be exploited to efficiently utilize the computer memory and storage space in analyzing structural systems that have very large stiffness matrices.

Diagonal matrices have only nonzero terms on the diagonal, whereas all the off-diagonal terms are zero. We can see that diagonal matrices are symmetric; for a diagonal matrix **A**, we have $a_{ij} = a_{ji} = 0$ for $i \neq j$. A special diagonal matrix is the identity matrix, denoted by **I**; all the diagonal terms of the identity matrix are equal to 1.

The *transpose* of an n × m matrix **A**, denoted as $\mathbf{A}^T$, is the m × n matrix obtained by interchanging the rows and columns in **A**. That is, if matrix $\mathbf{A} = \left[a_{ij}\right] \in \mathbf{R}^{n\times m}$, then $\mathbf{A}^T = \left[a_{ji}\right] \in \mathbf{R}^{m\times n}$. We call the summation of all diagonal terms of square matrix **A** the *trace* of **A**:

$$\operatorname{tr}(\mathbf{A}) = \sum_{j=1}^{n} a_{jj} \tag{1.34}$$

1.4.1 Rank of a matrix

The rank of a matrix **A**, rank(**A**), is defined as the number of linearly independent row vectors of the matrix. If matrix **A** is an n × n matrix, then rank(**A**) ≤ n. It can be proved that the rank of matrix **A** can be defined equivalently in terms of column vectors. If the rank of an n × n matrix is lower than n, then we say matrix **A** has rank deficiency. Rank deficiency can be discussed in terms of column rank deficiency and row rank deficiency. If a square matrix is rank deficient, then it is a singular matrix. The concept of the rank of a matrix is very important in determining the existence and uniqueness of solutions to a system of linear equations, which will be discussed in Chapter 3.

1.4.2 Inverse of a matrix

The inverse of an n × n matrix **A** is denoted by $\mathbf{A}^{-1}$ such that $\mathbf{A}\mathbf{A}^{-1} = \mathbf{A}^{-1}\mathbf{A} = \mathbf{I}$, where **I** is the identity matrix. If an n × n matrix **A** has an inverse, then **A** is called nonsingular; otherwise, **A** is singular. If we consider matrix **A** as representing the linear mapping from vector **x** into **y**, $\mathbf{y} = \mathbf{A}\mathbf{x}$, then the inverse represents the inverse mapping from vector **y** into vector **x**, $\mathbf{x} = \mathbf{A}^{-1}\mathbf{y}$.

For a nonsingular matrix, the inverse is unique. In general, for an n × n matrix **A**, the inverse exists if and only if rank(**A**) = n. It is obvious that columns of a nonsingular matrix are a linearly independent set of vectors. For a matrix with rank deficiency, there exists no inverse matrix. We can prove that the inverse of a product of two matrices is equal to the product of inverses of the matrices in the reverse order. That is, $(\mathbf{AB})^{-1} = \mathbf{B}^{-1}\mathbf{A}^{-1}$.

1.4.3 Orthogonal matrices

An orthogonal matrix **A** is defined as a real matrix with its inverse equal to its transpose: $\mathbf{A}^{-1} = \mathbf{A}^T$. This implies that if a matrix **A** is orthogonal, then $\mathbf{AA}^T = \mathbf{A}^T\mathbf{A} = \mathbf{I}$. The columns of an orthogonal matrix form a set of orthogonal vectors, which are linearly independent.

1.4.4 Adjoint and self-adjoint matrices

In numerical analysis, it is not uncommon to come across matrices that are designated as "adjoint." This is a very interesting mathematical concept. Let **A** be an $n \times n$ matrix with real or complex elements. We know that the conjugate of $\mathbf{A} = [a_{ij}]$ is an $n \times n$ matrix $\bar{\mathbf{A}} = [\bar{a}_{ij}]$. If $a_{ij} = a + i\,b$, then $\bar{a}_{ij} = a - i\,b$, $(i = \sqrt{-1})$. Let $\mathbf{A}^H$ be the conjugate transpose of **A**, that is, $\mathbf{A}^H = \bar{\mathbf{A}}^T$. Then we call the matrix $\mathbf{A}^H$ the *adjoint* matrix.

For a special class of matrices, $\mathbf{A}^H = \mathbf{A}$, and we say that **A** is a Hermitian matrix. A Hermitian matrix is equal to its adjoint; thereby, it is called *self-adjoint*. We can easily see that a real Hermitian matrix is symmetric, and a purely imaginary Hermitian matrix is skew symmetric. In structural mechanics, we usually deal with real Hermitian matrices or symmetric matrices.

1.4.5 Matrix product

Let **A** be an $m \times q$ matrix and **B** a $q \times n$ matrix; then the product of these two matrices is an $m \times n$ matrix $\mathbf{C} = \mathbf{AB}$. The elements of the **C** matrix are determined as follows:

$$c_{ij} = a_{i1}b_{1j} + a_{i2}b_{2j} + \cdots + a_{iq}b_{qj} = \sum_{j=1}^{q} a_{ik}\ b_{kj} \tag{1.35}$$

where $i = 1, \ldots, m$ and $j = 1, \ldots, n$. Multiplication of two matrices is possible if, and only if, the number of columns of matrix **A** is equal to the number of rows of matrix **B**.

1.4.6 Determinants

The concept of a determinant is introduced in the solution of a system of linear equations. With reference to a 2×2 square matrix **A**, a determinant of second order can be written in the following form:

$$\det(\mathbf{A}) = |\mathbf{A}| = \begin{vmatrix} a_{11} & a_{12} \\ a_{21} & a_{22} \end{vmatrix} = a_{11}\ a_{22} - a_{21}\ a_{12} \tag{1.36}$$

We can extend this notion to denote a determinant of order n as a square array of $n \times n$ elements enclosed between two vertical bars:

$$\det(\mathbf{A}) = \begin{vmatrix} a_{11} & \cdots & a_{1n} \\ \vdots & & \vdots \\ a_{n1} & \cdots & a_{nn} \end{vmatrix} \tag{1.37}$$

Similarly, we can use some terms defined for matrix, such as rows and columns, in describing a determinant. To facilitate the calculation of a determinant, we need to introduce the concepts of *minor* and *cofactor*. The *minor* of an element a_{ij}, denoted by M_{ij}, is the determinant of a matrix obtained by deleting the ith row and the jth column from the original matrix. The *cofactor* of a_{ij}, denoted by C_{ij}, is related to the minor as follows: $C_{ij} = (-1)^{i+j} M_{ij}$. It can be seen that the determinant of matrix **A** is equal to the sum of the products of the elements of any row or column and their respective cofactors:

$$\begin{aligned} \det(\mathbf{A}) &= \sum_{k=1}^{n} (-1)^{i+k} a_{ik} M_{ik} \; ; \text{for } i = 1, 2, \ldots, n \\ &= \sum_{i=1}^{n} (-1)^{i+k} a_{ik} M_{ik} \; ; \text{for } k = 1, 2, \ldots, n \end{aligned} \tag{1.38}$$

It can be verified that the interchange of any two rows or columns results in a change of sign of the determinant. Thus, if the corresponding elements within any two rows or columns are proportional to each other, then the determinant becomes zero. The following are useful formulae for determinant of matrices:

$$\begin{cases} \det(\mathbf{AB}) = \det(\mathbf{A}) \det(\mathbf{B}) \\ \det(\mathbf{A}^T) = \det(\mathbf{A}) \\ \det(\alpha\mathbf{A}) = \alpha^n \det(\mathbf{A}) \\ \det(\mathbf{A}^{-1}) = \dfrac{1}{\det(\mathbf{A})} \end{cases} \tag{1.39}$$

For triangular or diagonal matrices, the determinant is equal to the product of the diagonal terms. Determinant of identity matrix, $\det(\mathbf{I}) = 1$.

With the introduction of determinant, we can state that the inverse of an $n \times n$ matrix **A** exists if and only if $\det(\mathbf{A}) \neq 0$. If a matrix is singular, then $\det(\mathbf{A}) = 0$.

1.4.7 Cramer's rule

The Cramer's rule presents a method of obtaining the solution of a system of linear equations using determinants. For a system of n linear equations, we have the following matrix form:

$$\mathbf{Ax} = \mathbf{b} \tag{1.40}$$

If $\det(\mathbf{A})$ is nonzero, then the system has a unique solution. The terms of the solution vector can be computed by the following equation:

$$x_i = \frac{\det(\mathbf{A}_i)}{\det(\mathbf{A})} \tag{1.41}$$

The matrix $\mathbf{A}_i$ is obtained by replacing the ith column of **A** with vector **b**.

Calculating the determinant of a large matrix is computationally intensive, which makes the use of the Cramer's rule for solving a system of linear equations computationally costly.

However, the Cramer's rule is an important concept and of theoretical value in studying the properties of systems of linear equations.

1.4.8 Bilinear and quadratic forms

If we have vectors $\mathbf{x} \in \mathbf{R}^n$ and $\mathbf{y} \in \mathbf{R}^n$ and an $n \times n$ matrix $\mathbf{A} = [a_{ij}]$, then the following expression is called a *bilinear* form:

$$\mathbf{x}^T\mathbf{A}\mathbf{y} = \sum_{i=1}^{n}\sum_{j=1}^{n} a_{ij}\, x_i\, y_j \tag{1.42}$$

The matrix **A** in this expression is called the *coefficient matrix*. If $\mathbf{x} = \mathbf{y}$, then $\mathbf{x}^T\mathbf{A}\mathbf{x}$ is called a *quadratic* form in the n variables **x**. If the coefficient matrix **A** is real and symmetric, the quadratic form is referred to as the symmetric quadratic form in structural applications, specifically when the matrix **A** represents the structural stiffness matrix.

If a real symmetric quadratic form $\mathbf{x}^T\mathbf{A}\mathbf{x} > 0$ for all nonzero vectors **x**, then the symmetric coefficient matrix **A** is *positive definite*. If $\mathbf{x}^T\mathbf{A}\mathbf{x} \geq 0$ for all nonzero vectors **x**, then matrix **A** is *positive semidefinite*. Accordingly, matrix **A** is negative definite if $\mathbf{x}^T\mathbf{A}\mathbf{x} < 0$ or negative semidefinite if $\mathbf{x}^T\mathbf{A}\mathbf{x} \leq 0$. These last two types of matrices are seldom encountered in computational mechanics. We will see later that the positive definiteness of a symmetric matrix determines the signs of eigenvalues of the matrix. All eigenvalues of positive definite matrices are positive. Obviously, all eigenvalues of negative definite matrices are negative.

1.5 EIGENVALUES AND EIGENVECTORS OF A MATRIX

1.5.1 Standard eigenvalue problem

A square matrix **A** has eigenvalues λ and eigenvectors **x** as given in the following equation:

$$\mathbf{A}\mathbf{x} = \lambda\mathbf{x} \tag{1.43}$$

The set of eigenvalues constitutes the *spectrum* of **A**, $\lambda(\mathbf{A})$, and the largest eigenvalue in terms of its absolute quantity is called the *spectral radius* of **A**, $\rho(\mathbf{A})$.

The *standard* eigenvalue problem for an $n \times n$ matrix is in the following form:

$$\mathbf{A}\mathbf{x}_j = \lambda_j\mathbf{x}_j;\ j = 1,\ \dots,\ n \tag{1.44}$$

There are n eigenvalues of matrix **A**, λ_j, $j = 1, \dots, n$, and the corresponding n eigenvectors $\mathbf{x}_j$, $j = 1, \dots, n$.

Referring to Equation 1.44, we can write the vector equation in the following form:

$$(\mathbf{A} - \lambda\mathbf{I})\mathbf{x} = 0 \tag{1.45}$$

According to the Cramer's theorem, this homogeneous system of linear equations has a nontrivial solution if and only if $\det(\mathbf{A} - \lambda\mathbf{I}) = 0$, which results in a polynomial of nth degree in λ, called the characteristic equation:

$$\alpha_0 + \alpha_1\,\lambda + \alpha_2\,\lambda^2 + \dots + \alpha_n\,\lambda^n = 0 \tag{1.46}$$

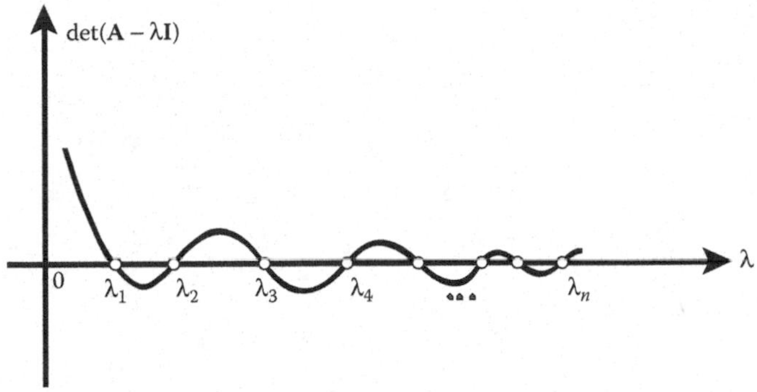

Figure 1.6 Eigenvalues as roots of the characteristic equation, det(**A** − λ**I**) = 0.

The first term in this equation is $\alpha_0 = \det(\mathbf{A})$. The eigenvalues are thus the roots of the characteristic Equation 1.46, which are illustrated in Figure 1.6.

1.5.2 Generalized eigenvalue problem

In structural mechanics, especially in structural dynamics, we often deal with the *generalized* eigenvalue problems in the following form:

$$\mathbf{A}\mathbf{x}_j = \lambda_j \mathbf{B}\mathbf{x}_j;\ j = 1, \ldots, n \tag{1.47}$$

The matrix **A** could be the stiffness matrix and **B** the mass matrix in a structural dynamic problem. The generalized eigenvalue problem is not solved directly. It is usually transformed to the standard form and solved with numerical methods, which will be discussed in Chapter 7.

If we group all eigenvalues and eigenvectors into matrices, we can write the standard and generalized eigenvalue problems in the following matrix form:

$$\begin{cases} \mathbf{A}\mathbf{X} = \mathbf{X}\Lambda \\ \mathbf{A}\mathbf{X} = \mathbf{B}\mathbf{X}\Lambda \end{cases}$$
$$\mathbf{X} = [\mathbf{x}_1, \ldots, \mathbf{x}_n] \tag{1.48}$$
$$\Lambda = \mathrm{diag}[\lambda_1, \ldots, \lambda_n]$$

For a *symmetric* matrix **A**, columns of **X** are *eigenvectors* that are both *linearly independent* and mutually *orthogonal.*

Usually, for each eigenvalue there is a corresponding eigenvector. However, it is common to have multiple eigenvalues. It can be shown that in some cases several linearly independent eigenvectors correspond to a single eigenvalue, and these eigenvectors form a subspace. If we denote the *spectrum* of a matrix **A** as λ(**A**), then there are many interesting properties of the spectrum:

$$
\begin{cases}
\mathrm{tr}(\mathbf{A}) = \sum_{j=1}^{n} \lambda_j \\
\lambda(\mathbf{A}^T) = \lambda(\mathbf{A}) \\
\lambda(\mathbf{A}^{-1}) = \dfrac{1}{\lambda_1}, \ \ldots, \ \dfrac{1}{\lambda_n} \\
\lambda(\beta\mathbf{A}) = \beta\lambda_1, \ \ldots, \ \beta\lambda_n \\
\lambda(\mathbf{A}^m) = \lambda_1^m, \ \ldots, \ \lambda_n^m
\end{cases}
\tag{1.49}
$$

If $\mathbf{A}$ is a triangular matrix, the elements of the principal diagonal are the eigenvalues of $\mathbf{A}$. The matrix $(\mathbf{A} - \kappa\mathbf{I})$ is called a *spectral shift* as it has the shifted eigenvalues of $\lambda_j - \kappa$. This is an important property in solving eigenvalue problems in computational mechanics.

1.5.3 Spectral decomposition

In a standard eigenvalue problem represented in the matrix form, $\mathbf{AX} = \mathbf{X\Lambda}$, the columns in the eigenvector matrix $\mathbf{X}$ are linearly independent. Therefore, $\mathbf{X}$ is nonsingular and its inverse, $\mathbf{X}^{-1}$, always exists. The matrix $\mathbf{A}$ can then be written as $\mathbf{A} = \mathbf{X\Lambda X}^{-1}$, which is known as *spectral decomposition* of $\mathbf{A}$. In addition, the eigenvalue matrix can be expressed as $\Lambda = \mathbf{X}^{-1}\mathbf{AX}$.

If matrix $\mathbf{A}$ is symmetric, then $\mathbf{X}$ is orthogonal, which means that $\mathbf{XX}^T = \mathbf{I}$ or $\mathbf{X}^{-1} = \mathbf{X}^T$. Therefore, the spectral decomposition of symmetric matrices is of the form $\mathbf{A} = \mathbf{X\Lambda X}^T$ and the eigenvalues are $\Lambda = \mathbf{X}^T\mathbf{AX}$.

Using spectral decomposition, we can verify properties of eigenvalue spectrum of matrix $\mathbf{A}$ discussed earlier; for example, we can determine the eigenvalues and eigenvectors of powers of a matrix $\mathbf{A}$:

$$
\mathbf{A}^k = \mathbf{A}\cdots\mathbf{A} = \left(\mathbf{X\Lambda X}^{-1}\right)\cdots\left(\mathbf{X\Lambda X}^{-1}\right) = \mathbf{X\Lambda}^k\mathbf{X}^{-1}
\tag{1.50}
$$

This means that the eigenvalues of kth power of a matrix are the kth power of the eigenvalues of the original matrix $\mathbf{A}$. The eigenvectors remain unchanged. If matrix $\mathbf{A}$ is positive definite, then the kth power of $\mathbf{A}$ can be expressed as $\mathbf{A}^k = \mathbf{X\Lambda}^k\mathbf{X}^T$.

We can show that eigenvalues of a positive definite matrix are also positive. By definition, if $\mathbf{A}$ is positive definite, for any vector $\mathbf{z}$, there should be $\mathbf{z}^T\mathbf{Az} > 0$. Since the columns of matrix $\mathbf{X}$ or the set of eigenvectors of $\mathbf{A}$ forms a basis in a vector space $\mathbf{R}^n$, the vector $\mathbf{z}$ can be expressed as a linear combination of the basis vectors:

$$
\mathbf{z} = \sum_{j=1}^{n} \alpha_j \mathbf{x}_j = \mathbf{X}\,\alpha
\tag{1.51}
$$

Then we calculate the quadratic form and enforce the positive definiteness of $\mathbf{A}$:

$$
\begin{aligned}
\mathbf{z}^T\mathbf{Az} &= \alpha^T\mathbf{X}^T\mathbf{AX}\alpha \\
&= \alpha^T\Lambda\alpha \\
&= \sum_{j=1}^{n} \lambda_j\,\alpha_j^2 > 0
\end{aligned}
\tag{1.52}
$$

We note that Equation 1.52 is valid for any vector **z**, including the cases when **z** is equal to the individual eigenvectors of matrix **A**. This implies that if **A** is positive definite, then all its eigenvalues are positive.

1.5.4 Principal minors of positive definite matrices

Let us consider an n × n square matrix $\mathbf{A} = [a_{ij}]$. The *principal minors* of **A** are principal submatrices generated along the diagonal line in the following form:

$$\mathbf{A}_{(1)} = [a_{11}];\ \mathbf{A}_{(2)} = \begin{bmatrix} a_{11} & a_{12} \\ a_{21} & a_{22} \end{bmatrix};\ \mathbf{A}_{(3)} = \begin{bmatrix} a_{11} & a_{12} & a_{13} \\ a_{21} & a_{22} & a_{23} \\ a_{31} & a_{32} & a_{33} \end{bmatrix};\ \cdots \tag{1.53}$$

It can be proved that if **A** is a positive definite matrix, then all its principal minors are also positive definite, $\det(\mathbf{A}_{(k)}) > 0$ for k = 1, …, n − 1. The proof can be accomplished by noticing that for matrix **A** to be positive definite, the quadratic form $\mathbf{z}^T\mathbf{A}\mathbf{z} > 0$ for any vector **z**. Thus, we can design the vector **z** by selecting the elements in it in such a way that the last (n − k) elements are all zero. Then $\mathbf{z}^T\mathbf{A}_{(k)}\mathbf{z} > 0$ is a quadratic form in the k variables.

In structural mechanics, if we consider matrix **A** as the global stiffness matrix of a structural system, then the above property regarding positive definiteness of principal minors has an important physical significance. A positive definite stiffness matrix corresponds to a stable system. The principal minors can be obtained successively by eliminating one row and corresponding column from the matrix **A** at a time in an ascending direction along the principal diagonal. If this operation is performed on a structural stiffness matrix, it corresponds to introducing constraints into the structural system. Since the principal minors are positive definite, it means that the successively constrained systems are also stable. This complies with the behavior of physical structural systems. If a structural system is stable, then introducing constraints into the system does not affect the stability of the system.

1.5.5 Similarity transformations

An n × n matrix $\bar{\mathbf{A}}$ is called *similar* to an n × n matrix **A** if $\bar{\mathbf{A}} = \mathbf{Q}^{-1}\mathbf{A}\mathbf{Q}$ for any nonsingular matrix **Q**. The process of transforming **A** to $\bar{\mathbf{A}}$ is called the *similarity transformation*. We can start by looking at the eigenvalues of these two matrices and the role of the similarity transformation:

$$\begin{aligned} &\mathbf{A}\mathbf{X} = \mathbf{X}\mathbf{\Lambda} \\ &\bar{\mathbf{A}}\bar{\mathbf{X}} = \bar{\mathbf{X}}\bar{\mathbf{\Lambda}} \\ &(\mathbf{Q}^{-1}\mathbf{A}\mathbf{Q})\bar{\mathbf{X}} = \bar{\mathbf{X}}\bar{\mathbf{\Lambda}} \ \Rightarrow\ \mathbf{A}(\mathbf{Q}\bar{\mathbf{X}}) = (\mathbf{Q}\bar{\mathbf{X}})\bar{\mathbf{\Lambda}} \end{aligned} \tag{1.54}$$

We observe that through similarity transformation, the eigenvalues remain unchanged, but the eigenvectors are transformed:

$$\begin{cases} \bar{\mathbf{\Lambda}} = \mathbf{\Lambda} \\ \bar{\mathbf{X}} = \mathbf{Q}^{-1}\mathbf{X} \end{cases} \tag{1.55}$$

Obviously, since the eigenvalues do not change in a similarity transformation, $\det(\bar{\mathbf{A}}) = \det(\mathbf{A})$. If $\mathbf{y} = \mathbf{A}\mathbf{x}$ and we set $\bar{\mathbf{x}} = \mathbf{Q}\mathbf{x}$ and $\bar{\mathbf{y}} = \mathbf{Q}^{-1}\mathbf{y}$, then $\bar{\mathbf{y}} = \bar{\mathbf{A}}\bar{\mathbf{x}}$.

1.5.6 Orthogonal similarity transformations

If matrix $\mathbf{Q}$ is an orthogonal matrix, $\mathbf{Q}^{-1} = \mathbf{Q}^T$, then the similarity transformation $\bar{\mathbf{A}} = \mathbf{Q}^T\mathbf{A}\mathbf{Q}$ is an *orthogonal similarity transformation*. Similarly, we can verify that through an orthogonal similarity transformation, the eigenvalues remain unchanged, but the eigenvectors are transformed:

$$\begin{cases} \bar{\Lambda} = \Lambda \\ \bar{\mathbf{X}} = \mathbf{Q}^T\mathbf{X} \end{cases} \tag{1.56}$$

There is a very important property associated with orthogonal transformation that has practical implications. For any two vectors $\mathbf{x}$ and $\mathbf{y}$, after orthogonal transformation, their inner product, or norm, remains unchanged. In a three-dimensional Euclidean space, an orthogonal transformation preserves the norm of any vector and the angle between any two vectors. This orthogonal transformation matrix is usually the rotation matrix.

For orthogonal matrices, we have $\mathbf{Q}^T\mathbf{Q} = \mathbf{I}$. This implies that $\det(\mathbf{Q}) = 1$. This relation can be verified by noting the following:

$$\det(\mathbf{Q}^T\mathbf{Q}) = [\det(\mathbf{Q})]^2 = \det(\mathbf{I}) = 1 \tag{1.57}$$

1.5.7 Gram–Schmidt orthogonalization

From our earlier discussions, we observe that sets of orthogonal vectors play a useful role in some numerical methods. A set of independent vectors $\mathbf{x}_1, \ldots, \mathbf{x}_n$ can be transformed into an orthogonal set $\mathbf{u}_1, \ldots, \mathbf{u}_n$ using the Gram–Schmidt orthogonalization method. We start by setting the first vector $\mathbf{u}_1$ equal to $\mathbf{x}_1$ and normalize it. Next we remove the component of $\mathbf{u}_1$ on $\mathbf{x}_2$ from $\mathbf{x}_2$ and normalize it and set it equal to $\mathbf{u}_2$. We continue this process and determine all the orthogonal vectors as follows:

$$\begin{cases} \bar{\mathbf{u}}_1 = \mathbf{x}_1 & \mathbf{u}_1 = \dfrac{\bar{\mathbf{u}}_1}{|\bar{\mathbf{u}}_1|} \\ \bar{\mathbf{u}}_2 = \mathbf{x}_2 - (\mathbf{x}_2 \cdot \mathbf{u}_1)\mathbf{u}_1 & \mathbf{u}_2 = \dfrac{\bar{\mathbf{u}}_2}{|\bar{\mathbf{u}}_2|} \\ \bar{\mathbf{u}}_3 = \mathbf{x}_3 - (\mathbf{x}_3 \cdot \mathbf{u}_1)\mathbf{u}_1 - (\mathbf{x}_3 \cdot \mathbf{u}_2)\mathbf{u}_2 & \mathbf{u}_3 = \dfrac{\bar{\mathbf{u}}_3}{|\bar{\mathbf{u}}_3|} \\ \quad\vdots & \quad\vdots \\ \bar{\mathbf{u}}_n = \mathbf{x}_n - \sum_{j=1}^{n-1} (\mathbf{x}_n \cdot \mathbf{u}_j)\mathbf{u}_j & \mathbf{u}_n = \dfrac{\bar{\mathbf{u}}_n}{|\bar{\mathbf{u}}_n|} \end{cases} \tag{1.58}$$

This process is known as the Gram–Schmidt orthogonalization process.

Because of the way the orthogonal vectors are computed in finite precision arithmetic, this orthogonalization process, if implemented in the current mathematical form, is not numerically stable. As the computation progresses, because of the propagation of a round-off error, the resulting vectors obtained from this process may be far from being orthogonal. In the worst case, if the original set of vectors is not perfectly independent, then the vectors generated may not be orthogonal at all. In practice, a mathematically equivalent but numerically stable version of the algorithm is adopted. This practical process is called the *modified Gram–Schmidt* process. We observe that in the original Gram–Schmidt process, the orthogonal vectors are computed in successive steps without altering the original independent vector set. In the modified Gram–Schmidt process, the basic idea is that when computing each orthogonal vector, the remaining original vectors are also modified by making them orthogonal to the vector being computed, as well as to all the previously determined orthonormal vectors. Therefore, the process proceeds as follows:

$$\begin{cases} \mathbf{u}_1 = \dfrac{\mathbf{x}_1}{|\mathbf{x}_1|} & \mathbf{x}_i^{(1)} = \mathbf{x}_i - (\mathbf{x}_i \cdot \mathbf{u}_1)\mathbf{u}_1; \quad i = 2, \ldots, n \\ \mathbf{u}_2 = \dfrac{\mathbf{x}_2^{(1)}}{|\mathbf{x}_2^{(1)}|} & \mathbf{x}_i^{(2)} = \mathbf{x}_i^{(1)} - (\mathbf{x}_i^{(1)} \cdot \mathbf{u}_2)\mathbf{u}_2; \quad i = 3, \ldots, n \\ & \vdots \\ \mathbf{u}_{n-1} = \dfrac{\mathbf{x}_{n-1}^{(n-2)}}{|\mathbf{x}_{n-1}^{(n-2)}|} & \mathbf{x}_n^{(n-1)} = \mathbf{x}_n^{(n-2)} - (\mathbf{x}_n^{(n-2)} \cdot \mathbf{u}_{n-1})\mathbf{u}_{n-1} \\ \mathbf{u}_n = \dfrac{\mathbf{x}_n^{(n-1)}}{|\mathbf{x}_n^{(n-1)}|} & \end{cases} \tag{1.59}$$

We observe that computing the vector $\mathbf{u}_j$ using the modified method does not involve the computation of summation of many inner products, so this algorithm is numerically stable. In practice, if the original set of vectors are independent, with the use of double precision, both methods yield the same results. However, when the vectors are nearly independent, the standard version of the Gram–Schmidt process fails to produce an orthogonal set of vectors, but the modified Gram–Schmidt process yields a set of nearly orthogonal vectors.

1.6 ELEMENTARY MATRICES

There are many types of elementary matrices for performing elementary operations on a matrix, including the Gaussian elementary matrices, projection matrices, reflection matrices, and rotation matrices. These elementary matrices are very important and form the building blocks for many numerical methods in the solution of a system of linear equations and eigenvalue problems.

1.6.1 Gaussian elementary matrices

In solving a system of linear equations, we often use elementary operations such as interchange of any two equations, multiplication of an equation with a constant, and addition of an equation multiplied by a constant to another equation. These operations when applied to the manipulation of matrices in the row order are called elementary row operations for matrices. Matrices that accomplish these elementary operations are called Gaussian elementary matrices.

The three kinds of elementary matrix transformations are as follows: (1) exchange of any two rows or two columns; (2) multiplication of a row or a column by a nonzero constant; and (3) addition of a row multiplied by a nonzero constant to another row or addition of a column multiplied by a nonzero constant to another column. These transformations can be accomplished by either premultiplying or postmultiplying matrix **A** with an elementary transformation matrix.

The elementary matrix for performing the exchange of any two rows or two columns can be expressed as $\mathbf{I}_{pq}$. Starting with an identity matrix, we set the diagonal terms in rows p and q to zero and introduce 1s in row p, column q and row q, and column p as follows:

$$\mathbf{I}_{pq} = \begin{bmatrix} 1 &&&&&&&&&& \\ & \ddots &&&&&&&&& \\ && 1 &&&&&&&& \\ &&& 0 && \cdots && 1 &&& \\ &&&& 1 &&&&&& \\ &&& \vdots && \ddots && \vdots &&& \\ &&&&&& 1 &&&& \\ &&& 1 && \cdots && 0 &&& \\ &&&&&&&& 1 && \\ &&&&&&&&& \ddots & \\ &&&&&&&&&& 1 \end{bmatrix} \tag{1.60}$$

Premultiplying matrix **A** by the exchange elementary matrix $\mathbf{I}_{pq}$ accomplishes the exchange of row p with row q, and postmultiplication of **A** by $\mathbf{I}_{pq}$ results in the exchange of column p with column q.

The elementary multiplication matrix $\mathbf{M}_p(\alpha)$ that performs multiplication of a row p or a column p with a nonzero constant, α, can be obtained by replacing the diagonal element at row p with α in an identity matrix:

$$\mathbf{M}_p(\alpha) = \begin{bmatrix} 1 &&&&&&&&&& \\ & \ddots &&&&&&&&& \\ && 1 &&&&&&&& \\ &&& \alpha &&&&&&& \\ &&&& 1 &&&&&& \\ &&&&& 1 &&&&& \\ &&&&&& 1 &&&& \\ &&&&&&& 1 &&& \\ &&&&&&&& 1 && \\ &&&&&&&&& \ddots & \\ &&&&&&&&&& 1 \end{bmatrix} \tag{1.61}$$

Row multiplication is accomplished by premultiplying matrix **A** with matrix $\mathbf{M}_p(\alpha)$, and column multiplication by postmultiplication of matrix **A** with the elementary multiplication matrix.

The third Gaussian elementary matrix is $\mathbf{M}_{pq}(\alpha)$. It is formed by adding a constant α to an identity matrix at row p and column q:

$$\mathbf{M}_{pq} = \begin{bmatrix} 1 & & & & & & & & & & \\ & \ddots & & & & & & & & & \\ & & 1 & & & & & & & & \\ & & & 1 & & & & & & & \\ & & & & 1 & & & & & & \\ & & & \vdots & & \ddots & & & & & \\ & & & & & & 1 & & & & \\ & & & \alpha & & \cdots & & 1 & & & \\ & & & & & & & & 1 & & \\ & & & & & & & & & \ddots & \\ & & & & & & & & & & 1 \end{bmatrix} \tag{1.62}$$

Premultiplying matrix **A** with matrix $\mathbf{M}_{pq}(\alpha)$ results in the multiplication of row q by α and its addition to row p. Postmultiplying with matrix $\mathbf{M}_{pq}^T(\alpha)$ performs the same task on the columns.

1.6.2 Elementary projection matrices

Considering two vectors **a** and **u**, we can determine the projection of vector **a** onto **u** as a vector **b** such that $\mathbf{b} = \mathbf{P}_u\mathbf{a}$, where $\mathbf{P}_u$ is called a projection matrix. Referring to Figure 1.7, we first determine a unit directional vector along **u**:

$$\bar{\mathbf{u}} = \frac{1}{\left(\mathbf{u}^T\mathbf{u}\right)^{1/2}}\mathbf{u} \tag{1.63}$$

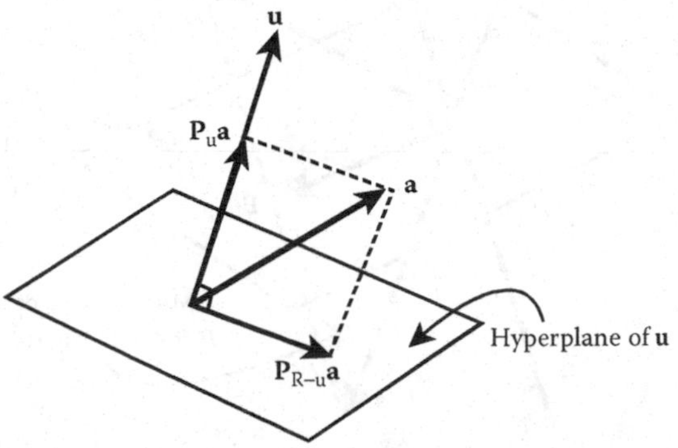

Figure 1.7 Determination of projection matrices in a three-dimensional space.

Vector **a** is projected onto the unit vector along **u**:

$$\mathbf{b}=\left(\bar{\mathbf{u}}^T\mathbf{a}\right)\bar{\mathbf{u}}=\frac{1}{\mathbf{u}^T\mathbf{u}}\ \mathbf{u}\mathbf{u}^T\mathbf{a} \tag{1.64}$$

$$\mathbf{P}_u=\bar{\mathbf{u}}\bar{\mathbf{u}}^T=\frac{1}{\mathbf{u}^T\mathbf{u}}\ \mathbf{u}\mathbf{u}^T \tag{1.65}$$

We can calculate that rank($\mathbf{P}_u$) = 1 and the projection matrix is always singular such that det($\mathbf{P}_u$) = 0. This is because the transformation from projection vector to its original vector is nonunique; different vectors can have the same projection onto a vector. Since self-projection does not change the image of the projection, the projection matrix is *idempotent*, which means that multiplying by itself does not change it $(\mathbf{P}_u)^k=\mathbf{P}_u$. In other words, once a vector has been projected, it cannot be projected again onto itself.

Referring to Figure 1.7, the projection of a vector **a** onto the hyperplane of **u**, which is a subspace orthogonal to the vector **u**, is vector **c** = **a** – **b**:

$$\mathbf{c}=\mathbf{a}-\mathbf{P}_u\mathbf{a}=(\mathbf{I}-\mathbf{P}_u)\mathbf{a}=\mathbf{P}_{R-u}\mathbf{a} \tag{1.66}$$

$$\mathbf{P}_{R-u}=\mathbf{I}-\frac{1}{\mathbf{u}^T\mathbf{u}}\ \mathbf{u}\mathbf{u}^T \tag{1.67}$$

It can be shown that this projection matrix $\mathbf{P}_{R-u}$ is singular and idempotent. It is also orthogonal to the projection matrix $\mathbf{P}_u$.

1.6.3 Reflection matrices

Referring to Figure 1.8, we consider two vectors **a** and **u** and calculate the reflection of vector **a** with respect to the hyperplane of **u**. Let the reflection vector be **d**:

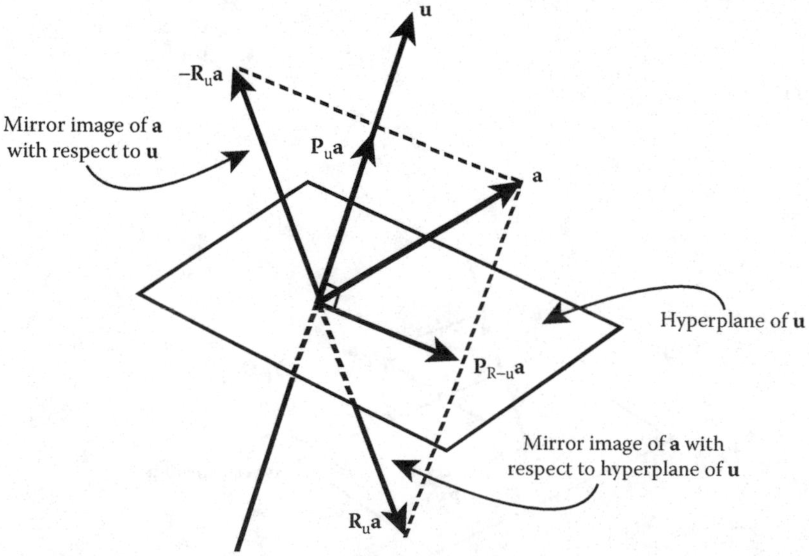

Figure 1.8 Determination of reflection matrices in a three-dimensional space.

$$\mathbf{d} = \mathbf{a} - 2\mathbf{P}_u\mathbf{a} = (\mathbf{I} - 2\mathbf{P}_u)\mathbf{a} = \mathbf{R}_u\mathbf{a} \tag{1.68}$$

$$\mathbf{R}_u = \mathbf{I} - \frac{2}{\mathbf{u}^T\mathbf{u}}\mathbf{u}\mathbf{u}^T \tag{1.69}$$

Multiplying any vector **a** by $\mathbf{R}_u$ will result in the mirror image of **a** with respect to the hyperplane of **u**. It is interesting to note that multiplying by $-\mathbf{R}_u$ will reflect vector a with respect to **u**:

$$\begin{aligned} -\mathbf{R}_u\mathbf{a} &= -(\mathbf{I} - 2\mathbf{P}_u)\mathbf{a} \\ &= \mathbf{a} - 2(\mathbf{I} - \mathbf{P}_u)\mathbf{a} \\ &= \mathbf{a} - 2\mathbf{P}_{R-u}\mathbf{a} \end{aligned} \tag{1.70}$$

It can be verified that the reflection matrix is nonsingular, $\det(\mathbf{R}_u) \neq 0$. This means that from the reflected vector we can reflect back to the original vector. This also implies that the reflection of a vector with respect to a vector **u** or its hyperplane is unique. The inverse of the reflection matrix transforms from the reflected vector to its original vector:

$$\begin{cases} \mathbf{R}_u^{-1} = \mathbf{R}_u \\ \mathbf{R}_u^2 = \mathbf{I} \end{cases} \tag{1.71}$$

The first equation indicates that forward and backward reflections are the same. The second equation tells us that reflecting twice will get us back to the original vector.

Reflection matrices are used in eigenvalue problems and in Householder decomposition. They are the most numerically stable matrices.

To illustrate the calculation of projection and reflection matrices, we consider a simple example in a three-dimensional Euclidean space as shown in Figure 1.9. We assume that we have a unit vector $\bar{\mathbf{u}}$ and vector **a** as follows:

$$\bar{\mathbf{u}} = \begin{Bmatrix} -1/\sqrt{2} \\ 1/\sqrt{2} \\ 0 \end{Bmatrix}; \quad \mathbf{a} = \begin{Bmatrix} 0 \\ 1 \\ 0 \end{Bmatrix}; \quad |\bar{\mathbf{u}}| = |\mathbf{a}| = 1 \tag{1.72}$$

The projection matrix can then be calculated as follows:

$$\mathbf{P}_u = \bar{\mathbf{u}}\bar{\mathbf{u}}^T = \begin{bmatrix} 1/2 & -1/2 & 0 \\ -1/2 & 1/2 & 0 \\ 0 & 0 & 0 \end{bmatrix}; \quad \mathbf{P}_{R-u} = \mathbf{I} - \mathbf{P}_u = \begin{bmatrix} 1/2 & 1/2 & 0 \\ 1/2 & 1/2 & 0 \\ 0 & 0 & 1 \end{bmatrix} \tag{1.73}$$

We can verify that $\det(\mathbf{P}_u) = \det(\mathbf{P}_{R-u}) = 0$, indicating that they are both singular matrices:

$$\mathbf{P}_u\mathbf{a} = \begin{Bmatrix} -1/2 \\ 1/2 \\ 0 \end{Bmatrix}; \quad \mathbf{P}_{R-u}\mathbf{a} = \begin{Bmatrix} 1/2 \\ 1/2 \\ 0 \end{Bmatrix} \tag{1.74}$$

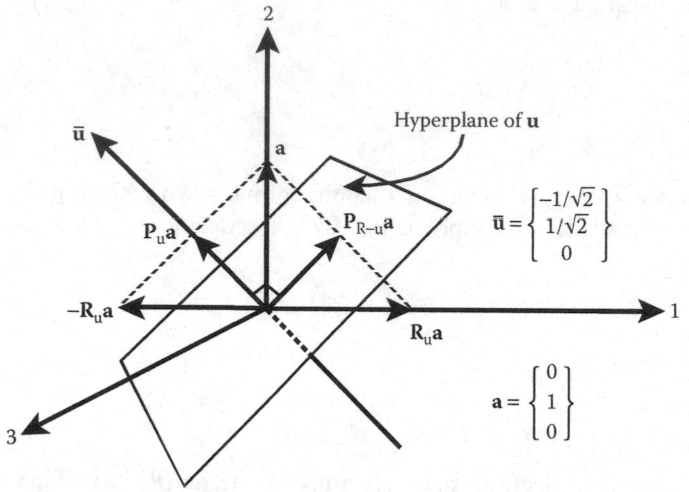

Figure 1.9 Example of calculating projection and reflection matrices.

The refection matrix and reflections of vector a are given as follows:

$$\mathbf{R}_u = \mathbf{I} - 2\mathbf{P}_u = \begin{bmatrix} 0 & 1 & 0 \\ 1 & 0 & 0 \\ 0 & 0 & 1 \end{bmatrix} \tag{1.75}$$

$$\mathbf{R}_u\mathbf{a} = \begin{Bmatrix} 1 \\ 0 \\ 0 \end{Bmatrix}; \quad -\mathbf{R}_u\mathbf{a} = \begin{Bmatrix} -1 \\ 0 \\ 0 \end{Bmatrix} \tag{1.76}$$

The above calculated reflection and projection vectors are schematically illustrated in Figure 1.9.

1.6.4 Plane rotation matrices

In a three-dimensional Euclidean space, a plane rotation matrix in the x–y plane is defined by the following relation:

$$\begin{Bmatrix} \bar{x} \\ \bar{y} \\ \bar{z} \end{Bmatrix} = \begin{bmatrix} \cos\theta & \sin\theta & 0 \\ -\sin\theta & \cos\theta & 0 \\ 0 & 0 & 1 \end{bmatrix} \begin{Bmatrix} x \\ y \\ z \end{Bmatrix} \tag{1.77}$$

In the multidimensional space, a general plane rotation matrix is of the following form:

$$\mathbf{R} = \begin{bmatrix} 1 & & & & & & & & \\ & \ddots & & & & & & & \\ & & 1 & & & & & & \\ & & & c & \cdots & s & & & \\ & & & 1 & & & & & \\ & & & \vdots & \ddots & \vdots & & & \\ & & & & 1 & & & & \\ & & & -s & \cdots & c & & & \\ & & & & & & 1 & & \\ & & & & & & & \ddots & \\ & & & & & & & & 1 \end{bmatrix} \tag{1.78}$$

$c = \cos\theta;\ s = \sin\theta$

The rotation matrix **R** is an orthogonal matrix. Because of its orthogonality, multiplying a vector by **R** does not change the length of the vector.

1.6.5 Projections and reflections on subspaces

So far, we have discussed the projection and reflection with respect to a vector **u** in a subspace S, which is one-dimensional. For $\mathbf{u} \in S$, the projection matrix $\mathbf{P}_u$ projects a vector a onto S, and $\mathbf{P}_{R-u}$ projects the vector onto $S^{\perp}$, which is the orthogonal complement of S. As for the reflection matrices, $-\mathbf{R}_u$ reflects a vector with respect to S, whereas $\mathbf{R}_u$ reflects with respect to $S^{\perp}$.

Now we can extend the discussion to more general cases such that **S** is an m-dimensional subspace $\mathbf{S} \in \mathbf{R}^m$. For example, the subspace **S** can be a plane in the three-dimensional Euclidean space. This subspace is thus spanned by columns of $n \times m$ matrix $\mathbf{U} \in \mathbf{R}^{n\times m}$, $m < n$. The projection matrix that projects a vector onto the subspace spanned by columns of **U** is defined as follows:

$$\mathbf{P}_u = \mathbf{U}\left(\mathbf{U}^T\mathbf{U}\right)^{-1}\mathbf{U}^T \tag{1.79}$$

Similarly, the projection matrix that projects a vector on the hyperspace that is the orthogonal complement of the subspace spanned by columns of **U** can be determined as follows:

$$\mathbf{P}_{R-u} = \mathbf{I} - \mathbf{U}\left(\mathbf{U}^T\mathbf{U}\right)^{-1}\mathbf{U}^T \tag{1.80}$$

The reflection matrix that reflects a vector with respect to the hyperspace that is orthogonal complement of the subspace spanned by columns of **U** is as follows:

$$\mathbf{R}_u = \mathbf{I} - 2\mathbf{U}\left(\mathbf{U}^T\mathbf{U}\right)^{-1}\mathbf{U}^T \tag{1.81}$$

For the special case of $m = 1$, **u** is a vector instead of a matrix, and we can arrive at the projection and reflection matrices obtained earlier for the one-dimensional case.

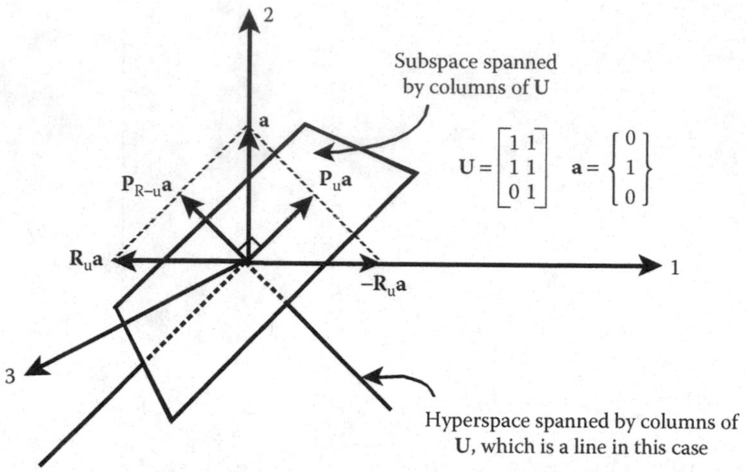

Figure 1.10 Example of calculating projection and reflection matrices on a multidimensional subspace.

Now, we work out a simple example to illustrate the calculation of these matrices in a multidimensional subspace. In a three-dimensional Euclidean space, suppose $\mathbf{U}$ consists of two vectors, the space diagonal and a vector along the line that bisects the angle between axes 1 and 2 as shown in Figure 1.10:

$$\mathbf{U} = \begin{bmatrix} 1 & 1 \\ 1 & 1 \\ 0 & 1 \end{bmatrix} \tag{1.82}$$

$$\left(\mathbf{U}^T\mathbf{U}\right)^{-1} = \begin{bmatrix} 2 & 2 \\ 2 & 3 \end{bmatrix}^{-1} = \frac{1}{2}\begin{bmatrix} 3 & -2 \\ -2 & 2 \end{bmatrix} \tag{1.83}$$

$$\mathbf{P}_u = \mathbf{U}\left(\mathbf{U}^T\mathbf{U}\right)^{-1}\mathbf{U}^T = \begin{bmatrix} 1/2 & 1/2 & 0 \\ 1/2 & 1/2 & 0 \\ 0 & 0 & 1 \end{bmatrix} \tag{1.84}$$

$$\mathbf{P}_{R-u} = \mathbf{I} - \mathbf{P}_u = \begin{bmatrix} 1/2 & -1/2 & 0 \\ -1/2 & 1/2 & 0 \\ 0 & 0 & 1 \end{bmatrix} \tag{1.85}$$

$$\mathbf{R}_u = \mathbf{I} - 2\mathbf{P}_u = \begin{bmatrix} 0 & -1 & 0 \\ -1 & 0 & 0 \\ 0 & 0 & 1 \end{bmatrix} \tag{1.86}$$

If vector a is along axis 2, that is, a = ⟨0, 1, 0⟩, then the projections and reflections of this vector are as follows:

$$\mathbf{P}_u\mathbf{a} = \begin{Bmatrix} 1/2 \\ 1/2 \\ 0 \end{Bmatrix}; \ \mathbf{P}_{R-u}\mathbf{a} = \begin{Bmatrix} -1/2 \\ 1/2 \\ 0 \end{Bmatrix}; \ \mathbf{R}_u\mathbf{a} = \begin{Bmatrix} -1 \\ 0 \\ 0 \end{Bmatrix}; \ -\mathbf{R}_u\mathbf{a} = \begin{Bmatrix} 1 \\ 0 \\ 0 \end{Bmatrix} \tag{1.87}$$

These vectors are also shown in Figure 1.10.

1.7 RAYLEIGH QUOTIENT

The *Rayleigh quotient* for an n × n symmetric matrix **A** and a real vector $\mathbf{x} \in \mathbf{R}^n$ is defined as follows:

$$\begin{aligned} \rho(\mathbf{x}) &= \frac{\mathbf{x}^T\mathbf{A}\mathbf{x}}{\mathbf{x}^T\mathbf{x}} \\ &= \left(\frac{\mathbf{x}}{|\mathbf{x}|}\right)^T \mathbf{A} \left(\frac{\mathbf{x}}{|\mathbf{x}|}\right) \end{aligned} \tag{1.88}$$

The Rayleigh quotient can be considered as a normalized quadratic form. Suppose matrix **A** represents the finite element stiffness matrix of a structural system, then we may ask the following: what is the physical interpretation of this quotient? We note that if vector **x** represents the displacement vector of the structural system, then from the equilibrium equation, we have **Ax** = **p**, where **p** is the load vector. Therefore, the numerator, $\mathbf{x}^T\mathbf{A}\mathbf{x} = \mathbf{x}^T\mathbf{p}$, is the external work that is equal to the strain energy stored in the system. This indicates that the Rayleigh quotient actually represents the strain energy stored in the system per unit displacement within a structural system. Because of this physical connection, the quantity of the Rayleigh quotient must be bounded.

1.7.1 Bounds of the Rayleigh quotient

To determine the bounds, we consider a standard eigenvalue problem:

$$\mathbf{A}\varphi_j = \lambda_j\varphi_j; \ j = 1, \ldots, n \tag{1.89}$$

Eigenvectors φ_j form the basis for the vector space, and all eigenvalues are distinct and ordered such that $\lambda_1 < \lambda_2 < \cdots < \lambda_n$. Since vector **x** is in the space spanned by the eigenvectors, we can express it as a linear combination of basis vectors:

$$\mathbf{x} = \sum_{i=1}^{n} y_i\varphi_i = \mathbf{\Phi}\mathbf{y} \tag{1.90}$$

Since **A** is a stiffness matrix that is symmetric and positive definite, its eigenvectors are orthogonal, that is, $\mathbf{\Phi}^T\mathbf{\Phi} = \mathbf{I}$. By using spectral decomposition, we can determine the numerator and the denominator of the the Rayleigh quotient:

$$\mathbf{x}^T\mathbf{A}\mathbf{x} = \mathbf{y}^T\mathbf{\Phi}^T\mathbf{A}\mathbf{\Phi}\mathbf{y} = \mathbf{y}^T\mathbf{\Lambda}\mathbf{y} = \sum_{j=1}^{n} \lambda_j y_j^2 \tag{1.91}$$

$$\mathbf{x}^T\mathbf{x} = \mathbf{y}^T\Phi^T\Phi\mathbf{y} = \mathbf{y}^T\mathbf{y} = \sum_{j=1}^{n} y_j^2 \tag{1.92}$$

$$\rho(\mathbf{x}) = \frac{\sum_{j=1}^{n}\lambda_j y_j^2}{\sum_{j=1}^{n} y_j^2} = \frac{\lambda_1\left(y_1^2 + \sum_{j=2}^{n}\frac{\lambda_j}{\lambda_1}y_j^2\right)}{\sum_{j=1}^{n} y_j^2} = \frac{\lambda_n\left(y_n^2 + \sum_{j=1}^{n-1}\frac{\lambda_j}{\lambda_n}y_j^2\right)}{\sum_{j=1}^{n} y_j^2} \tag{1.93}$$

From this equation, we can determine the lower and upper bounds for the Rayleigh quotient: Lower bound is λ_1 since $\lambda_j/\lambda_1 > 1$ for $j = 2,\ldots, n$, and upper bound is λ_n since $\lambda_j/\lambda_n < 1$ for $j = 1,\ldots, n-1$:

$$\lambda_1 < \rho(\mathbf{x}) < \lambda_n \tag{1.94}$$

If matrix **A** represents the stiffness matrix of a linear structural system discretized with finite elements and there are rigid body modes present, the smallest eigenvalue is $\lambda_1 = 0$. This corresponds to zero strain energy in the system. However, for every structural system, there is always an upper bound on the amount of strain energy that can be absorbed by the system.

The concept of the Rayleigh quotient is important for solving eigenvalue problems. An important characteristic of the Rayleigh quotient in the estimation of eigenvalues is that if vector **x** is close to an eigenvector φ_j, then $\rho(\mathbf{x})$ is second order close to λ_j:

$$\text{If } \mathbf{x} = \varphi_j + \varepsilon\,\mathbf{a}, \text{ for } \varepsilon \ll 1, \text{ then } \rho(\mathbf{x}) = \lambda_j + O(\varepsilon^2) \tag{1.95}$$

This implies that in eigenvalue problems, if we can arrive at a vector **x** that is close to an eigenvector, then we can estimate the corresponding eigenvalue with a higher degree of accuracy using the Rayleigh quotient.

1.8 NORMS OF VECTORS AND MATRICES

Norms for vectors and matrices are a measure of their size. They play an important role in measuring the accuracy and convergence in numerical solution methods.

1.8.1 Vector norms

In a three-dimensional Euclidean space, we are familiar with the length of a vector that characterizes its size. However, in a multidimensional space, we need to introduce a general concept of norm to provide a measure of the length of a vector in a broad sense.

In general, a norm is a nonnegative real function defined in a linear vector space. Let V be a linear vector space, and a vector $\mathbf{x} \in V$. If there exists a nonnegative real number $\|\mathbf{x}\|$ corresponding to vector **x**, then we call the function the norm of vector **x** if it satisfies the following three conditions:

1. $\|\mathbf{x}\| > 0$ if $\mathbf{x} \neq 0$ and $\|\mathbf{x}\| = 0$ if $\mathbf{x} = 0$

2. $\|\alpha\mathbf{x}\| = |\alpha|\,\|\mathbf{x}\|$ for any scalar quantity α
3. $\|\mathbf{x} + \mathbf{y}\| \le \|\mathbf{x}\| + \|\mathbf{y}\|$ for any vectors $\mathbf{x}, \mathbf{y} \in V$

If we denote $\|\mathbf{x}\|_m$ as the mth norm of vector $\mathbf{x}$, then we have the following norms:

$$\begin{cases} \|\mathbf{x}\|_1 = \sum_{j=1}^{n} |x_j| & \text{Sum norm} \\ \|\mathbf{x}\|_2 = \left(\sum_{j=1}^{n} x_j^2\right)^{1/2} & \text{Euclidean norm} \\ \|\mathbf{x}\|_m = \left(\sum_{j=1}^{n} x_j^m\right)^{1/m} & \text{mth norm} \\ \|\mathbf{x}\|_\infty = \max_j |x_j| & \text{Max norm} \end{cases} \tag{1.96}$$

The Euclidean norm is similar to the length of a vector in the Euclidean space. All these norms can be proven to satisfy the three conditions listed above. For example, to prove that $\|\mathbf{x}\|_1$ is a norm, we can verify the following:

1. When $\mathbf{x} \neq 0$, obviously, $\|\mathbf{x}\|_1 = \sum_{j=1}^{n} |x_j| > 0$, and when $\mathbf{x} = 0$, which means that all its components are zero, we have $|\mathbf{x}|_1 = 0$.
2. For any scalar quantity α, we have

$$\|\alpha\mathbf{x}\|_1 = \sum_{j=1}^{n} |\alpha x_j| = |\alpha| \sum_{j=1}^{n} |x_j| = |\alpha|\,\|\mathbf{x}\|_1.$$

3. For any vectors $\mathbf{x}, \mathbf{y} \in \mathbf{R}^n$, we have

$$\begin{aligned} \|\mathbf{x} + \mathbf{y}\|_1 &= \sum_{j=1}^{n} |x_j + y_j| \le \sum_{j=1}^{n} \left(|x_j| + |y_j|\right) \\ &= \sum_{j=1}^{n} |x_j| + \sum_{j=1}^{n} |y_j| = \|\mathbf{x}\|_1 + \|\mathbf{y}\|_1 \end{aligned} \tag{1.97}$$

Since $\|\mathbf{x}\|_1$ satisfies all the three conditions, it is a norm defined in the vector space $\mathbf{R}^n$. We can verify all the other norms by going through a similar procedure.

It can be proven that any norm defined in a finite-dimensional linear space is equivalent to any other norm defined similarly. In other words, if we define two norms $\|\mathbf{x}\|_a$ and $\|\mathbf{x}\|_b$ in a finite-dimensional vector space V, then there exist two independent positive constants α and β such that the following relations are satisfied:

$$\|\mathbf{x}\|_a \le \alpha \|\mathbf{x}\|_b, \quad \|\mathbf{x}\|_b \le \beta \|\mathbf{x}\|_a, \quad \forall \mathbf{x} \in V \tag{1.98}$$

Any two norms satisfying the above relations are called equivalent. It can also be proven that any vector $\mathbf{x}^{(k)}$ converges to vector $\mathbf{x}$ if and only if for any vector norm, the sequence $\{\|\mathbf{x}^{(k)} - \mathbf{x}\|\}$ converges to zero.

1.8.2 Matrix norms

An n × n matrix can be considered as an n × n–dimensional vector. Therefore, we can define matrix norms similar to vector norms. Suppose we define a nonnegative real function $\|\mathbf{A}\|$ in the set **M** consisting of all n × n square matrices. We call the function the norm of n × n matrices if for any matrices $\mathbf{A}, \mathbf{B} \in \mathbf{M}$ the following four conditions are satisfied:

1. $\|\mathbf{A}\| > 0$ if $\mathbf{A} \neq 0$, and $\|\mathbf{A}\| = 0$ if $\mathbf{A} = 0$
2. $\|\alpha\mathbf{A}\| = |\alpha|\, \|\mathbf{A}\|$ for any scalar quantity α
3. $\|\mathbf{A} + \mathbf{B}\| \leq \|\mathbf{A}\| + \|\mathbf{B}\|$
4. $\|\mathbf{A}\mathbf{B}\| \leq \|\mathbf{A}\|\, \|\mathbf{B}\|$

Some of the common matrix norms are as follows:

$$\begin{cases} \|\mathbf{A}\|_1 = \max\limits_{k} \sum\limits_{j=1}^{n} |a_{jk}| & \text{maximum absolute column sum} \\ \|\mathbf{A}\|_2 = \left[\lambda_n\left(\mathbf{A}^T\mathbf{A}\right)\right]^{1/2} & \text{spectral norm} \\ \|\mathbf{A}\|_\infty = \max\limits_{j} \sum\limits_{k=1}^{n} |a_{jk}| & \text{maximum absolute row sum} \end{cases} \tag{1.99}$$

For symmetric matrices, $\|\mathbf{A}\|_2 = \lambda_n$. In many cases, we can mix the use of matrix norms with that of vector norms because a matrix is a linear transformation of one vector to another. We need to make sure that the matrix norms are *compatible* with the vector norms. Consider an n × n square matrix **A** and a vector $\mathbf{x} \in \mathbf{R}^n$. The matrix norm $\|\mathbf{A}\|$ and vector norm $\|\mathbf{x}\|$ are compatible if the following condition is satisfied:

$$\|\mathbf{A}\mathbf{x}\| \leq \|\mathbf{A}\|\|\mathbf{x}\| \tag{1.100}$$

With the compatibility condition satisfied, we can use the maximum of vector norm as the norm for matrix **A**:

$$\|\mathbf{A}\| = \max_{|\mathbf{x}|=1} \|\mathbf{A}\mathbf{x}\| \tag{1.101}$$

The vector **x** is the set of all vectors with a unit norm. For example, for the spectral norm, we have the following:

$$\begin{aligned} \|\mathbf{A}\|_2 &= \max_{|\mathbf{x}|_2=1} \|\mathbf{A}\mathbf{x}\| \\ &= \max_{|\mathbf{x}|_2=1} \left[\mathbf{x}^T\left(\mathbf{A}^T\mathbf{A}\right)\mathbf{x}\right]^{1/2} \\ &= \max \left[\rho\left(\mathbf{A}^T\mathbf{A}\right)\right]^{1/2} \\ &= \left[\lambda_n\left(\mathbf{A}^T\mathbf{A}\right)\right]^{1/2} \end{aligned} \tag{1.102}$$

In practice, for any n × n matrix **A**, we also use the Frobenius norm, which is defined as follows:

$$\|\mathbf{A}\|_F = \left(\sum_{i=1}^{n} \sum_{j=1}^{n} a_{ij}^2 \right)^{1/2} \tag{1.103}$$

We note that for an n × n orthogonal matrix **Q**, the Frobenius norm of its multiplication with another matrix **A** is an invariant:

$$\begin{cases} \|\mathbf{A}\|_F = \|\mathbf{QA}\|_F = \|\mathbf{AQ}\|_F \\ \|\mathbf{A}\|_F = \|\mathbf{Q}^T\mathbf{AQ}\|_F \end{cases} \tag{1.104}$$

The Frobenius norm can also be computed from the following equation:

$$\|\mathbf{A}\|_F = \left[\mathrm{trace}\left(\mathbf{A}^T\mathbf{A} \right) \right]^{1/2} \tag{1.105}$$

Chapter 2

Review of methods of analysis in structural mechanics

2.1 INTRODUCTION

The finite element method, initially developed in 1960s, is the most widely used discretization method in computational mechanics. There are other methods also in use in computational continuum mechanics; the finite difference method has been in use long before the introduction of the finite element method, and meshless method has been introduced more recently. The finite element discretization approach is a general method that is used to solve linear and nonlinear problems in many fields, including structural mechanics, heat transfer, fluid dynamics, and electromagnetic field problems. In this chapter, we will briefly review the application of the finite element method in structural solid mechanics. We will first describe some of the most commonly used types of finite elements. We will follow this with a discussion of shape functions leading to the development of element stiffness matrices. The presentation in this chapter is a brief review and it is by no means comprehensive. We start by the assumption that the reader is already familiar with the finite element method and here we are just presenting a brief review.

Finite element discretization is used in the analysis of continuum solids as well as flexural structural systems. In continuum solids, we include one-, two-, and three-dimensional systems. In one-dimensional problems, we include both axial strain problems such as half-space subjected to uniform loading and axial stress problems such as bar elements commonly used in truss structures. For two-dimensional problems, we include two-dimensional plane strain and plane stress problems under the category of solids. Structural systems where flexural behavior is dominant include beams, columns, frames, plates, and shells.

2.2 A BRIEF REVIEW OF THE FINITE ELEMENT METHOD

A continuum is discretized into a finite number of elements. The elements are connected to each other at the nodes located on their boundaries. The structural system has degrees of freedom at the nodes, and there are forces and displacements at each degree of freedom, referring to a common global coordinate system, which is usually a Cartesian coordinate system. The element degrees of freedom are at the nodes that may be at the corners, along the sides, or within the element, depending on the element type. Some of the more commonly used elements are shown in Figure 2.1. With finite element discretization, the displacements

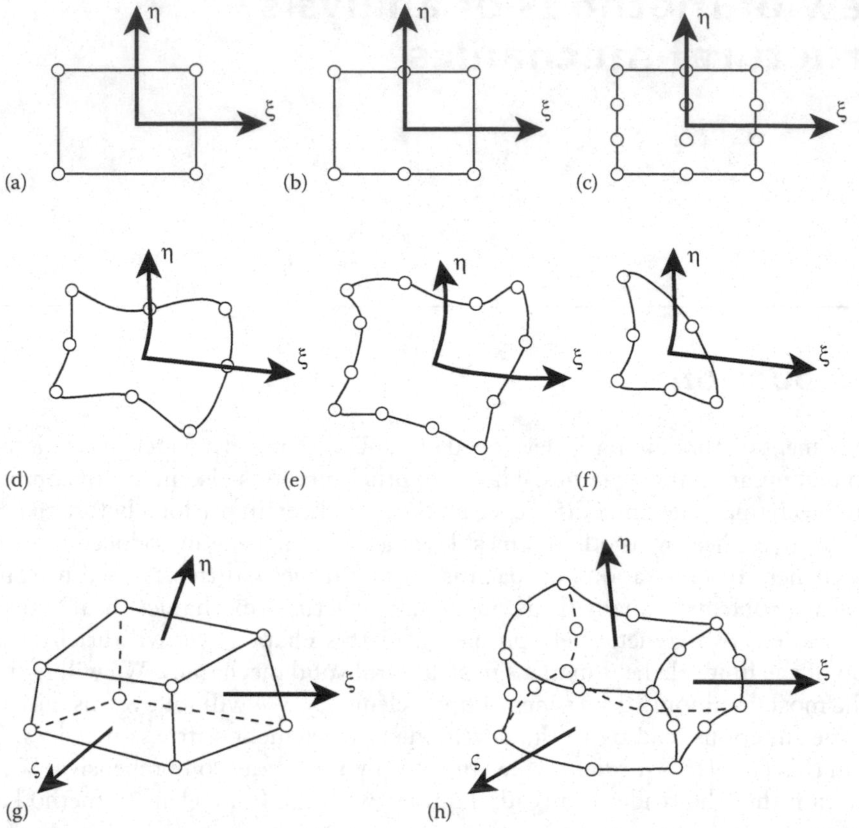

Figure 2.1 Samples of different types of Lagrangian and isoparametric plane and solid elements: (a) linear; (b) linear–quadratic; (c) quadratic–cubic; (d) quadratic plane; (e) cubic; (f) triangular quadratic; (g) linear solid; and (h) quadratic solid.

within a general three-dimensional solid element are related to nodal displacements through *shape functions* or *interpolation functions*:

$$\begin{cases} u_x = \mathbf{N}\ \mathbf{U}_x \\ u_y = \mathbf{N}\ \mathbf{U}_y \\ u_z = \mathbf{N}\ \mathbf{U}_z \end{cases} \tag{2.1}$$

In this equation, u_j, j = x, y, z, are the displacement components in the interior of each element; **N** is the matrix of shape functions; and $\mathbf{U}_j$, j = x, y, z, are components of nodal displacements. Element displacements are interpolated from the element nodal displacements using shape functions, which are usually polynomials.

For an element with n nodes, the matrix of shape functions contains n individual shape functions. The shape functions can be expressed in terms of the common Cartesian coordinates (x, y, z):

$$\mathbf{N}(x, y, z) = [N_1(x, y, z), N_2(x, y, z), \ldots, N_n(x, y, z)] \tag{2.2}$$

or in terms of natural coordinates:

$$N(\xi, \eta, \zeta) = [N_1(\xi, \eta, \zeta), N_2(\xi, \eta, \zeta), \ldots, N_n(\xi, \eta, \zeta)] \tag{2.3}$$

where (ξ, η, ζ) are natural coordinates.

For example, in a two-dimensional continuum, we consider the finite element spatial discretization using a simple linear Lagrangian element as shown in Figure 2.2. The displacement components in the x- and y-directions in the interior of the element, u_x and u_y, are related to nodal displacements as follows:

$$\begin{cases} u_x = \mathbf{N}\,\mathbf{U}_x \\ u_y = \mathbf{N}\,\mathbf{U}_y \end{cases} \tag{2.4}$$

where $\mathbf{N} = [N_1, N_2, N_3, N_4]$ is the shape function matrix. With four nodes within the element, the components of all nodal displacements are as follows:

$$\mathbf{U}_x = \begin{Bmatrix} U_{x1} \\ \vdots \\ U_{x4} \end{Bmatrix};\ \mathbf{U}_y = \begin{Bmatrix} U_{y1} \\ \vdots \\ U_{y4} \end{Bmatrix} \tag{2.5}$$

Referring to Figure 2.2, the Lagrangian shape functions for this element can be obtained as follows:

$$\begin{cases} N_1 = (b - x)(a - y)/(4ab) \\ N_2 = (b + x)(a - y)/(4ab) \\ N_3 = (b + x)(a + y)/(4ab) \\ N_4 = (b - x)(a + y)/(4ab) \end{cases} \tag{2.6}$$

We note that the size of the element is $2a \times 2b$.

It is important to note that the shape functions are defined in terms of *material* (convected) coordinates x_i in a Lagrangian formulation. If they are defined in terms of spatial

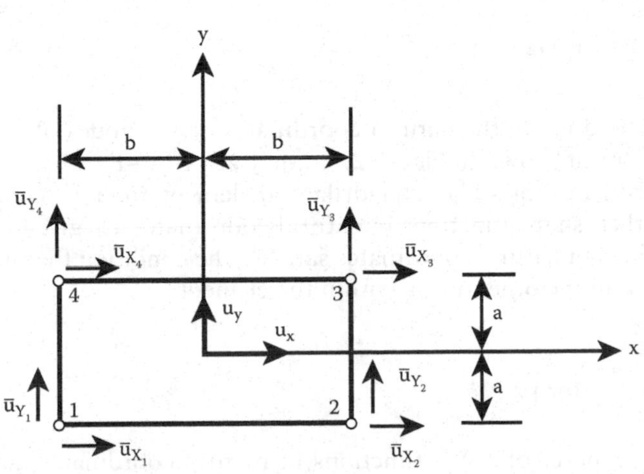

Figure 2.2 Interpolation of displacements within a linear element (Lagrangian shape functions).

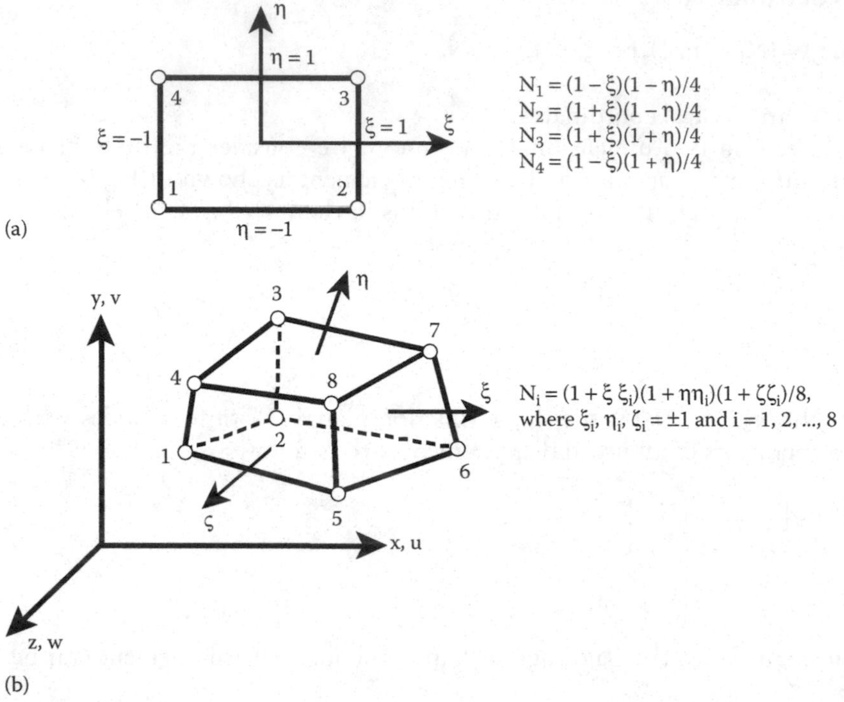

Figure 2.3 Natural coordinates and shape functions of some elements: (a) plane linear element and (b) eight-node brick element.

coordinates, they will change with the deformation of the continuum. The shape functions are usually polynomials of order n – 1 (n is the number nodes in each direction). However, in some special cases, they may be functions other than polynomials. Moreover, they are often defined in terms of *natural coordinates*, which can be related to common global coordinate system through the global coordinates of the nodes. For a general four-node linear element as shown in Figure 2.3a, in terms of natural coordinates, the shape functions are as follows:

$$N_i = \frac{1}{4}(1 + \xi_i\ \xi)(1 + \eta_i\ \eta);\ i = 1, \ldots, 4 \tag{2.7}$$

In this equation, ξ_i and η_i are the natural coordinates of the nodes of the element. In the two-dimensional element shown in Figure 2.3a, they are 1 or –1.

The natural coordinates for a plane quadrilateral element and a three-dimensional brick element as well as their shape functions in natural coordinates are given in Figure 2.3. The shape function defined in natural coordinates satisfies the condition that its value is unity at node i and is zero at all the other nodes within the element:

$$\begin{cases} N_i(\xi_i, \eta_i, \zeta_i) = 1 \\ N_i(\xi_j, \eta_j, \zeta_j) = 0, \quad \text{for } j \neq i \end{cases} \tag{2.8}$$

This fundamental property of shape functions in natural coordinates makes a systematic determination of shape functions possible.

Often, we use *isoparametric* elements to model continua in two- and three-dimensional finite element analysis. An element is isoparametric if the displacements and the global coordinates in the interior of the element are interpolated with the same set of shape functions defined in a natural coordinate system. That is, the element displacements and coordinates of any point in the interior of the element are interpolated as in the following equations:

$$\begin{cases} \mathbf{u} = \mathbf{N}\,\mathbf{U} \\ \mathbf{x} = \bar{\mathbf{N}}\,\mathbf{X} \end{cases} \tag{2.9}$$

In this equation, **X** is the vector of nodal coordinates. Obviously, for an isoparametric element, $\bar{\mathbf{N}} = \mathbf{N}$. The shape functions are polynomials of order (n – 1). In case the shape functions used for coordinate interpolation are of a higher order than those for displacement interpolation, the resulting element is referred to as superparametric. If the shape functions for coordinate interpolation are of a lower order than their counterparts in displacement interpolation, the element is called subparametric.

From a mapping point of view, we use shape functions based on the natural coordinate system of a *parent element*, which is usually a square element, to define the geometry of the *actual element* in physical coordinates. This is called the isoparametric coordinate mapping. The advantage of using an isoparametric formulation is that the element matrices corresponding to the element local degrees of freedom can be directly obtained, and the framework can be routinely extended to any type of elements, especially for those with non-rectangular or curved boundaries.

With the element displacements defined, the *linear strain–nodal displacement relations* for a particular solid element are used to obtain the element generalized strains as follows:

$$\boldsymbol{\varepsilon} = \mathbf{B}\mathbf{U} \tag{2.10}$$

In this equation, **ε** is the element strain vector and **B** is the strain–nodal displacement matrix. We note that in finite element analysis, the stress and strain tensors are represented as stress and strain vectors. For a four-node linear element shown in Figure 2.3a, we have the following:

$$\begin{Bmatrix} \varepsilon_{xx} \\ \varepsilon_{yy} \\ \varepsilon_{xy} \end{Bmatrix} = \begin{bmatrix} N_{,x} & 0 \\ 0 & N_{,y} \\ N_{,y} & N_{,x} \end{bmatrix} \begin{Bmatrix} U_x \\ U_y \end{Bmatrix} \tag{2.11}$$

With the strains and displacements defined, we can compute the internal and external virtual work for the element. The virtual work for the entire continuum can then be assembled from element contributions. By expressing all the externally applied loads as **P**, we obtain the following relationship from the virtual work principle:

$$\delta\mathbf{u}^T \left[\sum \int_V \mathbf{B}^T \boldsymbol{\sigma}\, dV - \mathbf{P} \right] = 0 \tag{2.12}$$

In this equation, the symbol $\sum$ denotes the direct assembly process that will be described in the next Section 2.3. Since the virtual displacement $\delta\mathbf{u}$ is an arbitrary but kinematically admissible displacement, we obtain the following system of equilibrium equations:

$$\sum \int_V \mathbf{B}^T \boldsymbol{\sigma}\, dV - \mathbf{P} = 0 \tag{2.13}$$

If the element material behavior is *elastic*, element stresses and strains are related by a constitutive relation:

$$\sigma = C\ \varepsilon \tag{2.14}$$

In this equation, C is the constitutive matrix, which is a constant for linear elastic materials. Substituting the element strain–displacement relation in the constitutive relation yields the stress–displacement relation:

$$\sigma = C\ B\ U \tag{2.15}$$

Combining these equations, we arrive at the system of equilibrium equations in terms of system stiffness matrix **K**:

$$K\ U = P \tag{2.16}$$

$$K = \sum \int_V B^T C\ B\ dV \tag{2.17}$$

The linear responses of the structure in terms of nodal displacements are obtained by solving the above system of equilibrium equations.

It is important to note that for other types of field problems, a similar coefficient matrix can also be derived.

Up to this point, we have assumed linear elastic material properties (Equation 2.14) and linear strain–displacement relations based on infinitesimal deformations (Equation 2.10). Elastoplastic material nonlinearity and/or large deformation geometric nonlinearity can be present in a finite element simulation. The resulting equilibrium equations will obviously be nonlinear. When there is only geometric nonlinearity, the total Lagrangian method will lead to the secant stiffness matrix:

$$K_s(U)\ U = P \tag{2.18}$$

The total Lagrangian with the secant stiffness matrix is not used very often in computational simulations. An incremental equilibrium equation with a tangent stiffness matrix is most commonly used in nonlinear finite element simulations. In the presence of geometric and material nonlinearity, the updated Lagrangian method is used to arrive at the tangent stiffness matrix:

$$K_t(U)\ \Delta U = \Delta P \tag{2.19}$$

The detailed formulation of nonlinear finite element problems is presented in another book by the same authors.*

2.3 DIRECT STIFFNESS ASSEMBLY

In Equation 2.12 we used the symbol $\sum$ to denote the direct assembly process. One of the key procedures in any finite element analysis is to assemble the stiffness matrix for the structural system from element stiffness matrices. We will describe later a number of specific numerical tasks that are performed to arrive at the stiffness matrix for the structural system. It is important to point out that those numerical tasks are based on satisfying the two conditions of equilibrium and compatibility at the structural system's degrees of freedom. These two

* J. Ghaboussi, D.A. Pecknold and S.X. Wu (in press). *Nonlinear Computational Solid Mechanics*, CRC Press, Boca Raton, FL.

conditions relate the forces and displacements at element degrees of freedom to those at the structural system nodal degrees of freedom. They can be expressed in matrix form as follows:

$$\begin{cases} \mathbf{u} = \mathbf{A}\,\mathbf{U} \\ \mathbf{P} = \mathbf{A}^T\,\mathbf{p} \end{cases} \tag{2.20}$$

The vectors **P** and **U** contain the forces and displacements at the structural degrees of freedom, and **p** and **u** contain the forces and displacements at the degrees of freedom of all the elements. The element and structural stiffness matrices relation for the force and displacement vectors are as follows:

$$\begin{cases} \bar{\mathbf{k}}\,\mathbf{u} = \mathbf{p} \\ \mathbf{K}\,\mathbf{U} = \mathbf{P} \end{cases} \tag{2.21}$$

Combining the three relations of equilibrium, compatibility, and element force–displacement relation results in the following structural stiffness matrix:

$$\mathbf{P} = \left(\mathbf{A}^T\,\bar{\mathbf{k}}\,\mathbf{A}\right)\mathbf{U} \tag{2.22}$$

$$\mathbf{K} = \mathbf{A}^T\,\bar{\mathbf{k}}\,\mathbf{A} \tag{2.23}$$

To illustrate the structural assembly process, we develop the structural stiffness matrix for the continuous beam shown in Figure 2.4. Assume that EI = 10,000. The structural degrees of freedom and member degrees of freedom are also illustrated in the figure. In this case, the member stiffness matrix can be calculated using the following equations:

$$\begin{Bmatrix} p_1 \\ p_2 \end{Bmatrix} = \frac{2EI}{L}\begin{bmatrix} 2 & 1 \\ 1 & 2 \end{bmatrix}\begin{Bmatrix} u_1 \\ u_2 \end{Bmatrix} \tag{2.24}$$

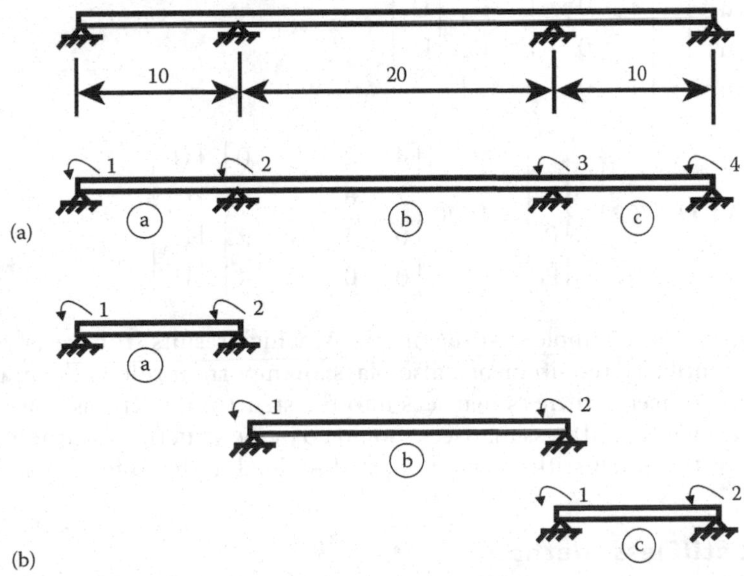

Figure 2.4 Determination of structural stiffness matrix for the continuous beam: (a) structural degrees of freedom and (b) member degrees of freedom.

$$\mathbf{k}^{(a)} = \mathbf{k}^{(c)} = 2000\begin{bmatrix}2 & 1\\ 1 & 2\end{bmatrix};\ \mathbf{k}^{(b)} = 1000\begin{bmatrix}2 & 1\\ 1 & 2\end{bmatrix} \tag{2.25}$$

We can combine the element stiffness matrices into a single 6 × 6 stiffness matrix that relates the total number of element force and displacement degrees of freedom without connecting them to each other:

$$\mathbf{p} = \bar{\mathbf{k}}\ \mathbf{u};\quad \begin{Bmatrix} p_1^{(a)}\\ p_2^{(a)}\\ p_1^{(b)}\\ p_2^{(b)}\\ p_1^{(c)}\\ p_2^{(c)} \end{Bmatrix} = 1000\begin{bmatrix} 4 & 2 & & & & \\ 2 & 4 & & & & \\ & & 2 & 1 & & \\ & & 1 & 2 & & \\ & & & & 4 & 2\\ & & & & 2 & 4 \end{bmatrix}\begin{Bmatrix} u_1^{(a)}\\ u_2^{(a)}\\ u_1^{(b)}\\ u_2^{(b)}\\ u_1^{(c)}\\ u_2^{(c)} \end{Bmatrix} \tag{2.26}$$

The equilibrium and compatibility matrices are given as follows:

$$\mathbf{P} = \mathbf{A}^T\ \mathbf{p};\quad \begin{Bmatrix} P_1\\ P_2\\ P_3\\ P_4 \end{Bmatrix} = \begin{bmatrix} 1 & 0 & 0 & 0 & 0 & 0\\ 0 & 1 & 1 & 0 & 0 & 0\\ 0 & 0 & 0 & 1 & 1 & 0\\ 0 & 0 & 0 & 0 & 0 & 1 \end{bmatrix}\begin{Bmatrix} p_1^{(a)}\\ p_2^{(a)}\\ p_1^{(b)}\\ p_2^{(b)}\\ p_1^{(c)}\\ p_2^{(c)} \end{Bmatrix} \tag{2.27}$$

$$\mathbf{u} = \mathbf{A}\ \mathbf{U};\quad \begin{Bmatrix} u_1^{(a)}\\ u_2^{(a)}\\ u_1^{(b)}\\ u_2^{(b)}\\ u_1^{(c)}\\ u_2^{(c)} \end{Bmatrix} = \begin{bmatrix} 1 & 0 & 0 & 0\\ 0 & 1 & 0 & 0\\ 0 & 1 & 0 & 0\\ 0 & 0 & 1 & 0\\ 0 & 0 & 1 & 0\\ 0 & 0 & 0 & 1 \end{bmatrix}\begin{Bmatrix} U_1\\ U_2\\ U_3\\ U_4 \end{Bmatrix} \tag{2.28}$$

$$\mathbf{P} = \left(\mathbf{A}^T\ \bar{\mathbf{k}}\ \mathbf{A}\right)\ \mathbf{U} = \mathbf{K}\ \mathbf{U};\quad \begin{Bmatrix} P_1\\ P_2\\ P_3\\ P_4 \end{Bmatrix} = 1000\begin{bmatrix} 4 & 2 & 0 & 0\\ 2 & 6 & 1 & 0\\ 0 & 1 & 6 & 2\\ 0 & 0 & 2 & 4 \end{bmatrix}\begin{Bmatrix} U_1\\ U_2\\ U_3\\ U_4 \end{Bmatrix} \tag{2.29}$$

It can be seen from this example that the matrix **A**, which results from satisfying the equilibrium and compatibility requirements, also plays an important role in the practical aspect of assembling the element stiffness matrices into the structural stiffness matrix. Matrix **A** directs where the element stiffness matrices should go in the structural stiffness matrix. This forms the basis of the direct stiffness method, as described in the following section.

2.3.1 Direct stiffness method

As a result of satisfying the equilibrium and compatibility requirements, a simple rule develops for directly assembling the structural stiffness matrix from the member stiffness matrices. This assembly process is known as direct stiffness method. In this method, the member stiffness matrices are first formed and transformed to a common global coordinate system. The correspondence between the member degrees of freedom and the structural degrees of freedom is established by *destination arrays* (DAs) for each element. With the aid of the DAs, the terms of member stiffness matrices are directly assembled into the structural stiffness matrix. DAs for the example in Figure 2.4 are given as follows:

$$DA_{(a)} = \begin{Bmatrix} 1 \\ 2 \end{Bmatrix}; \; DA_{(b)} = \begin{Bmatrix} 2 \\ 3 \end{Bmatrix}; \; DA_{(c)} = \begin{Bmatrix} 3 \\ 4 \end{Bmatrix} \tag{2.30}$$

These DAs direct the terms of the element stiffness matrices to specific rows and columns of the structural stiffness matrix.

In any finite element analysis, using the direct stiffness method, we typically go through the following computational steps:

1. Form element stiffness matrices and load vectors for all members.
2. In some cases, the element degrees of freedom are in local coordinate system. Element stiffness matrices and load vectors need to be rotated to coincide with the structural degrees of freedom in global coordinate system.
3. Form the DAs for all the elements.
4. Assemble the structural stiffness matrix and the structural load vector.
5. Impose the necessary boundary conditions.
6. Solve the system of equations to compute the nodal displacements. This is the computationally most intensive step.
7. Determine the element displacements in local coordinate systems from structural nodal displacement vector.
8. Compute the strains and stresses within each element.

2.4 INTERNAL RESISTING FORCE VECTOR

In finite element analysis, specifically in nonlinear problems, we often need to calculate the internal resisting force vector of a structural system in response to an imposed displacement vector. The simplest way of looking at the internal resisting force vector is that it is the result of multiplication of the stiffness matrix by the displacement vector. However, in most situations, this is not possible. In those cases, the internal resisting force vector is computed by direct assembly of element contributions. For a structural displacement vector $\mathbf{x}$, the internal resisting force vector is $\mathbf{I}(\mathbf{x})$:

$$\mathbf{I}(\mathbf{x}) = \mathbf{K}\,\mathbf{x} = \sum \int_V \mathbf{B}\,\sigma\,dv \tag{2.31}$$

Later on, in our discussion of semi-iterative methods, such as the conjugate gradient method in Chapter 5, we need only to compute the internal resisting force vector of a system rather than forming the global stiffness matrix of the system. Algorithms that do not require the explicit formulation of the stiffness matrix are well suited for solving very large problems.

To illustrate the calculation of the internal resisting force vector of a system, we consider an example of a simple one-dimensional structure acted on by a single horizontal load at the right end, as shown in Figure 2.5. The structure is modeled with four bar elements.

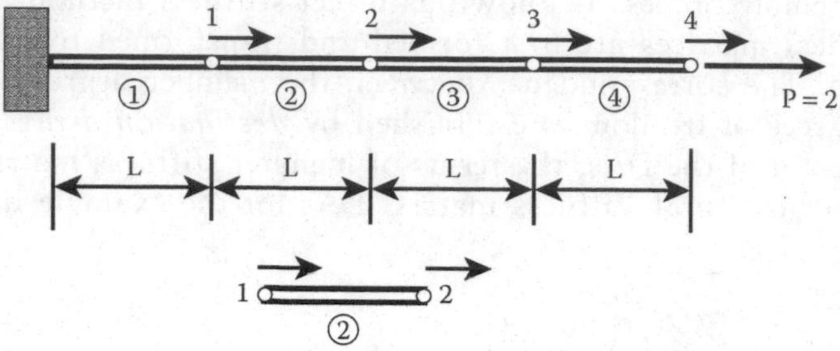

Figure 2.5 Determination of internal resisting force vector for a simple structure.

Assume that the element properties are the following: EA = 10, L = 1, and the load P = 2. The strain–displacement matrix and element stiffness matrices are as follows:

$$\mathbf{B} = \frac{1}{L}\begin{bmatrix}-1 & 1\end{bmatrix} \tag{2.32}$$

$$\mathbf{k}_{(j)} = \frac{EA}{L}\begin{bmatrix}1 & -1\\ -1 & 1\end{bmatrix} = 10\begin{bmatrix}1 & -1\\ -1 & 1\end{bmatrix} \tag{2.33}$$

After assembling the structural stiffness matrix, we can apply the load and solve the equilibrium equation to determine the structural displacement vector. We said earlier that one way of determining the internal resisting force vector is to multiply the stiffness matrix and the displacement vector. This path is shown in the following equation:

$$10\begin{bmatrix}2 & -1 & & \\ -1 & 2 & -1 & \\ & -1 & 2 & -1\\ & & -1 & 1\end{bmatrix}\begin{Bmatrix}u_1\\ u_2\\ u_3\\ u_4\end{Bmatrix} = \begin{Bmatrix}0\\ 0\\ 0\\ 2\end{Bmatrix} \Rightarrow \mathbf{u} = \begin{Bmatrix}0.2\\ 0.4\\ 0.6\\ 0.8\end{Bmatrix} \tag{2.34}$$

$$\mathbf{I} = 10\begin{bmatrix}2 & -1 & & \\ -1 & 2 & -1 & \\ & -1 & 2 & -1\\ & & -1 & 1\end{bmatrix}\begin{Bmatrix}0.2\\ 0.4\\ 0.6\\ 0.8\end{Bmatrix} = \begin{Bmatrix}0\\ 0\\ 0\\ 2\end{Bmatrix} \tag{2.35}$$

In this case, the internal resisting force vector is equal to the external force vector.

Now we compute the internal resisting force vector through direct assembly of element contributions:

$$\begin{aligned}\mathbf{I} &= \sum \int_V \mathbf{B}^T \sigma \, dA \, dx\\ &= \sum \int_0^{L=1} \mathbf{B}^T p \, dx\end{aligned} \tag{2.36}$$

Since all the four elements are the same and have the same axial stress, the contribution from each element is the same. Using the DAs, we can directly assemble the internal resisting force vector:

$$\int_0^1 B^T p \, dx = \begin{bmatrix} 1 \\ -1 \end{bmatrix} \times 2 = \begin{Bmatrix} 2 \\ -2 \end{Bmatrix} \tag{2.37}$$

$$DA_{(1)} = \begin{Bmatrix} - \\ 1 \end{Bmatrix}; DA_{(2)} = \begin{Bmatrix} 1 \\ 2 \end{Bmatrix}; DA_{(3)} = \begin{Bmatrix} 2 \\ 3 \end{Bmatrix}; DA_{(4)} = \begin{Bmatrix} 3 \\ 4 \end{Bmatrix} \tag{2.38}$$

$$I = \begin{Bmatrix} 2-2 \\ 2-2 \\ 2-2 \\ 2 \end{Bmatrix} = \begin{Bmatrix} 0 \\ 0 \\ 0 \\ 2 \end{Bmatrix} \tag{2.39}$$

This is exactly what we obtained from direct multiplication of stiffness matrix with the displacement vector. As we will see in later chapters, for example in Chapter 5, direct assembly of the internal resisting force vector from the element contribution is often the most practical method. In most nonlinear problems, the first method of directly computing the internal resisting force vector from the structural stiffness matrix is not possible.

2.5 MAJOR NUMERICAL TASKS IN COMPUTATIONAL MECHANICS

There are three major computational tasks in most computational mechanics problems that we will discuss in the next chapters.

Solving a system of equations to determine the structural displacement vector is the most time-consuming part of most computational simulations with the finite element method. If the structural system is linear, this task is performed once. In nonlinear problems with incremental and/or iterative formulation, the system of equations has to be solved many times.

In some cases, we have to determine some of the eigenvalues and eigenvectors of the structural stiffness matrix. This is another major task that is computationally intensive. Most often we perform eigenanalysis in the context of structural dynamics. Eigenanalysis is also performed in structural stability problems.

The third major computational task involves numerical integration in time of the linear or nonlinear dynamic equation of motion of structural systems. There are many methods of performing this task. There are important issues of accuracy and numerical stability associated with each of these methods.

These are the important issues in any computational mechanics problem that we will discuss in detail in the remaining chapters of this book.

MATHEMATICAL TASKS IN COMPUTATIONAL MECHANICS

Chapter 3

Solution of systems of linear equations

3.1 INTRODUCTION

In the finite element analysis of linear systems, we obtain the system of equilibrium equations in the form of

$$KU = P \tag{3.1}$$

where:

K is the system stiffness matrix
P is the global load vector
U is the unknown displacement

The stiffness matrix is usually symmetric, and positive definite. In nonlinear finite element analysis, at the nth load increment, we have the following incremental form of system of equilibrium equations:

$$K_t \Delta U_n = P - I_{n-1} \tag{3.2}$$

In this equation, K_t is the tangential stiffness matrix, P is the load vector, and I_{n-1} is the internal resisting force vector at the end of the previous increment. These systems of linear equations are usually written in the following generic form:

$$Ax = b \tag{3.3}$$

The displacement (solution) vector **x** is computed by solving the system of linear equations. In general, there are three classes of methods for solving the system of linear equations:

1. Direct methods
2. Iterative methods
3. Semi-iterative methods

A direct method solves the system of equations to arrive at the solution vector in a finite number of computational operations. An iterative method approaches the solution vector of the system in a number of iterations. If the iterative method is convergent, then as the number of iterations increases, the solution vector will converge to the exact solution. However, the number of iterations to reach convergence cannot be determined in advance. A semi-iterative method is a direct method in exact mathematical operations, but it is implemented as an iterative method in practice. Because of their unique numerical characteristics, iterative and semi-iterative methods have been increasingly used in solving very large systems, often on massively parallel, distributed or vector computers. Formation of the actual structural system stiffness matrix for very large systems may not be efficient and practical.

In this chapter, we will discuss some commonly used direct solution methods, which include Gauss-elimination, Cholesky, Givens, and Householder methods. The Gauss-elimination method is the most commonly used direct method in solving a system of linear equations. As the name implies, the Gauss-elimination method and its variant for symmetric systems, the Cholesky method, use Gaussian elementary matrices for performing triangular decomposition of the original coefficient matrix. Givens and Householder methods are based on performing orthogonal decomposition of the coefficient matrix using plane rotation matrix and reflection matrix, respectively. For iterative methods, which will be discussed in Chapter 4, we will discuss the RF method, Gauss–Seidel, Jacobi, and steepest descent methods. In semi-iterative methods, we will discuss the conjugate gradient method and the preconditioned conjugate gradient method for linear and nonlinear problems. With each solution method, we will discuss the basic concept of the method, the mathematical algorithm, and the numerical algorithm (procedure) that lead to the computer implementation of the algorithm. In many cases, we will illustrate the performance of the algorithm by solving some representative problems in computational mechanics.

From a purely mathematical point of view, it seems obvious that for a system described in Equation 3.3, we can directly compute the inverse of the coefficient matrix **A** and then obtain the displacement vector **x**:

$$\mathbf{x} = \mathbf{A}^{-1}\mathbf{b} \tag{3.4}$$

This process appears straightforward and simple and involves only computing the inverse of matrix **A** and performing a matrix–vector multiplication operation. Although the matrix–vector operation can be easily carried out, it is not difficult to see that this procedure works only for a system of small number of equations, which corresponds to systems with a few degrees of freedom. For larger matrices, computing the direct inverse of a matrix is computationally expensive. For a system with a large number of degrees of freedom, which we often encounter in engineering problems, it is practically implausible to compute the direct inverse of a full matrix because of the prohibitive computational expenses and the length of time required to complete the task. This suggests that in solving the system of linear equations in Equation 3.3 within a reasonable time frame, the computing of the inverse of the matrix should be avoided. To get around the problem, a practical procedure is to decompose the coefficient matrix **A** as a product of two matrices with the expectation that the solution process can be carried out much faster with the decomposed matrices:

$$\mathbf{A} = \mathbf{L}\mathbf{U} \tag{3.5}$$

With this decomposition of the coefficient matrix, the original system of linear equations becomes

$$\mathbf{L}\mathbf{U}\mathbf{x} = \mathbf{b} \tag{3.6}$$

This decomposed system can then be solved by calculating first the inverse of matrix **L**. Subsequently the solution vector **x** can be obtained by solving the following system:

$$\mathbf{U}\mathbf{x} = \mathbf{L}^{-1}\mathbf{b} \tag{3.7}$$

This is the basic idea behind most of the direct solution methods—decomposing the coefficient matrix.

Obviously, there are an infinite number of ways to decompose the coefficient matrix **A**. However, in general, any decomposition scheme should satisfy the following two criteria: (1) the inverse of matrix **L** must be inexpensive to compute and (2) the solution of the system of equations with **U** as the coefficient matrix should also be inexpensive. To satisfy the first criterion, matrix **L** should be a special matrix. The possibilities are that it is the product of either Gaussian elementary matrices or orthogonal matrices as the inverse of these matrices can be readily obtained. To satisfy the second criterion, the **U** matrix should at most be a triangular matrix, which can be solved easily. With the constraints of these two criteria, which are very reasonable, the possibilities for decomposing the coefficient matrix **A** are reduced to these two operations. In the following, we will discuss the Gauss-elimination method, which is the most widely used method, using these basic ideas for solving system of linear equations.

3.2 GAUSS-ELIMINATION METHOD

The Gauss-elimination method is a direct method for solving a system of linear equations. The method basically consists of two operations. First, the coefficient matrix is reduced to a triangular matrix, often an upper triangular matrix, through a series of elementary row operations: multiplication of a row with a constant and its addition to other rows of the matrix. This step is called *triangular decomposition* and it is usually accompanied by vector reduction, which determines the effect of the row operations on the right-hand side of the equation, the vector of the known loads. The solution of the system of equations with an upper triangular matrix can be readily obtained. This step is called *back-substitution*, which means that the system of linear equations with an upper triangular matrix is solved by starting with the last equation first and proceeding backward until the first equation is solved.

In the triangular decomposition process, the coefficient matrix **A** is decomposed into the product of matrices **L** and **U**:

$$\mathbf{A} = \mathbf{LU} \tag{3.8}$$

where:
- **L** is a lower triangular matrix
- **U** is an upper triangular matrix

The original system is transformed into the following form:

$$(\mathbf{LU})\mathbf{x} = \mathbf{b} \tag{3.9}$$

For vector reduction, we solve for vector **y** from the system of linear equations with lower triangular matrix:

$$\mathbf{Ly} = \mathbf{b} \tag{3.10}$$

Next, through back-substitution, we solve for the displacement vector **x** from the following equation with the upper triangular matrix:

$$\mathbf{Ux} = \mathbf{y} \tag{3.11}$$

To illustrate the basic steps involved in the solution of a system of linear equations with the Gauss-elimination method, we consider a simple one-dimensional structural example. The example structure consists of a number of one-dimensional bar elements, assembled as shown in Figure 3.1, and subjected to an end force P = 1. The structure has four degrees of freedom, the axial displacements at the four nodes. We assume that the member properties are EA = 2 and L = 1. We can form the structural stiffness matrix and obtain the system of equilibrium equations:

$$\begin{bmatrix} 3 & -2 & -1 & 0 \\ -2 & 5 & -2 & -1 \\ -1 & -2 & 5 & -2 \\ 0 & -1 & -2 & 5 \end{bmatrix} \begin{Bmatrix} x_1 \\ x_2 \\ x_3 \\ x_4 \end{Bmatrix} = \begin{Bmatrix} 1 \\ 0 \\ 0 \\ 0 \end{Bmatrix} \tag{3.12}$$

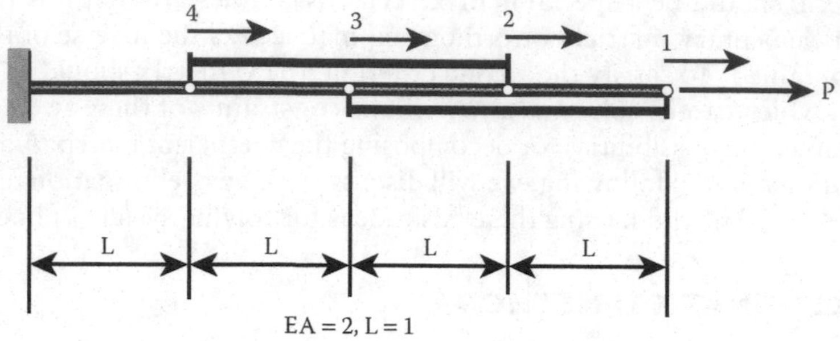

Figure 3.1 Example structure modeled with one-dimensional bar elements.

To illustrate the transformation of the coefficient matrix and the load vector, we denote the original matrix as $\mathbf{A}^{(1)}$ and the original load vector as $\mathbf{b}^{(1)}$. Now we carry out elementary row operations on the original system of equations to reduce the terms in successive columns below diagonal to zero.

To reduce the terms below the diagonal of the first column to zero, we perform row operations by multiplying the first row with a constant and then adding it to the following row:

$$\begin{cases} \text{Row } 2-\left(\dfrac{-2}{3}\right)\text{row } 1 \\ \text{Row } 3-\left(\dfrac{-1}{3}\right)\text{row } 1 \\ \text{Row } 4-\left(\dfrac{0}{3}\right)\text{row } 1 \end{cases} \tag{3.13}$$

These row operations reduce the original system to the following:

$$\begin{bmatrix} 3 & -2 & -1 & 0 \\ 0 & \dfrac{11}{3} & \dfrac{-8}{3} & -1 \\ 0 & \dfrac{-8}{3} & \dfrac{14}{3} & -2 \\ 0 & -1 & -2 & 5 \end{bmatrix} \begin{Bmatrix} x_1 \\ x_2 \\ x_3 \\ x_4 \end{Bmatrix} = \begin{Bmatrix} 1 \\ \dfrac{2}{3} \\ \dfrac{1}{3} \\ 0 \end{Bmatrix} \Rightarrow \mathbf{A}^{(2)}\mathbf{x} = \mathbf{b}^{(2)} \tag{3.14}$$

The effect of these row operations is equivalent to premultiplying the original matrix equation by a Gaussian elementary matrix. That is, we can obtain the above system of equations by carrying out the following computation:

$$\mathbf{L}_{(1)}\mathbf{A}^{(1)}\mathbf{x} = \mathbf{L}_{(1)}\mathbf{b}^{(1)} \tag{3.15}$$

As has been discussed in Chapter 1, this Gaussian matrix is a unit lower triangular matrix and has nonzero off-diagonal terms only in the first column:

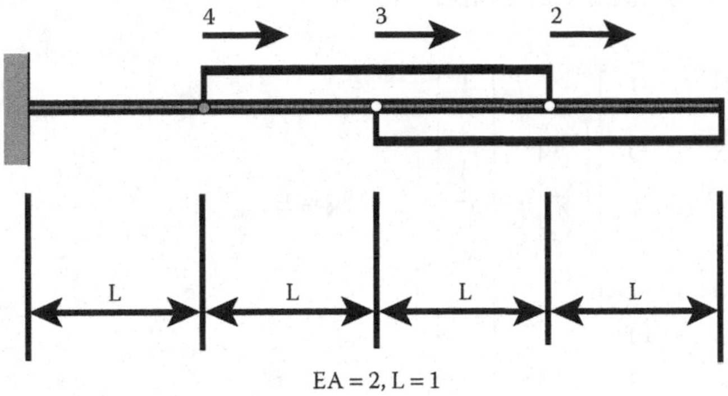

Figure 3.2 Substructure after condensing out the degree of freedom 1.

$$\mathbf{L}_{(1)} = \begin{bmatrix} 1 & & & \\ \frac{2}{3} & 1 & & \\ \frac{1}{3} & 0 & 1 & \\ 0 & 0 & 0 & 1 \end{bmatrix} \tag{3.16}$$

If we eliminate the first row and column from the matrix $\mathbf{A}^{(2)}$, the remaining matrix is still symmetric. This is because by reducing the terms below the diagonal of the first column to zero, we have in fact condensed out the degree of freedom (1) from the original structural system. The symmetric submatrix is the stiffness matrix corresponding to the condensed structure as shown in Figure 3.2. The stiffness matrix for this substructure can be determined as follows:

$$\mathbf{K}^{(2)} = \begin{bmatrix} \frac{11}{3} & \frac{-8}{3} & -1 \\ \frac{-8}{3} & \frac{14}{3} & -2 \\ -1 & -2 & 5 \end{bmatrix} \tag{3.17}$$

Because of the conservation of symmetry after performing Gauss elimination, in practice, we can work on half of the stiffness matrix.

Similarly, the following row operations reduce the terms below the diagonal of the second column to zero:

$$\begin{cases} \text{row } 3 - \left(\dfrac{-8/3}{11/3}\right) \text{row } 2 \\ \text{row } 4 - \left(\dfrac{-1}{11/3}\right) \text{row } 2 \end{cases} \tag{3.18}$$

The resulting system of linear equations becomes

$$\begin{bmatrix} 3 & -2 & -1 & 0 \\ 0 & \frac{11}{3} & \frac{-8}{3} & -1 \\ 0 & 0 & \frac{30}{11} & \frac{-30}{11} \\ 0 & 0 & \frac{-30}{11} & \frac{52}{11} \end{bmatrix} \begin{Bmatrix} x_1 \\ x_2 \\ x_3 \\ x_4 \end{Bmatrix} = \begin{Bmatrix} 1 \\ \frac{2}{3} \\ \frac{27}{33} \\ \frac{6}{33} \end{Bmatrix} \Rightarrow \mathbf{A}^{(3)}\mathbf{x} = \mathbf{b}^{(3)} \tag{3.19}$$

The effect of these row operations is again equivalent to premultiplying the matrix equation by a Gaussian elementary matrix with nonzero off-diagonal terms in the second column:

$$\mathbf{L}_{(2)}\mathbf{A}^{(2)}\mathbf{x} = \mathbf{L}_{(2)}\mathbf{b}^{(2)} \tag{3.20}$$

$$\mathbf{L}_{(2)} = \begin{bmatrix} 1 & & & \\ 0 & 1 & & \\ 0 & \frac{8}{11} & 1 & \\ 0 & \frac{3}{11} & 0 & 1 \end{bmatrix} \tag{3.21}$$

Similarly, the effect of this operation is also equivalent to performing static condensation on the structure by condensing out the degree of freedom (2).

We perform the following row operation to reduce the terms below the diagonal of the third column in matrix to zero:

$$\text{row } 4 - \left(\frac{-30/11}{30/11}\right)\text{row } 3 \tag{3.22}$$

The system of linear equations after the above row operation becomes

$$\begin{bmatrix} 3 & -2 & -1 & 0 \\ 0 & \frac{11}{3} & \frac{-8}{3} & -1 \\ 0 & 0 & \frac{30}{11} & \frac{-30}{11} \\ 0 & 0 & 0 & 2 \end{bmatrix} \begin{Bmatrix} x_1 \\ x_2 \\ x_3 \\ x_4 \end{Bmatrix} = \begin{Bmatrix} 1 \\ \frac{2}{3} \\ \frac{27}{33} \\ 1 \end{Bmatrix} \Rightarrow \mathbf{A}^{(4)}\mathbf{x} = \mathbf{b}^{(4)} \tag{3.23}$$

The effect of these row operations is equivalent to premultiplying the matrix equation by a Gaussian elementary matrix with nonzero off-diagonal terms in the third column:

$$\mathbf{L}_{(3)}\mathbf{A}^{(3)}\mathbf{x} = \mathbf{L}_{(3)}\mathbf{b}^{(3)} \tag{3.24}$$

$$\mathbf{L}_{(3)} = \begin{bmatrix} 1 & & & \\ 0 & 1 & & \\ 0 & 0 & 1 & \\ 0 & 0 & 1 & 1 \end{bmatrix} \tag{3.25}$$

As we have seen in this example, the system of equations with the triangular matrix is obtained after a series of row operations, in order to reduce all the terms below the diagonal to zero. The same result can also be achieved by premultiplying the matrix equation by a series of Gaussian elementary matrices:

$$(\mathbf{L}_{(3)}\mathbf{L}_{(2)}\mathbf{L}_{(1)})\mathbf{A}\mathbf{x} = (\mathbf{L}_{(3)}\mathbf{L}_{(2)}\mathbf{L}_{(1)})\mathbf{b} \tag{3.26}$$

$$\mathbf{U} = (\mathbf{L}_{(3)}\mathbf{L}_{(2)}\mathbf{L}_{(1)})\mathbf{A} \tag{3.27}$$

3.3 TRIANGULAR DECOMPOSITION

At this point, we concentrate on the triangular decomposition of the coefficient (stiffness) matrix. When the coefficient matrix is premultiplied by the appropriate Gaussian elementary matrices, the result is an upper triangular matrix:

$$(\mathbf{L}_{(n-1)}\cdots\mathbf{L}_{(2)}\mathbf{L}_{(1)})\mathbf{A} = \mathbf{U} \tag{3.28}$$

$$\mathbf{L}_{(j)} = \begin{bmatrix} 1 & & & & & & \\ & \ddots & & & & & \\ & & 1 & & & & \\ & & & 1 & & & \\ & & & -l_{j+1,j} & 1 & & \\ & & & -l_{j+2,j} & & 1 & \\ & & & \vdots & & & \ddots & \\ & & & -l_{n,j} & & & & 1 \end{bmatrix} \tag{3.29}$$

The Gaussian elementary matrices are unit lower triangular matrix as shown. The subscript (j) indicates that the jth column has the nonzero terms. The nonzero elements in this matrix are computed from the partially decomposed matrix $\mathbf{A}^{(j-1)}$:

$$l_{j+k,j} = \frac{a_{j+k,j}^{(j-1)}}{a_{jj}^{(j-1)}} \tag{3.30}$$

The inverse of the Gaussian elementary matrix is also a unit lower triangular matrix, which can be readily obtained by reversing the sign of the nonzero elements below diagonal in the jth column:

$$\mathbf{L}_{(j)}^{-1} = \begin{bmatrix} 1 & & & & & & \\ & \ddots & & & & & \\ & & 1 & & & & \\ & & & 1 & & & \\ & & & l_{j+1,j} & 1 & & \\ & & & l_{j+2,j} & & 1 & \\ & & & \vdots & & & \ddots \\ & & & l_{n,j} & & & & 1 \end{bmatrix} \tag{3.31}$$

After determining all the Gaussian elementary matrices, we can obtain the triangular decomposition for **A**:

$$\mathbf{A} = (\mathbf{L}_{(1)}^{-1}\mathbf{L}_{(2)}^{-1}\cdots\mathbf{L}_{(n-1)}^{-1})\mathbf{U} = \mathbf{LU} \tag{3.32}$$

L is a unit lower triangular matrix because the multiplication of unit lower triangular matrices results in another unit lower triangular matrix:

$$\mathbf{L} = \begin{bmatrix} 1 & & & & \\ l_{21} & 1 & & & \\ l_{31} & l_{32} & 1 & & \\ \vdots & \vdots & \ddots & \ddots & \\ l_{n1} & l_{n2} & \cdots & l_{n,n-1} & 1 \end{bmatrix} \tag{3.33}$$

For the example problem discussed earlier, the triangular decomposition process decomposes the coefficient (stiffness) matrix **A** into the product of the following unit lower triangular matrix and upper triangular matrix:

$$\mathbf{A} = \mathbf{LU} \tag{3.34}$$

$$\mathbf{L} = \mathbf{L}_{(1)}^{-1}\mathbf{L}_{(2)}^{-1}\mathbf{L}_{(3)}^{-1} = \begin{bmatrix} 1 & & & \\ \frac{-2}{3} & 1 & & \\ \frac{-1}{3} & \frac{-8}{11} & 1 & \\ 0 & \frac{-3}{11} & -1 & 1 \end{bmatrix} \tag{3.35}$$

$$\mathbf{U} = \begin{bmatrix} 3 & -2 & -1 & 0 \\ 0 & \frac{11}{3} & \frac{-8}{3} & -1 \\ 0 & 0 & \frac{30}{11} & \frac{-30}{11} \\ 0 & 0 & 0 & 2 \end{bmatrix} \tag{3.36}$$

3.3.1 Triangular decomposition of symmetric matrices

Any upper triangular matrices can be further decomposed into a product of a diagonal matrix D and the transpose of the unit lower triangular matrix:

$$\mathbf{U} = \mathbf{D}\bar{\mathbf{L}} \tag{3.37}$$

If matrix **A** is symmetric, then we have the following triangular decomposition:

$$\begin{aligned} \mathbf{A} &= \mathbf{L}\mathbf{D}\bar{\mathbf{L}} \\ \mathbf{A}^{T} &= \bar{\mathbf{L}}^{T}\mathbf{D}\mathbf{L}^{T} \end{aligned} \tag{3.38}$$

Considering the symmetry of matrix **A**, we have $\bar{\mathbf{L}} = \mathbf{L}^{T}$:

$$\mathbf{A} = \mathbf{L}\mathbf{D}\mathbf{L}^{T} \tag{3.39}$$

In addition, since matrix **L** is a unit lower triangular matrix, its determinant is equal to 1:

$$\det(\mathbf{A}) = \det(\mathbf{L})\det(\mathbf{D})\det(\mathbf{L}^{T}) = \det(\mathbf{D}) \tag{3.40}$$

For the example problem, we can obtain the following symmetric decomposition:

$$\mathbf{A} = \begin{bmatrix} 1 & & & \\ \frac{-2}{3} & 1 & & \\ \frac{-1}{3} & \frac{-8}{11} & 1 & \\ 0 & \frac{-3}{11} & -1 & 1 \end{bmatrix} \begin{bmatrix} 3 & & & \\ & \frac{11}{3} & & \\ & & \frac{30}{11} & \\ & & & 2 \end{bmatrix} \begin{bmatrix} 1 & \frac{-2}{3} & \frac{-1}{3} & 0 \\ & 1 & \frac{-8}{11} & \frac{-3}{11} \\ & & 1 & -1 \\ & & & 1 \end{bmatrix} \tag{3.41}$$

We can also verify that det(**A**) = det(**D**) = 60.

3.3.2 Computational expense: number of operations

To measure the computational expense of a solution algorithm, we usually count the number of operations incurred with the execution of an algorithm. In numerical analysis, by convention, one operation is defined as the process of one multiplication followed by one addition. If we denote the total number of degrees of freedom of a system as n, then the number of operations involved in the Gauss-elimination process is as follows:

1. Triangular decomposition: $n^3 + [(n-2)^3/3] - n(n-1)(n-2)$, which is $\approx (n^3/3)$ when n becomes very large
2. Vector reduction: $n(n+1)^2/2 \approx n^2/2$
3. Back-substitution: same for vector reduction, $n(n+1)^2/2 \approx n^2/2$

It is observed that the computational expense incurred in triangularization is almost n times higher than the rest of the computations. Compared with the computational expense involved in triangularization, the operation expenses incurred in vector reduction and back-substitution can be practically neglected. Therefore, the total operations involved in the Gauss-elimination process can be estimated to be approximately αn^3, $0 < \alpha < 1$.

In practice, the structural stiffness matrix obtained from finite element analysis is always symmetric and banded. Under normal conditions, we seldom have a full stiffness matrix. Denoting the half bandwidth as m, we can observe that in most structural problems m is much smaller than n. Because of symmetry and bandedness of the stiffness matrix, the original n × n matrix can be compactly stored as an n × m matrix.

Using the compact storage scheme for stiffness matrix, the number of operations involved in the Gauss-elimination process is reduced to the following:

1. Triangular decomposition: $\alpha\, nm^2$, $0 < \alpha < 1$
2. Vector reduction: $\beta\, nm$, $0 < \beta < 1$
3. Back-substitution: $\beta\, nm$

Therefore, with banded storage of stiffness matrix, the total number of operations incurred in the Gauss-elimination process is approximately $\alpha\, nm^2$, for a large n.

3.4 PRACTICAL ALGORITHMS IN GAUSS-ELIMINATION PROCEDURE

We have seen in Section 3.3.1 that for a symmetric matrix there exists a symmetric decomposition:

$$\mathbf{A} = \mathbf{L}\mathbf{D}\mathbf{L}^T \tag{3.42}$$

Obviously, to complete the triangular decomposition process in any practical algorithm, we need to determine the elements in matrices **L** and **D** directly from the elements of matrix **A**. To start with, we consider the standard triangular decomposition, **A** = **LU**, as illustrated in Figure 3.3. From this figure, we can obtain the following relationship between elements on the left-hand and right-hand sides of the matrix equation:

$$a_{ij} = u_{ij} + \sum_{k=1}^{i-1} l_{ik} u_{kj} \tag{3.43}$$

From this equation, we can determine the terms of matrix **U**:

$$u_{ij} = a_{ij} - \sum_{k=1}^{i-1} l_{ik} u_{kj} \tag{3.44}$$

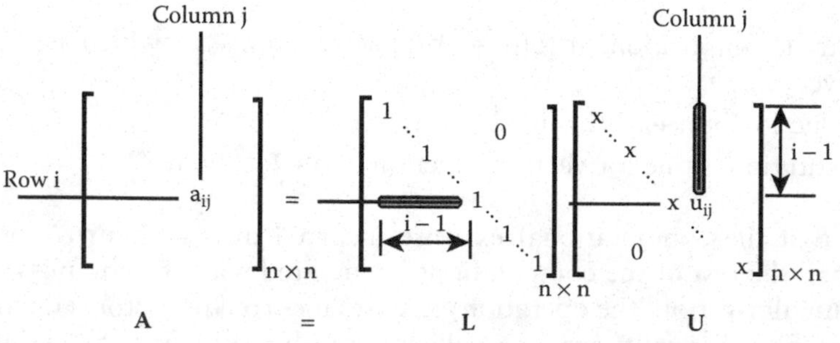

Figure 3.3 Triangular decomposition of coefficient matrix **A** = **LU**.

This equation means that the element in the upper triangular matrix can be determined once the elements in the blocked area shown in the figure have been computed. With this equation, we can determine the elements in matrices **L** and **U** directly from the elements in matrix **A**. As a matter of fact, this equation serves as the basis of all the Gauss-elimination methods. This basic procedure can be implemented in the following algorithm:

for $j = 1, \ldots, n$

$u_{1j} = a_{1j}$ first row of **U**

$l_{j1} = \frac{u_{1j}}{u_{11}}$ first column of **L**

for $i = 2, \ldots, n$

for $j = i, \ldots, n$

$$u_{ij} = a_{ij} - \sum_{k=1}^{i-1} l_{ik} u_{kj}$$

$$l_{ji} = \frac{u_{ij}}{u_{ii}}$$

This algorithm is the direct implementation of the basic equation and it works. However, it is not an efficient algorithm because we need to store all the three matrices **A**, **L**, and **U**. This can be a major drawback in analyzing very large structural systems.

A closer look at the symmetric decomposition in Equation 3.42 indicates that because of symmetry, $\mathbf{L}^T$ does not need to be stored. On the other hand, **D** is a diagonal matrix and **L** a unit lower triangular matrix. Therefore, we can store the elements of **D** matrix on the diagonal of the **L** matrix, and we need to store only the elements below the diagonal in the **L** matrix. This storage scheme is implemented in a symmetric decomposition procedure in the following algorithm:

Initialization

$l_{11} \leftarrow d_{11} = a_{11}$

for $j = 2, \ldots, n$

$u_1 \leftarrow a_{1j}$ **u** is a temporary storage vector

for $i = 2, \ldots, j-1$ (skip for $j = 2$)

$$u_i \leftarrow a_{ij} - \sum_{k=1}^{i-1} l_{ik} u_k$$

for $i = 1, \ldots, j-1$

$$l_{ji} \leftarrow \frac{u_i}{l_{ii}}$$

$$l_{jj} \leftarrow a_{jj} - \sum_{k=1}^{j-1} l_{jk} u_k$$

We note that vector **u** is only a temporary storage vector of size n; we are not calculating the terms of **u**, rather we store information in **u**. This should not be confused with the elements in the upper triangular matrix **U**, which is not directly computed in this algorithm.

A more practical algorithm is to store the **L** and **D** matrices back into **A** matrix itself. As the terms of **L** and **D** matrices are computed, they are stored in the upper part of the original matrix **A**. We can do this because those terms of the **A** matrix will not be needed for the remaining operations in decomposition. This algorithm replaces elements a_{ij}, $j > i + 1$ by l_{ij} and elements a_{ii} by d_{ii}:

Algorithm TD_1

for $i = 2, \ldots, n$

$$(1)\ t_k \leftarrow \frac{a_{ki}}{a_{kk}} \qquad ; k = 1, \ldots, i-1$$

$$(2)\ a_{ii} \leftarrow a_{ii} - \sum_{k=1}^{i-1} t_k\, a_{ki}$$

$$(3)\ a_{ki} \leftarrow t_k \qquad ; k = 1, \ldots, i-1$$

$$(4)\ a_{ij} \leftarrow a_{ij} - \sum_{k=1}^{i-1} a_{ki}\, a_{kj} \qquad ; j = i+1, \ldots, n$$

Vector **t** of size n is for temporary storage.

In Algorithm TD_1, step (1) changes the terms of the ith column of **U** to $\mathbf{L}^T$ and stores in the temporary vector **t**; step (2) computes the diagonal terms $u_{ii} = d_{ii}$ and store them in a_{ii}; step (3) replaces the terms in the ith column by **t**, that is, it changes from **U** to $\mathbf{L}^T$; and step (4) computes the terms of ith row of **U**. This algorithm (TD_1) is schematically illustrated in Figure 3.4.

A variation of this procedure can also be implemented in Algorithm TD_2 by storing the **L** and **D** matrices into **A** matrix itself. The following algorithm also replaces elements a_{ij}, $j > i + 1$ by l_{ij} and elements a_{ii} by d_{ii}:

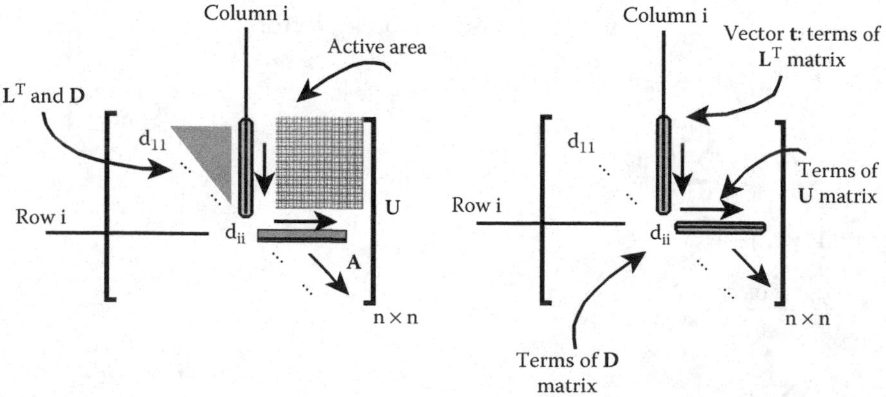

Figure 3.4 Illustration of active area and storage scheme in Algorithm TD_1.

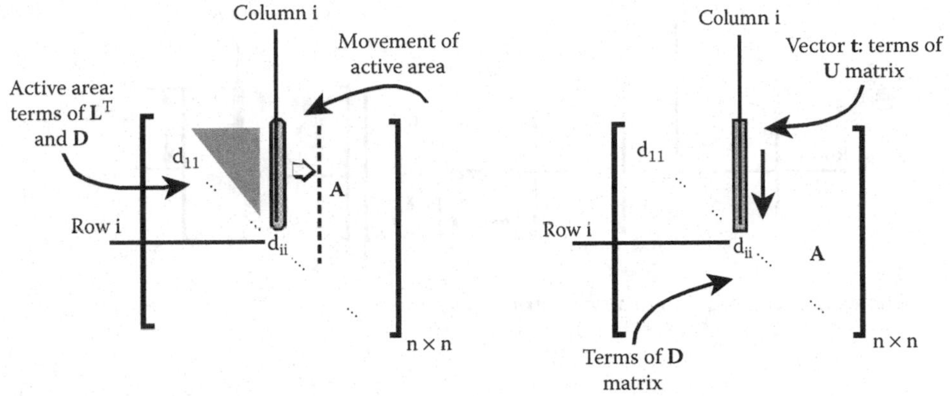

Figure 3.5 Illustration of active area and storage scheme in Algorithm TD_2.

Algorithm TD_2

for i = 2, ..., n

(1) $t_1 \leftarrow a_{1i}$

(2) $t_j \leftarrow a_{ji} - \sum_{k=1}^{j-1} a_{kj}\, t_k \qquad ; j = 2, \ldots, i-1$

(3) $a_{ji} \leftarrow \frac{t_j}{a_{jj}} \qquad ; j = 1, \ldots, i-1$

(4) $a_{ii} \leftarrow a_{ii} - \sum_{k=1}^{i-1} a_{ki}\, t_k$

Vector **t** is for temporary storage with at most n elements. In this algorithm, steps (1) and (2) compute the jth column of **U** and store in vector **t**, step (3) computes the ith row of **L** or the ith column of $\mathbf{L}^T$ and stores in the ith column of **A**, and finally step (4) computes the diagonal term d_{ii} and stores in a_{ii}. This algorithm is illustrated schematically in Figure 3.5.

3.5 CHOLESKY DECOMPOSITION

Cholesky decomposition is also called symmetric decomposition, which decomposes a symmetric matrix **A** into the following form:

$$\mathbf{A} = \bar{\mathbf{L}}\bar{\mathbf{L}}^T \tag{3.45}$$

$\bar{\mathbf{L}}$ is a lower triangular matrix, rather than a *unit* lower triangular matrix. This decomposition scheme applies to only symmetric matrices. However, Gauss elimination can be applied to any general matrix, that is, either positive definite or negative definite. We can derive the Cholesky decomposition from Gauss elimination:

$$\mathbf{A} = \mathbf{L}\mathbf{D}\mathbf{L}^T = (\mathbf{L}\mathbf{D}^{1/2})(\mathbf{D}^{1/2}\mathbf{L}^T) \tag{3.46}$$

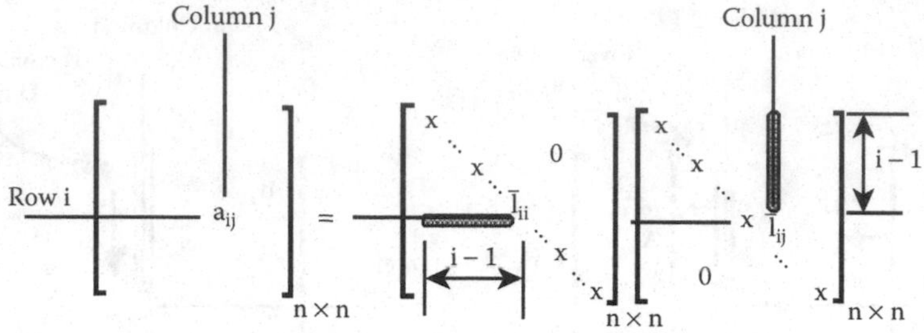

Figure 3.6 Cholesky decomposition of a symmetric matrix.

$$\bar{L} = LD^{1/2} \tag{3.47}$$

The basic equation for the algorithm can be derived from Equation 3.45 by considering the relationship between elements in the matrices on the left-hand and right-hand sides of the equation, as illustrated in Figure 3.6.

From matrix multiplication, by equating the term in the matrix on the left-hand side of the matrix equation with that on right-hand side, we have

$$a_{ij} = \sum_{k=1}^{i-1} \bar{l}_{ik}\bar{l}_{jk} + \bar{l}_{ij}\bar{l}_{ii} \tag{3.48}$$

Thus, the element in the lower triangular matrix can be computed in the following relation:

$$\bar{l}_{ij} = \frac{1}{\bar{l}_{ii}}\left[a_{ij} - \sum_{k=1}^{i-1} \bar{l}_{ik}\bar{l}_{jk}\right] \tag{3.49}$$

If we let i = j in the above equation, then we have the following:

$$\bar{l}_{ii} = \left[a_{ii} - \sum_{k=1}^{i-1} \bar{l}_{ik}^2\right]^{1/2} \tag{3.50}$$

This indicates that the Cholesky decomposition process will break down if there exists a zero eigenvalue in the coefficient matrix. In order to use Cholesky decomposition, the stiffness matrix must be symmetric and positive definite. In linear finite element analysis, this condition is usually satisfied. However, the satisfaction of this condition usually cannot be guaranteed in nonlinear finite element analysis, especially with structural stability problems such as buckling, snap-through, and bifurcation.

3.6 ORTHOGONAL DECOMPOSITION

In orthogonal decomposition, we basically premultiply the coefficient matrix with a series of orthogonal matrices in order to reduce it to an upper triangular matrix. Through this procedure, matrix **A** is decomposed into the product of an orthogonal matrix **Q** and an upper triangular matrix $\mathbf{U}_q$:

$$\mathbf{A} = \mathbf{Q}\mathbf{U}_q \tag{3.51}$$

The inverse of an orthogonal matrix Q is equal to its transpose. This allows us to reduce the system of linear equations, $\mathbf{Ax} = \mathbf{b}$, to an upper triangular matrix:

$$\mathbf{U}_q\mathbf{x} = \mathbf{Q}^T\mathbf{b} \tag{3.52}$$

The unknown vector **x** can now be determined through the standard back-substitution procedure.

3.6.1 Givens method

It has been observed that an off-diagonal term in a matrix can be reduced to zero by premultiplying the matrix with a plane rotation matrix. Therefore, if we premultiply the coefficient matrix **A** with a series of plane rotation matrices, then eventually matrix **A** will be reduced to an upper triangular matrix. This is the Givens method for orthogonal decomposition of matrix because the plane rotation matrices are orthogonal. For example, if we want to reduce the off-diagonal terms below the diagonal in the first column of matrix **A**, we can carry out the following operations:

$$(\mathbf{G}_{n,1}\ldots\mathbf{G}_{2,1})\mathbf{A} = \mathbf{G}_1\mathbf{A} \tag{3.53}$$

If we continue this procedure to the last column of the matrix, we have finally reduced matrix **A** to an upper triangular matrix:

$$(\mathbf{G}_{n,n-1})\ldots(\mathbf{G}_{n,2}\ldots\mathbf{G}_{3,2})(\mathbf{G}_{n,1}\ldots\mathbf{G}_{2,1})\mathbf{A} = \mathbf{U}_g \tag{3.54}$$

$$(\mathbf{G}_{n-1}\ldots\mathbf{G}_2\mathbf{G}_1)\mathbf{A} = \mathbf{U}_g \tag{3.55}$$

$$\mathbf{G}_j = (\mathbf{G}_{n,j}\mathbf{G}_{n-1,j}\ldots\mathbf{G}_{j+1,j}) \tag{3.56}$$

Premultiplying matrix **A** with the Givens rotation matrix $\mathbf{G}_j$ reduces all the terms below the diagonal in column j to zero.

With these operations, the matrix **A** is decomposed into the following form:

$$\begin{aligned}\mathbf{A} &= (\mathbf{G}_1^T\mathbf{G}_2^T\ldots\mathbf{G}_{n-1}^T)\mathbf{U}_g\\ &= \mathbf{G}\mathbf{U}_g\end{aligned} \tag{3.57}$$

where **G** is an orthogonal matrix. We note that the product of orthogonal matrices is also an orthogonal matrix.

To determine the plane rotation matrix in carrying out the orthogonal decomposition process, we consider the effect of premultiplying matrix **A** with a rotation matrix, as illustrated in Figure 3.7.

From matrix multiplication, we can obtain the following equations:

$$\begin{cases}\bar{a}_{jj} = ca_{jj} + sa_{ij}\\ \bar{a}_{ij} = -sa_{jj} + ca_{ij}\end{cases} \tag{3.58}$$

where $c = \cos\alpha$ and $s = \sin\alpha$. By imposing the condition that $\bar{a}_{ij} = 0$ and $c^2 + s^2 = 1$, we obtain the terms of the rotation matrix:

$$c = \frac{a_{jj}}{(a_{jj}^2 + a_{ij}^2)^{1/2}}, \quad s = \frac{a_{ij}}{(a_{jj}^2 + a_{ij}^2)^{1/2}} \tag{3.59}$$

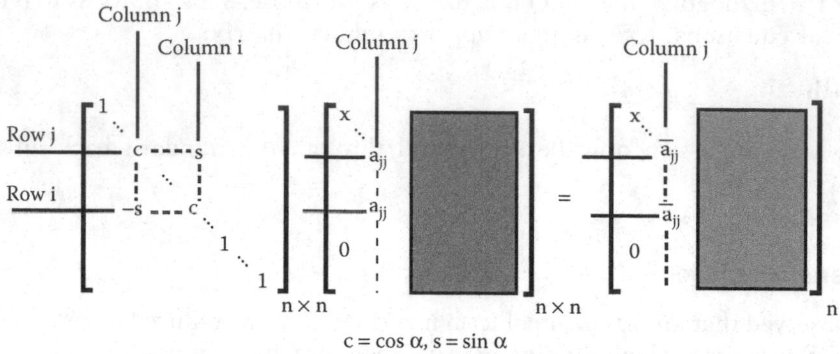

Figure 3.7 Determination of plane rotation matrix in Givens method.

Thus, the rotation matrix for performing Givens decomposition is determined directly from the elements in matrix **A**. We note that if $a_{jj}^2 + a_{ij}^2 = 0$, then no transformation needs to be performed, and we can use cos α = 1 and sin α = 0 in the plane rotation matrix.

3.6.2 Householder transformation

In Householder transformation (Householder, 1958), the matrix **A** is premultiplied with reflection matrices to reduce it to an upper triangular matrix $\mathbf{U}_h$:

$$(\mathbf{H}_{n-1} \ldots \mathbf{H}_2\mathbf{H}_1)\mathbf{A} = \mathbf{U}_h \tag{3.60}$$

where $\mathbf{H}_j$ is a Householder transformation matrix. The effect of premultiplying matrix **A** with a Householder matrix $\mathbf{H}_j$ reduces the terms below diagonal in the jth column of the matrix to zero.

Because the reflection matrices are orthogonal, their product is also orthogonal:

$$\mathbf{HA} = \mathbf{U}_h \tag{3.61}$$

$$\mathbf{H} = \mathbf{H}_{n-1} \ldots \mathbf{H}_2\mathbf{H}_1 \tag{3.62}$$

Through Householder transformation, we can decompose matrix **A** into the product of an orthogonal matrix and an upper triangular matrix:

$$\mathbf{A} = \mathbf{H}^T\mathbf{U}_h \tag{3.63}$$

To describe the transformation of matrix A after conducting a Householder transformation, we denote the succession of the intermediate steps in the following form:

$$\begin{aligned}
&\mathbf{H}_1\ \mathbf{A} = \mathbf{A}_{(1)} \\
&\mathbf{H}_2\ \mathbf{A}_{(1)} = \mathbf{A}_{(2)} \\
&\quad\vdots \\
&\mathbf{H}_k\ \mathbf{A}_{(k-1)} = \mathbf{A}_{(k)} \\
&\quad\vdots \\
&\mathbf{H}_{n-1}\ \mathbf{A}_{(n-2)} = \mathbf{A}_{(n-1)}
\end{aligned} \tag{3.64}$$

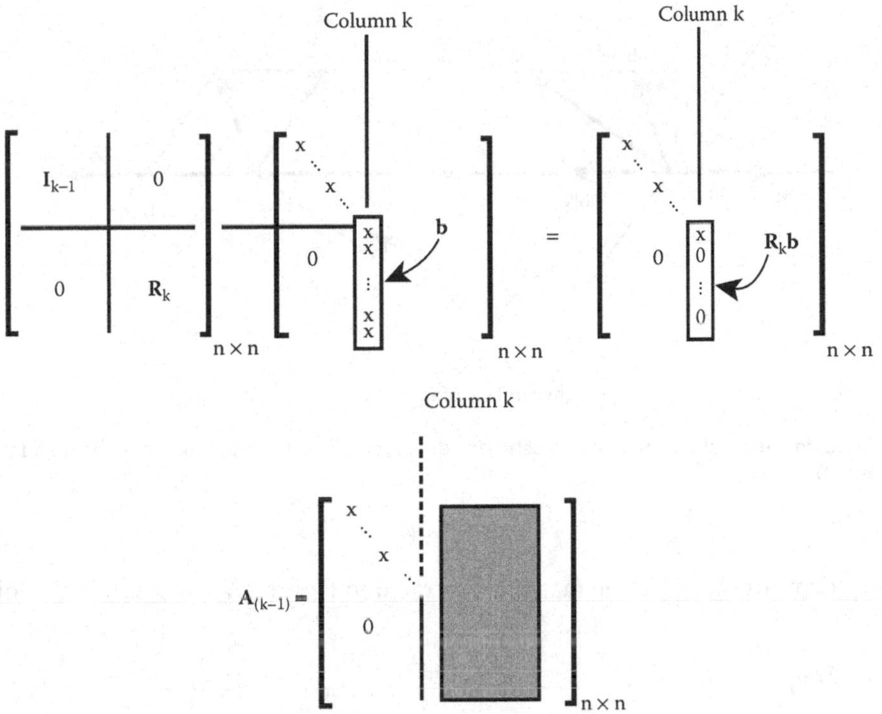

Figure 3.8 Householder reflection matrix to reduce the terms below the diagonal of column k to zero.

The multiplication of the Householder matrix $\mathbf{H}_k$ with $\mathbf{A}_{(k-1)}$ is to reduce all terms below the diagonal of column k to zero. The detailed computation process is illustrated in Figure 3.8.

$\mathbf{R}_k$ is a reflection submatrix of dimension $[n - k + 1] \times [n - k + 1]$ and $\mathbf{b}$ is a vector of the dimension $[n - k + 1]$. As illustrated, the reflection matrix $\mathbf{R}_k$ when operated on vector $\mathbf{b}$, the reflected vector has all terms below the first term equal to zero:

$$\mathbf{R}_k\ \mathbf{b} = |\mathbf{b}|\ \mathbf{e}_1 = \begin{Bmatrix} |\mathbf{b}| \\ 0 \\ \vdots \\ 0 \end{Bmatrix} \tag{3.65}$$

We note that, as discussed in Chapter 1, the length of vector $\mathbf{b}$ is preserved through reflection. We also recall from Chapter 1 that the reflection matrix is of the following general form:

$$\mathbf{R}_k = \mathbf{I} - \frac{2}{\mathbf{u}^T\mathbf{u}}\mathbf{u}\mathbf{u}^T \tag{3.66}$$

Next, we need to determine the vector $\mathbf{u}$ such that the above transformation can be realized. We recall that the reflection matrix produces a mirror image of a vector with respect to the hyperplane of vector $\mathbf{u}$, or the reflecting plane is perpendicular to vector $\mathbf{u}$. The vector $\mathbf{u}$ can be determined in the following two ways, which are illustrated in Figure 3.9.

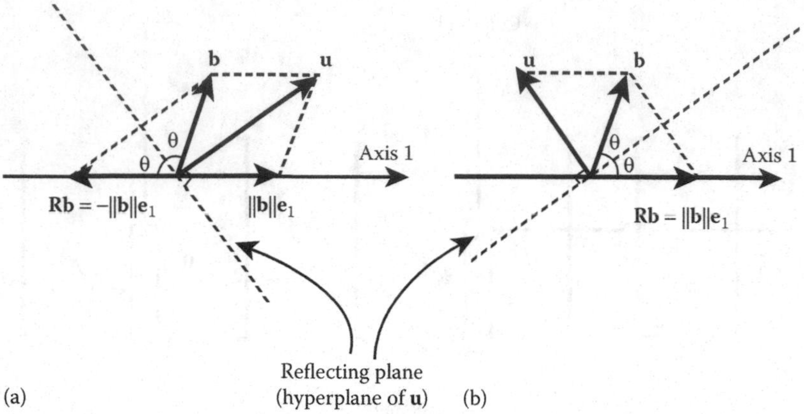

Figure 3.9 Determination of vector **u** in Householder decomposition: (a) case I, $\mathbf{u} = \mathbf{b} + \|\mathbf{b}\|\mathbf{e}_1$ and (b) case II, $\mathbf{u} = \mathbf{b} - \|\mathbf{b}\|\mathbf{e}_1$.

The two forms of the reflection matrix, illustrated in Figure 3.9, are given in the following equations:

$$\mathbf{u} = \mathbf{b} + \|\mathbf{b}\|\mathbf{e}_1 \tag{3.67}$$

$$\mathbf{u} = \mathbf{b} - \|\mathbf{b}\|\mathbf{e}_1 \tag{3.68}$$

Once the vector **u** is determined, the reflection matrix **R** can be readily computed using Equation 3.66. The two reflection matrices will produce the following results:

$$\mathbf{R}\mathbf{b} = -\|\mathbf{b}\|\mathbf{e}_1 \tag{3.69}$$

$$\mathbf{R}\mathbf{b} = \|\mathbf{b}\|\mathbf{e}_1 \tag{3.70}$$

The reflection matrices developed from vector **u** in Equations 3.67 and 3.68 may result in numerical problems. This may occur if the vector **b** is very close to a vector $|\mathbf{b}|\mathbf{e}_1$, as illustrated in Figure 3.10. To illustrate the problem, we assume the following form of the vector **b**:

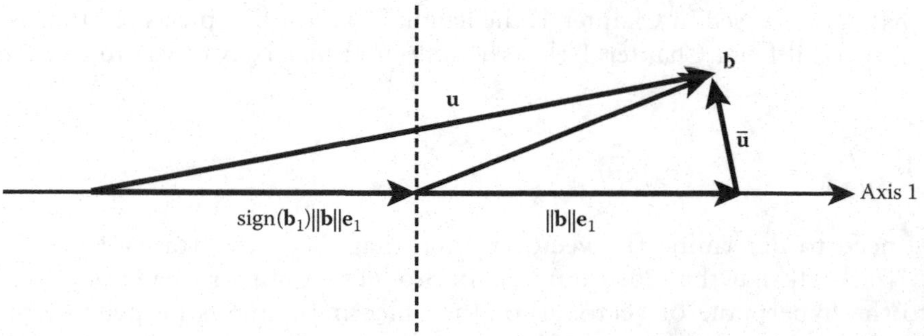

Figure 3.10 Determination of Householder decomposition matrix when vector **b** is very close to $|\mathbf{b}|\mathbf{e}_1$.

$$\mathbf{b} = \begin{Bmatrix} b_1 \\ \varepsilon_1 \\ \vdots \\ \varepsilon_{n-1} \end{Bmatrix} \tag{3.71}$$

In this vector $b_1 = |\mathbf{b}|$ and $\varepsilon_1, \ldots, \varepsilon_{n-1}$ are very small numbers. Then we calculate the vector **u** from the above two equations:

$$\mathbf{u} = \mathbf{b} + |\mathbf{b}|\ \mathbf{e}_1 = \begin{Bmatrix} |\mathbf{b}| \\ \varepsilon_1 \\ \vdots \\ \varepsilon_{n-1} \end{Bmatrix} + \begin{Bmatrix} |\mathbf{b}| \\ 0 \\ \vdots \\ 0 \end{Bmatrix} = \begin{Bmatrix} 2|\mathbf{b}| \\ \varepsilon_1 \\ \vdots \\ \varepsilon_{n-1} \end{Bmatrix} \tag{3.72}$$

$$\bar{\mathbf{u}} = \mathbf{b} - |\mathbf{b}|\ \mathbf{e}_1 = \begin{Bmatrix} |\mathbf{b}| \\ \varepsilon_1 \\ \vdots \\ \varepsilon_{n-1} \end{Bmatrix} - \begin{Bmatrix} |\mathbf{b}| \\ 0 \\ \vdots \\ 0 \end{Bmatrix} = \begin{Bmatrix} 0 \\ \varepsilon_1 \\ \vdots \\ \varepsilon_{n-1} \end{Bmatrix} \tag{3.73}$$

This indicates that vector **u** is very small, which may lead to numerical error if it is used in computing the reflection matrix **R**. To prevent this problem, we should calculate the **u** vector using the following equation:

$$\mathbf{u} = \mathbf{b} + \text{sign}(b_1)|\mathbf{b}|\ \mathbf{e}_1 \tag{3.74}$$

Here we have used the sign of the first term in the vector **b**.

3.6.3 Comparison of Givens and Householder decompositions

In previous Sections 3.6.1 and 3.6.2, we discussed Givens and Householder methods of decomposing a matrix into product of an orthogonal matrix and an upper triangular matrix. This may indicate that there is more than one way to arrive at an orthogonal matrix in the triangular decomposition. Here we will further explore this question. The following are the Givens and Householder decompositions of the same matrix:

$$\begin{cases} \mathbf{A} = \mathbf{G}\mathbf{U}_g \\ \mathbf{A} = \mathbf{H}\mathbf{U}_h \end{cases} \Rightarrow \mathbf{G}\mathbf{U}_g = \mathbf{H}\mathbf{U}_h \tag{3.75}$$

We note that the Householder orthogonal matrix in the above equation is the transpose of the matrix in Equation 3.63.

Premultiplying Equation 3.75 by $\mathbf{H}^T$ and postmultiplying by $\mathbf{U}_g^{-1}$, we arrive at the following:

$$\mathbf{H}^T\mathbf{G}\mathbf{U}_g\mathbf{U}_g^{-1} = \mathbf{H}^T\mathbf{H}\mathbf{U}_h\mathbf{U}_g^{-1} \tag{3.76}$$

$$\mathbf{H}^T\mathbf{G} = \mathbf{U}_h\mathbf{U}_g^{-1} \tag{3.77}$$

We know that the inverse of a triangular matrix is also a triangular matrix, and the product of two triangular matrices is a triangular matrix. Therefore, we have the following, where U is an upper triangular matrix:

$$\mathbf{H}^T\mathbf{G} = \mathbf{U}_h\mathbf{U}_g^{-1} = \mathbf{U} \tag{3.78}$$

From the following equation, we can see that the upper triangular matrix **U** is an orthogonal matrix:

$$\mathbf{U}^T\mathbf{U} = (\mathbf{G}^T\mathbf{H})\,(\mathbf{H}^T\mathbf{G}) = \mathbf{I} \tag{3.79}$$

Since $\mathbf{U}^T$ is a lower triangular matrix and its inverse should also be a lower triangular matrix, the only possibility suggested from the above relation is that the matrix **U** is a diagonal matrix with +1 or −1 elements in the diagonal or some alternating ±1 elements:

$$\mathbf{U} = (\pm)\mathbf{I} \tag{3.80}$$

This observation on the matrix **U** leads to the following conclusions:

$$\mathbf{U} = \mathbf{U}_h\mathbf{U}_g^{-1} = (\pm)\mathbf{I} \tag{3.81}$$

$$\mathbf{U}_h = (\pm)\mathbf{I}\mathbf{U}_g \tag{3.82}$$

This indicates that the absolute value of elements in the Householder triangular matrix and the Givens triangular matrix should be the same:

$$\left|U_{ij}\right|_g = (\pm)\left|U_{ij}\right|_h \tag{3.83}$$

We can thus conclude that orthogonal decomposition is basically unique.

3.7 ERROR ANALYSIS

3.7.1 Sources of numerical errors

In modeling of any engineering system, it is inevitable that we have to make assumptions, idealization, abstraction, and simplifications about the physical system so that the problem can be formulated within the realm of mathematical modeling tools. The deviation between the mathematical model and the physical system is generally referred to as *modeling error*. Only if we understand the physical problem and make reasonable assumption and abstraction in the construction of mathematical model can we expect to get reasonable results from any computational simulation. Obviously, in most cases it is very difficult, if not impossible, to quantify the modeling error. Therefore, in numerical analysis, we usually consider the mathematical model as a reasonable representation of the physical system.

Within any mathematical model, usually there are parameters and other physical entities that need to be determined from observations. For example, in predicting the linear behavior of a structural system, we need to determine the material properties such as Young's modulus and Poisson's ratio. No matter what kind of measurement instruments and methodology we use, the error present in the observation process is inherent. This kind of error is called *observation error*, which cannot be reliably quantified.

In this section, we will discuss only the numerical errors generated in using solution methods for solving the mathematical modeling problems. These include *cutoff* or methodology error, *round-off* error, and *initial value* error. When it is difficult to obtain an analytical solution to a mathematical modeling problem, we can resort to numerical methods to obtain an approximate solution. The cutoff error is the difference between the exact solution and the approximate solution. We can consider the remainder of a Taylor series expansion of a function as a measure of the cutoff error. Because numbers are represented as an incomplete finite set in any digital computer, depending on the precision of a machine, we cannot ignore numerical representation error or round-off error in carrying out any algebraic operations. For certain problems involving initial values, any error present in the set of initial values would affect the solution of the system. Therefore, when we use numerical methods to solve a system of linear equations obtained from finite element analysis, $\mathbf{A}\mathbf{x} = \mathbf{b}$, if we expect the solution vector $\mathbf{x}$ to be exact, then, in a sense, we will never be able to find it. Instead, we usually find a solution vector $\bar{\mathbf{x}} = \mathbf{x} + \delta\mathbf{x}$, which is in the neighborhood of the exact solution vector.

More precisely, in the solution of a system of linear equations, unless the coefficient matrix and load vector are integers, elements in the matrix $\mathbf{A}$ and vector $\mathbf{b}$ can never be represented exactly in a digital computer. In other words, matrix $\mathbf{A}$ and vector $\mathbf{b}$ will be represented in computers as $\mathbf{A} + \delta\mathbf{A}$ and $\mathbf{b} + \delta\mathbf{b}$. The precision on numerical representation depends on the architectural design of computers.

This leads to the following question: how much will be the error in the solution vector $\mathbf{x}$ if there are errors present in $\mathbf{A}$ and $\mathbf{b}$ of the original system from modeling and/or representation process? To carry out this kind of error analysis is called *forward* error analysis. However, a more interesting question is the following one: supposing we can obtain an exact solution to a problem, how close is the problem to the one we are solving? In other words, if we have error in the solution vector, then what will be the effect of this deviation on the matrix $\mathbf{A}$ and even vector $\mathbf{b}$ of the system? This kind of error analysis is called *backward* error analysis.

With the presence of modeling error and numerical representation error, in general, instead of solving a system as $\mathbf{A}\mathbf{x} = \mathbf{b}$, we are actually solving the following perturbed system of linear equations:

$$(\mathbf{A} + \delta\mathbf{A})(\mathbf{x} + \delta\mathbf{x}) = (\mathbf{b} + \delta\mathbf{b}) \tag{3.84}$$

In order to discuss forward and backward error analyses for solution methods, we first introduce some preliminaries on the representation of data within a digital computer and some basics in perturbation theory and error analysis.

3.7.2 Floating point representation of numbers

In mathematics, the set of real numbers used in calculus is a complete set, which is also infinite, densely distributed, and continuous. However, the set of real numbers that can be stored in a digital computer is a finite discrete set with nonuniform distributions. Because of this incompleteness in representing real numbers within a digital computer, it is possible that a solution method when implemented in computers will be demonstrating different characteristics from the mathematical algorithm itself. It is thus important to understand the way data are represented within a computer and how the associated operations are performed.

3.7.3 Representation of integer and real numbers

In a digital computer, a number is represented as a pattern in a string of n binary bits. For example, an integer value may be represented as a 32-bit pattern in computer. If we fix the left-end bit in the string to represent the sign of the number, then the remaining 31 bits are used to store integer values between 0 and $2^{31} - 1$, which is approximately 10^{10}. Some computers may use longer length of bits such as 60 bits to extend the range of integer numbers that can be represented. However, it is obvious that the range of integers is too small for reasonable engineering calculations. For example, the Young's modulus for steel is approximately 200 GPa = 200×10^9 Pa. Even if a longer string is used, it is not uncommon for the resulting integer value from calculations to become too big to be represented properly.

For a real number or very large integer number, we resort to store the number as a floating point number in a digital computer. This is done by expressing the real number in two parts: the exponent and the mantissa. The mantissa varies between 0 and 1. The form of exponent depends on the type of computers. For example, with a 32-bit machine, the format for a floating point representation consists of the following: 24 bits for mantissa, 6 bits for exponent, and 1 bit each for sign of exponent and sign of number. With this format, the exponent can contain a binary number between 0 and $2^6 - 1$. The mantissa can store a binary number between 0 and $2^{24} - 1$. The hardware then forms the real number as $\pm 16^{exp}[\text{mantissa}/(2^{24} - 1)]$. This implies that the range of floating point numbers is approximately $\pm 16^{\pm 63} \approx \pm 10^{\pm 75}$. The smallest recognizable change in the fraction is $[1/(2^{24} - 1)]$, which means that we have slightly better than seven-digit accuracy. If the calculation requires more than seven significant figures, then the computer will *underflow* and operations will be incorrect. However, if the operation results in a number larger than the maximum number representable within the number set of that computer, then the computer may *overflow* or the program terminates as a result.

In general, a number is represented as floating point data in digital computers in the following form:

$$\begin{cases} f = \pm 0.b_1 b_2 \ldots b_t x \beta^e \\ 0 \le b_j \le \beta \\ L \le e \le U \end{cases} \tag{3.85}$$

where β is the machine base, which can be 2, 10, or 16; t is the precision or the length of a word; L is the underflow limit; and U is the upper flow limit. This four-number set (β, t, L, U) varies with different types of computers and it defines the precision of a computer.

Let F be the set of floating point numbers stored in a computer. Since this set is a finite, discrete set, it is inevitable that a number, say x, be it initial data or an intermediate result, may not be a member in the set F. In this case, we can represent the number x only approximately with a floating point number $fl(x)$ within the set. Therefore, a floating point is represented in the machine as follows:

$$fl(x) = x(1 + \epsilon) \tag{3.86}$$

where $\in \le (1/2)\beta^{1-t}$ for *round-off* operation and $\in \le \beta^{1-t}$ for *chopped-off* operation. For example, suppose t = 4, e = 2, and β = 10; then with this computer, we have the following floating point representation of numbers:

$$fl(0.382874 \times 10^{15}) = 0.3829 \times 10^{15}$$

$$fl(0.875648 \times 10^{06}) = 0.8756 \times 10^{06}$$

This kind of operation in determining the floating point number $fl(x)$ is called *round-off* operation. We can easily show that the relative error in this operation is

$$\left|\frac{fl(x)-x}{x}\right| \leq 5\times 10^{t} \tag{3.87}$$

For a binary system, we can derive the relative error as follows:

$$\left|\frac{fl(x)-x}{x}\right| \leq 2^{t} \tag{3.88}$$

The original floating point representation of number in Equation 3.86 can be written as follows:

$$fl(x)-x = x\epsilon \tag{3.89}$$

$$\frac{fl(x)-x}{x} = \epsilon \tag{3.90}$$

Obviously, from the above equations, it can be observed that the quantity $x\epsilon$ is the *absolute error* and ϵ is the *relative error* in representing x with a floating point number $fl(x)$.

3.7.4 Algebraic operations with floating point numbers

Assume that a and b are floating point numbers; $a, b \in F$. If we perform floating point algebraic operations on the two numbers, then we have

$$fl(a \otimes b) = (a \otimes b)(1 + |\epsilon|) \tag{3.91}$$

where the symbol $\otimes$ denotes any algebraic operator. It can be easily verified that the exact result after performing algebraic operations on any two floating point numbers may not be a member in the original floating point number set F. For example, let $t = 4$ and $\beta = 10$, and let the two floating point numbers a and b be

$$a = 0.2256\times 10^{-6}; \quad b = 0.3832\times 10^{-4}$$

Then, after performing addition of the two floating point numbers, the exact algebraic result is

$$a + b = 0.002256 \times 10^{-4} + 0.3832 \times 10^{-4}$$
$$= 0.385456 \times 10^{-4}$$

which is apparently not in the set F. This result is in fact represented approximately as 0.3855×10^{-4}. This shows that the exact solution after performing algebraic operations on any two floating point data may not be in the original set of F.

The floating point operations apply to vectors and matrices in the same way as to individual numbers. Suppose we have two vectors $\mathbf{x}, \mathbf{y} \in \mathbf{R}^n$; then

$$|fl(\mathbf{x}^T\mathbf{y}) - (\mathbf{x}^T\mathbf{y})| \leq n\,|\epsilon|\,\|\mathbf{x}\|\,\|\mathbf{y}\| \tag{3.92}$$

where n is the size of the vector. For matrices, such as $\mathbf{A}, \mathbf{B} \in \mathbf{R}^{n\times n}$, then we have

$$\begin{cases} fl(\alpha \mathbf{A}) = (\alpha \mathbf{A}) + \mathbf{E}; & \|\mathbf{E}\| \le |\epsilon| \, \|\alpha \mathbf{A}\| \\ fl(\mathbf{A} + \mathbf{B}) = (\mathbf{A} + \mathbf{B}) + \mathbf{E}; & \|\mathbf{E}\| \le |\epsilon| \, \|\mathbf{A} + \mathbf{B}\| \\ fl(\mathbf{AB}) = (\mathbf{AB}) + \mathbf{E}; & \|\mathbf{E}\| \le |\epsilon| \, \|\mathbf{A}\| \, \|\mathbf{B}\| \end{cases} \tag{3.93}$$

3.7.5 Propagation of round-off error in algebraic operations

With the use of floating point numbers, in general, some algebraic operations such as the associativity law in addition and multiplication, and the distributivity law in multiplication may not work in the way as with exact mathematics. For example, in exact algebraic operations, the associativity law holds such that

$$a + b + c = (a + b) + c = a + (b + c) \tag{3.94}$$

However, in floating point operation, the results may be different. We give an example as follows:

$$\begin{aligned} fl[(a + b) + c] &= fl[(a + b)(1 + \epsilon_1) + c] \\ &= [(a + b)(1 + \epsilon_1) + c](1 + \epsilon_2) \\ &= (a + b + c)\left[1 + \frac{a + b}{a + b + c}\epsilon_1 \; (1 + \epsilon_2) + \epsilon_2\right] \end{aligned} \tag{3.95}$$

If we ignore the higher-order terms in the above relation, it becomes

$$\begin{aligned} fl[(a + b) + c] &= (a + b + c)\left[1 + \frac{a + b}{a + b + c}\epsilon_1 + \epsilon_2\right] \\ &= (a + b + c)(1 + \epsilon) \end{aligned} \tag{3.96}$$

where the relative error is

$$\epsilon = \frac{a + b}{a + b + c}\epsilon_1 + \epsilon_2 \tag{3.97}$$

In the same way, we can derive the following relationship:

$$fl[a + (b + c)] = (a + b + c)(1 + \bar{\epsilon}) \tag{3.98}$$

where $\bar{\epsilon}$ is the relative error in the form of

$$\bar{\epsilon} = \frac{b + c}{a + b + c}\bar{\epsilon}_1 + \bar{\epsilon}_2 \tag{3.99}$$

It can be seen that if $|a + b| < |b + c|$, then performing the first kind of associative operation will produce more accurate results, and vice versa.

3.7.6 Effect of representation error in initial data

At this juncture, we have discussed numerical error caused by round-off error in floating point operations. In conducting any computer-based analysis of engineering systems, we also need to deal with representation error in the initial data and its effect on the result of subsequent computations. Assume that **x** is an n-dimensional initial data vector:

$$\mathbf{x} = [x_1, \ldots, x_n]^T \tag{3.100}$$

This initial data vector will be used for computing an m-dimensional vector **y** according to the relation $y_i = f_i(\mathbf{x})$, i = 1, …, m:

$$\begin{aligned} \mathbf{y} &= [y_1, \ldots, y_m]^T \\ &= [f_1(\mathbf{x}), \ldots, f_m(\mathbf{x})]^T \end{aligned} \tag{3.101}$$

If $\bar{\mathbf{x}}$ is an approximation to the initial data vector **x**, then we can denote the absolute and the relative error of replacing **x** with $\bar{\mathbf{x}}$ as follows:

$$\Delta x_i = \bar{x}_i - x_i, \; i = 1, \ldots, n \tag{3.102}$$

$$\epsilon_{xi} = \frac{\Delta x_i}{x_i}, \text{ if } x_i \neq 0 \tag{3.103}$$

By ignoring the higher-order terms, we can estimate the error in vector **y** after replacing **x** with $\bar{\mathbf{x}}$ from the following relation:

$$\begin{aligned} \Delta y_i &= \bar{y}_i - y_i = f_i(\bar{\mathbf{x}}) - f_i(\mathbf{x}) \\ &= \sum_{j=1}^{n} \frac{\partial f_i(\mathbf{x})}{\partial x_j} \Delta x_j, \; i = 1, \ldots, m \end{aligned} \tag{3.104}$$

This relation can be written in the matrix form as follows:

$$\begin{aligned} \Delta \mathbf{y} = \begin{Bmatrix} \Delta y_1 \\ \vdots \\ \Delta y_m \end{Bmatrix} &= \begin{bmatrix} \frac{\partial f_1(\mathbf{x})}{\partial x_1} & \cdots & \frac{\partial f_1(\mathbf{x})}{\partial x_n} \\ \vdots & \ddots & \vdots \\ \frac{\partial f_m(\mathbf{x})}{\partial x_1} & \cdots & \frac{\partial f_m(\mathbf{x})}{\partial x_n} \end{bmatrix} \begin{Bmatrix} \Delta x_1 \\ \vdots \\ \Delta x_n \end{Bmatrix} a \\ &= J[f(\mathbf{x})] \, \Delta \mathbf{x} \end{aligned} \tag{3.105}$$

where J[f(**x**)] is the Jacobian matrix of function f(**x**) with respect to initial data vector **x**. We can directly obtain the relative error in **y**:

$$\begin{aligned} \epsilon_{yi} = \frac{\Delta y_i}{f_i(\mathbf{x})} &= \sum_{j=1}^{n} \frac{\partial f_i(\mathbf{x})}{\partial x_j} \frac{\Delta x_j}{f_i(\mathbf{x})} \\ &= \sum_{j=1}^{n} \left(\frac{x_j}{f_i(\mathbf{x})} \frac{\partial f_i(\mathbf{x})}{\partial x_j} \right) \epsilon_{xj} \end{aligned} \tag{3.106}$$

From this equation, we observe that the effect of error in initial data vector on resulting computation is directly influenced by the following parameter, which is referred to as the *condition number*:

$$\left(\frac{x_j}{f_i(\mathbf{x})} \frac{\partial f_i(\mathbf{x})}{\partial x_j} \right) \tag{3.107}$$

The magnitude of this condition number represents the degree of influence of representation error in initial data in subsequent computations. If the condition number is large, obviously, the error will get propagated and its adverse effect on subsequent computation may be amplified. We can verify, for simple algebraic operations, the following relationships between the error in initial data and the resulting error in subsequent operations:

$$\begin{cases} \epsilon_{x\pm y} = \dfrac{x}{x \pm y}\,\epsilon_x \pm \dfrac{y}{x \pm y}\,\epsilon_y \\ \epsilon_{xy} = \epsilon_x + \epsilon_y \\ \epsilon_{x/y} = \epsilon_x - \epsilon_y \\ \epsilon_{\sqrt{x}} = \dfrac{1}{2}\,\epsilon_x \end{cases} \tag{3.108}$$

The above relationships indicate that the condition numbers corresponding to performing multiplication, division, and square root operations are fixed and small. Therefore, initial error does not have much effect on subsequent results after performing these operations. However, for addition operation, if x and y have the opposite sign, especially when $x \approx -y$, then the condition number may become very large. With subtraction operation, if x and y have the same sign, especially when $x \approx y$, then the condition number would also become very large. For these extreme cases with possibly large condition numbers, any initial error present in data would lead to larger errors in the results after performing the operation.

From above discussion on the propagation of initial representation error and round-off error in floating point operations, we observe that a mathematical algorithm, if implemented differently, may lead to drastically different results, especially when a system is ill-conditioned. That a system is ill-conditioned means that in subsequent computations a system is very sensitive to any small deviations in the initial data vector. A system is considered well-conditioned if it is insensitive to any small perturbation on the initial data vector.

3.7.7 Convergent matrices

A matrix **A** is called convergent if it satisfies the following condition:

$$\lim_{k \to \infty} \left\| \mathbf{A}^k \right\| \to 0 \tag{3.109}$$

It can be demonstrated that for a matrix to be convergent, all its eigenvalues should be less than 1. For a symmetric matrix, since the second (or spectral) norm of **A** is its largest eigenvalue, $\|\mathbf{A}\|_2 = \lambda_{max}(\mathbf{A})$, for it to be convergent it must satisfy the condition $\lambda_{max}(\mathbf{A}) < 1$. If the matrix **A** is convergent, then the spectral radius $\rho(\mathbf{A}) < 1$. We can verify this property through solving the following standard eigenvalue problem:

$$\mathbf{A}\mathbf{\Phi} = \mathbf{\Phi}\mathbf{\Lambda} \tag{3.110}$$

where $\mathbf{\Phi}$ is the matrix of eigenvectors that are orthonormal and $\mathbf{\Lambda}$ is the diagonal matrix of eigenvalues of $\mathbf{A}$:

$$\mathbf{A} = \mathbf{\Phi\Lambda\Phi}^T \quad (3.111)$$

The kth power of this matrix can be obtained as follows:

$$\begin{aligned}\mathbf{A}^k &= (\mathbf{\Phi\Lambda\Phi}^T)^k \\ &= (\mathbf{\Phi\Lambda\Phi}^T)(\mathbf{\Phi\Lambda\Phi}^T)\ldots(\mathbf{\Phi\Lambda\Phi}^T) \\ &= \mathbf{\Phi\Lambda}^k\mathbf{\Phi}^T\end{aligned} \quad (3.112)$$

We note that $\mathbf{\Phi}^T\mathbf{\Phi} = \mathbf{I}$, where $\mathbf{I}$ is the identity matrix. The matrix $\mathbf{\Lambda}^k$ is of the following form:

$$\mathbf{\Lambda}^k = \begin{bmatrix} \lambda_1^k & & & \\ & \lambda_2^k & & \\ & & \ddots & \\ & & & \lambda_n^k \end{bmatrix} \quad (3.113)$$

Now if we assume that all the eigenvalues of $\mathbf{A}$ are less than 1, then we get

$$\lim_{k\to\infty} \mathbf{\Lambda}^k \to 0 \quad (3.114)$$

$$\lim_{k\to\infty} \mathbf{A}^k = \lim_{k\to\infty}(\mathbf{\Phi\Lambda}^k\mathbf{\Phi}^T) \to 0 \quad (3.115)$$

This shows that the eigenvalue properties of a matrix determine its convergence.

3.7.8 Neumann series

The Neumann series refers to a matrix series of the following form:

$$\mathbf{I} + \mathbf{A} + \mathbf{A}^2 + \cdots + \mathbf{A}^n + \cdots \quad (3.116)$$

This series converges to $(\mathbf{I} - \mathbf{A})^{-1}$ if and only if matrix $\mathbf{A}$ is convergent. This can be proven by premultiplying the partial summation of the series with $(\mathbf{I} - \mathbf{A})$:

$$\begin{aligned}&(\mathbf{I} - \mathbf{A})(\mathbf{I} + \mathbf{A} + \mathbf{A}^2 + \cdots + \mathbf{A}^m) \\ &\quad = \mathbf{I} + \mathbf{A} + \mathbf{A}^2 + \cdots + \mathbf{A}^m \\ &\qquad - \mathbf{A} - \mathbf{A}^2 - \cdots - \mathbf{A}^m - \mathbf{A}^{m+1} \\ &\quad = \mathbf{I} - \mathbf{A}^{m+1}\end{aligned} \quad (3.117)$$

If matrix $\mathbf{A}$ is convergent, then $\lim_{m\to\infty} \mathbf{A}^m \to 0$:

$$(\mathbf{I} - \mathbf{A})(\mathbf{I} + \mathbf{A} + \mathbf{A}^2 + \cdots + \mathbf{A}^m + \cdots) = \lim_{m\to\infty}(\mathbf{I} - \mathbf{A}^{m+1}) = \mathbf{I} \quad (3.118)$$

$$\mathbf{I} + \mathbf{A} + \mathbf{A}^2 + \cdots + \mathbf{A}^m + \cdots = (\mathbf{I} - \mathbf{A})^{-1} \quad (3.119)$$

It is interesting to note that the Neumann series has a scalar counterpart in the geometric series:

$$1 + a + a^2 + \cdots + a^m + \cdots \tag{3.120}$$

This can be shown to converge to 1/(1 – a) if and only if a < 1. The convergence property of the Neumann series serves as the basis in analyzing iterative solution methods, which will be discussed in later chapters.

3.7.9 Banach Lemma

For any matrix **A**, the Banach Lemma can be stated by the following relations:

$$\frac{1}{1+\|\mathbf{A}\|} \le \left\|(\mathbf{I} \pm \mathbf{A})^{-1}\right\| \le \frac{1}{1-\|\mathbf{A}\|} \tag{3.121}$$

We will show later that this lemma is very important in carrying out error analysis of solution methods for system of linear equations. This lemma can be proven by starting from the following identity:

$$\mathbf{I} = (\mathbf{I} \pm \mathbf{A})(\mathbf{I} \pm \mathbf{A})^{-1} \tag{3.122}$$

Some fundamental relations regarding norms of vectors and matrices are also needed. For example, we showed in Chapter 1 that for any two vectors, **a**, **b** ∈ $\mathbf{R}^n$, and for any two matrices, **A**, **B** ∈ $\mathbf{R}^{n\times n}$, we have the following relations:

$$\|\mathbf{a} \pm \mathbf{b}\| \le \|\mathbf{a}\| + \|\mathbf{b}\| \tag{3.123}$$

$$\begin{cases} \|\mathbf{A} \pm \mathbf{B}\| \le \|\mathbf{A}\| + \|\mathbf{B}\| \\ \|\mathbf{AB}\| \le \|\mathbf{A}\|\|\mathbf{B}\| \end{cases} \tag{3.124}$$

With these relations, we can consider the norm of the identity equation (Equation 3.122):

$$\begin{aligned} \|\mathbf{I}\| = 1 &= \left\|(\mathbf{I} \pm \mathbf{A})\,(\mathbf{I} \pm \mathbf{A})^{-1}\right\| \\ &\le \|(\mathbf{I} \pm \mathbf{A})\|\,\left\|(\mathbf{I} \pm \mathbf{A})^{-1}\right\| \\ &\le (\|\mathbf{I}\| + \|\mathbf{A}\|)\,\left\|(\mathbf{I} \pm \mathbf{A})^{-1}\right\| \\ &\le (1 + \|\mathbf{A}\|)\,\left\|(\mathbf{I} \pm \mathbf{A})^{-1}\right\| \end{aligned} \tag{3.125}$$

From the above equation, we can see that we have proven the left-hand side of the inequality in Equation 3.121.

Now we prove the right-hand side of the inequality in Banach Lemma by starting from the same identity as in Equation 3.122 and going through the following steps:

$$\begin{aligned} \mathbf{I} &= (\mathbf{I} \pm \mathbf{A})\,(\mathbf{I} \pm \mathbf{A})^{-1} \\ &= (\mathbf{I} \pm \mathbf{A})^{-1} \pm \mathbf{A}\,(\mathbf{I} \pm \mathbf{A})^{-1} \end{aligned} \tag{3.126}$$

$$(\mathbf{I} \pm \mathbf{A})^{-1} = \mathbf{I} \mp \mathbf{A}\,(\mathbf{I} \pm \mathbf{A})^{-1} \tag{3.127}$$

$$\begin{aligned} \left\|(\mathbf{I} \pm \mathbf{A})^{-1}\right\| &= \left\|\mathbf{I} \mp \mathbf{A}\,(\mathbf{I} \pm \mathbf{A})^{-1}\right\| \\ &\le \left\|\mathbf{I}\right\| + \left\|\mathbf{A}\,(\mathbf{I} \pm \mathbf{A})^{-1}\right\| \\ &\le 1 + \left\|\mathbf{A}\right\| \left\|(\mathbf{I} \pm \mathbf{A})^{-1}\right\| \end{aligned} \tag{3.128}$$

$$(1 - \left\|\mathbf{A}\right\|) \left\|(\mathbf{I} \pm \mathbf{A})^{-1}\right\| \le 1 \tag{3.129}$$

This equation proves the right-hand side of the inequality in Equation 3.121. Thus, the Banach Lemma is proven.

3.7.10 Error analysis in the solution of a system of equations

With the preliminary knowledge provided in the previous sections, we are now ready to study the propagation of different types of errors in the solution of a system of linear equations:

$$\mathbf{Ax} = \mathbf{b} \tag{3.130}$$

Suppose that there exist errors in the data (the stiffness matrix **A** and the load vector **b**), such that

$$\begin{cases} \mathbf{A} \leftarrow \mathbf{A} + \delta\mathbf{A} \\ \mathbf{b} \leftarrow \mathbf{b} + \delta\mathbf{b} \end{cases} \tag{3.131}$$

Then we can expect that the solution vector obtained would be changed to $\mathbf{x} + \delta\mathbf{x}$. We assume that the deviation on the stiffness matrix $\delta\mathbf{A}$ is convergent, which is a reasonable assumption because $\delta\mathbf{A}$ consists mainly of round-off errors. Then in any practical solution process, we are solving a perturbed system as follows:

$$(\mathbf{A} + \delta\mathbf{A})\,(\mathbf{x} + \delta\mathbf{x}) = (\mathbf{b} + \delta\mathbf{b}) \tag{3.132}$$

In *forward* error analysis, our objective is to establish an *a priori* bound on the error in the solution vector $\delta\mathbf{x}$ with the presence of error in data $\delta\mathbf{A}$ and $\delta\mathbf{b}$. It provides the answers to the following question: what are the largest errors possible in the solution? In backward error analysis, it tries to establish *a posteriori* bounds on deviations in the original data that define the system. Rather than solving a problem exactly, we solve that problem approximately with some errors present in the solution vector. This approximate solution is the exact solution of a different problem. This type of analysis tries to answer the following question: how different is the problem we have solved from the problem we intended to solve? As with perturbation analysis, it explores the problem's susceptibility to growth of round-off errors. It also determines the relations between the errors in data and the resulting errors in the results. In carrying out error analysis, it is natural to use the perturbation idea and consider perturbed systems.

3.7.11 Errors in load vector

First, we consider the case that there is error in the load vector **b** only and assume that the system coefficient matrix is exact. As a consequence, we are solving the following perturbed systems:

$$\mathbf{A}(\mathbf{x} + \delta\mathbf{x}) = (\mathbf{b} + \delta\mathbf{b}) \tag{3.133}$$

From the originally unperturbed system, we have

$$\|\mathbf{A}\mathbf{x}\| = \|\mathbf{b}\| \tag{3.134}$$

Since $\|\mathbf{A}\mathbf{x}\| \le \|\mathbf{A}\|\,\|\mathbf{x}\|$, we have

$$\|\mathbf{A}\|\,\|\mathbf{x}\| \ge \|\mathbf{A}\mathbf{x}\| = \|\mathbf{b}\| \tag{3.135}$$

Hence, we get

$$\|\mathbf{x}\| \ge \|\mathbf{A}\|^{-1}\|\mathbf{b}\| \tag{3.136}$$

The difference between Equations 3.130 and 3.133 yields the following:

$$\mathbf{A}\delta\mathbf{x} = \delta\mathbf{b} \tag{3.137}$$

$$\begin{aligned}\|\delta\mathbf{x}\| &= \|\mathbf{A}^{-1}\delta\mathbf{b}\| \\ &\le \|\mathbf{A}^{-1}\|\,\|\delta\mathbf{b}\|\end{aligned} \tag{3.138}$$

Subsequently, from Equations 3.136 and 3.138, we have the following:

$$\begin{aligned}\frac{\|\delta\mathbf{x}\|}{\|\mathbf{x}\|} &\le \frac{\|\mathbf{A}^{-1}\|}{\|\mathbf{A}\|^{-1}}\frac{\|\delta\mathbf{b}\|}{\|\mathbf{b}\|} \\ &= \left(\|\mathbf{A}^{-1}\|\,\|\mathbf{A}\|\right)\frac{\|\delta\mathbf{b}\|}{\|\mathbf{b}\|} \\ &= \kappa(\mathbf{A})\frac{\|\delta\mathbf{b}\|}{\|\mathbf{b}\|}\end{aligned} \tag{3.139}$$

The quantity $\kappa(\mathbf{A})$ is called the *condition number* of matrix **A**. We can rewrite Equation 3.139 in terms of the second norms:

$$\frac{\|\delta\mathbf{x}\|_2}{\|\mathbf{x}\|_2} \le \left(\|\mathbf{A}^{-1}\|_2\,\|\mathbf{A}\|_2\right)\frac{\|\delta\mathbf{b}\|_2}{\|\mathbf{b}\|_2} \tag{3.140}$$

The second norm of symmetric and positive definite matrices is equal to their maximum eigenvalue:

$$\begin{cases}\|\mathbf{A}\|_2 = \lambda_{max}(\mathbf{A}) \\ \|\mathbf{A}^{-1}\|_2 = \lambda_{max}(\mathbf{A}^{-1}) = \dfrac{1}{\lambda_{min}(\mathbf{A})}\end{cases} \tag{3.141}$$

From this equation, we can see that the condition number of matrix **A** is the ratio of its largest and smallest eigenvalues:

$$\kappa(\mathbf{A}) = \left(\|\mathbf{A}^{-1}\|\,\|\mathbf{A}\|\right) = \frac{\lambda_{max}}{\lambda_{min}} \tag{3.142}$$

Therefore, for matrices that are symmetric and positive definite, we have the following:

$$\frac{\|\delta \mathbf{x}\|_2}{\|\mathbf{x}\|_2} \le \left(\frac{\lambda_{\max}}{\lambda_{\min}}\right)\frac{\|\delta \mathbf{b}\|_2}{\|\mathbf{b}\|_2} \tag{3.143}$$

This relation shows that the condition number of the coefficient matrix sets the upper limit of the relation between the relative error in the solution vector and the error present in the load vector.

3.7.12 Errors in coefficient matrix and load vector

Now we consider a general perturbed system with errors in data, $\delta\mathbf{A}$ and $\delta\mathbf{b}$, present at the same time, and our objective in a forward error analysis is to derive a relationship of the form

$$\|\delta \mathbf{x}\| = f\left(\|\delta \mathbf{A}\|, \|\delta \mathbf{b}\|\right) \tag{3.144}$$

We start with the general perturbed system of equations as follows:

$$(\mathbf{A} + \delta\mathbf{A})\,(\mathbf{x} + \delta\mathbf{x}) = (\mathbf{b} + \delta\mathbf{b}) \tag{3.145}$$

We can premultiply this equation with $\mathbf{A}^{-1}$ and rearrange it through the following steps to arrive at an expression for $\delta\mathbf{x}$:

$$\mathbf{x} + \delta\mathbf{x} + \mathbf{A}^{-1}\delta\mathbf{A}\mathbf{x} + \mathbf{A}^{-1}\delta\mathbf{A}\delta\mathbf{x} = \mathbf{A}^{-1}\mathbf{b} + \mathbf{A}^{-1}\delta\mathbf{b} \tag{3.146}$$

$$\delta\mathbf{x} + \mathbf{A}^{-1}\delta\mathbf{A}\mathbf{x} + \mathbf{A}^{-1}\delta\mathbf{A}\delta\mathbf{x} = \mathbf{A}^{-1}\delta\mathbf{b} \tag{3.147}$$

$$(\mathbf{I} + \mathbf{A}^{-1}\delta\mathbf{A})\delta\mathbf{x} = \mathbf{A}^{-1}(\delta\mathbf{b} - \delta\mathbf{A}\mathbf{x}) \tag{3.148}$$

$$\delta\mathbf{x} = (\mathbf{I} + \mathbf{A}^{-1}\delta\mathbf{A})^{-1}\mathbf{A}^{-1}(\delta\mathbf{b} - \delta\mathbf{A}\mathbf{x}) \tag{3.149}$$

With a similar approach used in the first case, we take the norm of the above equation and apply triangular inequality relation and the Banach Lemma. The error in the solution vector is bounded by the following relation:

$$\begin{aligned}\|\delta\mathbf{x}\| &= \left\|(\mathbf{I} + \mathbf{A}^{-1}\delta\mathbf{A})^{-1}\mathbf{A}^{-1}(\delta\mathbf{b} - \delta\mathbf{A}\mathbf{x})\right\| \\ &\le \left\|(\mathbf{I} + \mathbf{A}^{-1}\delta\mathbf{A})^{-1}\right\|\,\left\|\mathbf{A}^{-1}\right\|\,\left(\|\delta\mathbf{b}\| + \|\delta\mathbf{A}\|\,\|\mathbf{x}\|\right)\end{aligned} \tag{3.150}$$

$$\begin{aligned}\frac{\|\delta\mathbf{x}\|}{\|\mathbf{x}\|} &\le \left\|(\mathbf{I} + \mathbf{A}^{-1}\delta\mathbf{A})^{-1}\right\|\,\left\|\mathbf{A}^{-1}\right\|\,\|\mathbf{A}\|\left(\frac{\|\delta\mathbf{b}\|}{\|\mathbf{A}\|\,\|\mathbf{x}\|} + \frac{\|\delta\mathbf{A}\|}{\|\mathbf{A}\|}\right) \\ &\le \frac{1}{1 - \left\|\mathbf{A}^{-1}\right\|\,\|\delta\mathbf{A}\|}\,\kappa(\mathbf{A})\left(\frac{\|\delta\mathbf{b}\|}{\|\mathbf{b}\|} + \frac{\|\delta\mathbf{A}\|}{\|\mathbf{A}\|}\right) \\ &\le \frac{\kappa(\mathbf{A})}{1 - \kappa(\mathbf{A})\dfrac{\|\delta\mathbf{A}\|}{\|\mathbf{A}\|}}\left(\frac{\|\delta\mathbf{b}\|}{\|\mathbf{b}\|} + \frac{\|\delta\mathbf{A}\|}{\|\mathbf{A}\|}\right)\end{aligned} \tag{3.151}$$

If we ignore the higher-order terms, we can simplify the right-hand side of the above equation as follows:

$$\frac{\|\delta \mathbf{x}\|}{\|\mathbf{x}\|} \leq \kappa(\mathbf{A})\left(\frac{\|\delta \mathbf{A}\|}{\|\mathbf{A}\|}+\frac{\|\delta \mathbf{b}\|}{\|\mathbf{b}\|}\right) \tag{3.152}$$

From the above equation, if we assume that the decimal precision of a computer representation in data is t, and the precision in computational output is α, then approximately we have

$$\begin{cases} \dfrac{\|\delta \mathbf{x}\|}{\|\mathbf{x}\|} = 10^{-\alpha} \\ \left(\dfrac{\|\delta \mathbf{A}\|}{\|\mathbf{A}\|}+\dfrac{\|\delta \mathbf{b}\|}{\|\mathbf{b}\|}\right) = 10^{-t} \end{cases} \tag{3.153}$$

$$10^{-\alpha} \leq \kappa(\mathbf{A})\, 10^{-t} \tag{3.154}$$

$$\alpha \geq t - \log[\kappa(\mathbf{A})] \tag{3.155}$$

We know that for symmetric and positive definite matrices, $\kappa(\mathbf{A}) = \lambda_n/\lambda_1 \geq 1$. If $\kappa(\mathbf{A}) = 1$, then there must be $\mathbf{A} = \mathbf{I}$. For this special case, we have $\alpha \geq t$, which means that there is no loss of accuracy during the computation process. If $\kappa(\mathbf{A})$ is a very large number, it is very likely that $\alpha < t$, resulting in loss of precision in computation.

From the above discussions, we observe that the condition number works like an amplification factor on the relative error in initial data. In other words, the condition number represents the sensitivity of solution vector $\mathbf{x}$ of the system of linear equations, $\mathbf{Ax} = \mathbf{b}$, with respect to error in the initial data. The larger the condition number, the more sensitive the solution process becomes. If the system is too sensitive, a very small initial error in the data, such as $\delta\mathbf{A}$ and $\delta\mathbf{b}$, would result in large relative error in the solution vector $\mathbf{x}$. These kinds of problems in which the solution vector is very sensitive to any small perturbation on the initial data are called "ill-conditioned."

In nonlinear analysis, for softening type of structures, the tangential stiffness matrix of the structure becomes ill-conditioned once the solution path is in the neighborhood of the limit point or bifurcation points.

3.7.13 Solution error estimation with residual vector

We can argue that if the solution vector obtained is not the exact solution for the original system, then it must be the exact solution for a perturbed system. Let us define a nonzero vector, $\mathbf{r}$, as follows:

$$\mathbf{r}\big|_{\mathbf{x}+\delta\mathbf{x}} = \mathbf{b} - \mathbf{A}(\mathbf{x}+\delta\mathbf{x}) \tag{3.156}$$

which is the residual vector corresponding to the approximate solution $\mathbf{x} + \delta\mathbf{x}$. The solution vector $\mathbf{x} + \delta\mathbf{x}$ is the exact solution of the following system of equations:

$$\mathbf{A}(\mathbf{x}+\delta\mathbf{x}) = \mathbf{b} - \mathbf{r}\big|_{\mathbf{x}+\delta\mathbf{x}} \tag{3.157}$$

$$\mathbf{x}+\delta\mathbf{x}=\mathbf{A}^{-1}(\mathbf{b}-\mathbf{r}|_{\mathbf{x}+\delta\mathbf{x}}) \tag{3.158}$$

$$\delta\mathbf{x}=-\mathbf{A}^{-1}\mathbf{r}|_{\mathbf{x}+\delta\mathbf{x}} \tag{3.159}$$

Therefore, round-off errors in solution vector are bounded in the following relation:

$$\|\delta\mathbf{x}\| \le \|\mathbf{A}^{-1}\| \, \|\mathbf{r}|_{\mathbf{x}+\delta\mathbf{x}}\| \tag{3.160}$$

When there is error in the initial coefficient matrix, we can also prove that if

$$\|\mathbf{I}-(\mathbf{A}+\delta\mathbf{A})^{-1}\|<1 \tag{3.161}$$

then

$$\|\delta\mathbf{x}\| \le \frac{\|(\mathbf{A}+\delta\mathbf{A})^{-1}\|}{1-\|\mathbf{I}-(\mathbf{A}+\delta\mathbf{A})^{-1}\mathbf{A}\|}\|\mathbf{r}|_{\mathbf{x}+\delta\mathbf{x}}\| \tag{3.162}$$

At this juncture, we have seen from proceeding discussions that the condition number represents the sensitivity of the solution of the system to initial errors. Now the question is whether the magnitude of the residual vector represents the accuracy of the approximate solution vector. To answer this question, we need to consider the stability of solution algorithms.

3.7.14 Stability of solution algorithms

Suppose that we intend to solve the following system of linear equations:

$$\mathbf{A}_0\mathbf{x}=\mathbf{b}_0 \tag{3.163}$$

where $\mathbf{A}_0$ and $\mathbf{b}_0$ are initial data with certain sampling or representation errors. Suppose we have obtained a solution vector $\bar{\mathbf{x}}$ by solving the above system with certain algorithms such as the Gauss-elimination method. Because of presence of round-off error, obviously, $\bar{\mathbf{x}}$ is not the exact solution of the original system; rather it is the exact solution of a perturbed system in the neighborhood of the original system:

$$(\mathbf{A}_0+\delta\mathbf{A}_0)\,\bar{\mathbf{x}}=\mathbf{b}_0+\delta\mathbf{b} \tag{3.164}$$

If the magnitudes of resulting perturbations on system data $\|\delta\mathbf{A}_0\|$ and $\|\delta\mathbf{b}\|$ are small, then we can say that the solution algorithm is "stable." If the original system is not ill-conditioned, we can expect to obtain fairly accurate solutions using a stable algorithm. We note that the concept of stability of an algorithm is different from the concept of sensitivity of the solution vector to errors in original data or the conditioning of the coefficient matrix. Conditioning is an inherent property of a system, whereas stability varies with different algorithms. By conventional wisdom, a solution vector obtained using a stable algorithm should usually be acceptable.

For example, we consider the back-substitution procedure in the Gauss-elimination method:

$$\mathbf{U}\mathbf{x}=\mathbf{y} \tag{3.165}$$

where $\mathbf{U}$ is a $n \times n$ nonsingular upper triangular matrix. Let $\bar{\mathbf{x}}$ be the solution vector obtained from the standard back-substitution procedure conducted in a digital computer. If we use single-precision operations, it can be verified that the solution vector obtained is the exact solution for the following perturbed system:

$$(\mathbf{U}+\mathbf{E})\,\bar{\mathbf{x}}=\mathbf{y} \tag{3.166}$$

where elements in the perturbation matrix E satisfy the following relation:

$$|e_{ij}| \le (n+1)\,\eta\,|u_{ij}|\,10^{t} \tag{3.167}$$

where:

t is the machine precision
η is a parameter corresponding to algebraic operations, which is on the order of 1
n is the order of matrix **U**

It is obvious that if n is not very large, the magnitude of the perturbation matrix **E** is very close to the round-off error in representing matrix **U**. This shows that the back-substitution procedure is stable.

3.7.15 Set of neighboring system of equations

Given $\mathbf{A}_0\mathbf{x}=\mathbf{b}_0$ as the original system of equations, assume that δ**A** and δ**b** are nonnegative matrix and vector, respectively. If any matrix **A** and vector **b** satisfy the following relations:

$$|a_{ij}-a_{ij}^{0}| \le \delta a_{ij}; \quad |b_k-b_k^{0}| \le \delta b_k \tag{3.168}$$

then we say that matrix **A** forms a matrix set $\{\mathbf{A}_0, \delta\mathbf{A}\}$ in the neighborhood of $\mathbf{A}_0$, and vector **b** forms a vector set $\{\mathbf{b}_0, \delta\mathbf{b}\}$ in the neighborhood of $\mathbf{b}_0$. Correspondingly, the family of system of equations determined from any matrix **A** and vector **b** satisfying the above relations forms a set of system of equations in the neighborhood of $\{\delta\mathbf{A}, \delta\mathbf{b}\}$ with respect to $\mathbf{A}_0\mathbf{x}=\mathbf{b}_0$. Obviously, for a nonsingular matrix $\mathbf{A}_0$, if δ**A** is sufficiently small, then any matrix in the matrix set $\{\mathbf{A}_0, \delta\mathbf{A}\}$ is also nonsingular.

Now we can discuss the Oettli–Prager theorem (Oettli and Prager, 1964). It states that for any matrix $\mathbf{A} \in \{\mathbf{A}_0, \delta\mathbf{A}\}$ and for any vector $\mathbf{b} \in \{\mathbf{b}_0, \delta\mathbf{b}\}$, if the exact solution for the neighboring system of equations $\mathbf{A}\mathbf{x}=\mathbf{b}$ is $\bar{\mathbf{x}}$, then the residual vector $\mathbf{r}(\bar{\mathbf{x}})=\mathbf{b}_0-\mathbf{A}_0\bar{\mathbf{x}}$ satisfies the following relation:

$$\mathbf{r}(\bar{\mathbf{x}}) \le \delta\mathbf{A}\,|\bar{\mathbf{x}}| + \delta\mathbf{b} \tag{3.169}$$

On the other hand, if the above relation is satisfied, then there exist a neighboring matrix **A** and a vector **b** such that $\mathbf{A}\bar{\mathbf{x}}=\mathbf{b}$. We note that the above relation compares elements in vectors from both sides of the equation. We can prove the first part of the theorem in a straightforward manner.

Since $\bar{\mathbf{x}}$ is the exact solution for $\mathbf{A}\bar{\mathbf{x}}=\mathbf{b}$, and $|\mathbf{A}-\mathbf{A}_0| \le \delta\mathbf{A}$, $|\mathbf{b}-\mathbf{b}_0| \le \delta\mathbf{b}$ by definition, we have

$$\begin{aligned}
\mathbf{r}(\bar{\mathbf{x}}) &= |\mathbf{b}_0-\mathbf{A}_0\bar{\mathbf{x}}| \\
&= |\mathbf{b}-(\mathbf{b}-\mathbf{b}_0)-[\mathbf{A}-(\mathbf{A}-\mathbf{A}_0)]\bar{\mathbf{x}}| \\
&= |-(\mathbf{b}-\mathbf{b}_0)+(\mathbf{A}-\mathbf{A}_0)\bar{\mathbf{x}}| \\
&\le |\mathbf{b}-\mathbf{b}_0| + |\mathbf{A}-\mathbf{A}_0|\;|\bar{\mathbf{x}}| \\
&\le \delta\mathbf{b} + \delta\mathbf{A}\;|\bar{\mathbf{x}}|
\end{aligned} \tag{3.170}$$

Hence, the first part of the theorem is proved. This relation also refers to elements on the left-hand and right-hand sides of the equation. The above theorem gives the relation between the solution vector and its residual vector under the condition that the system is not ill-conditioned. From this theorem, we state that if the relative error for **A** and **b** is ϵ, that is, $\delta\mathbf{A} = \epsilon|\mathbf{A}_0|$, $\delta\mathbf{b} = \epsilon|\mathbf{b}_0|$, then when the residual vector satisfies the following relation:

$$\mathbf{r}(\bar{\mathbf{x}}) = |\mathbf{b}_0 - \mathbf{A}_0\bar{\mathbf{x}}| \le (|\mathbf{b}_0| - |\mathbf{A}_0||\bar{\mathbf{x}}|) \tag{3.171}$$

we can say that the approximate solution is the exact solution of the following neighboring system of equations:

$$\mathbf{A}\bar{\mathbf{x}} = \mathbf{b}, \quad \mathbf{A} \in \{\mathbf{A}_0, \delta\mathbf{A}\}, \ \mathbf{b} \in \{\mathbf{b}_0, \delta\mathbf{b}\} \tag{3.172}$$

However, it should be noted that when the original system is *ill-conditioned*, the results from this theorem may not be applied and could be misleading. In other words, with ill-conditioning, a small residual norm does not necessarily correspond to an accurate solution vector, and an accurate solution vector may have a sizable residual norm. To illustrate this point, we consider a simple system that is obviously ill-conditioned:

$$\begin{bmatrix} 1.000 & 1.001 \\ 1.000 & 1.000 \end{bmatrix} \begin{Bmatrix} x_1 \\ x_2 \end{Bmatrix} = \begin{Bmatrix} 1 \\ 0 \end{Bmatrix} \tag{3.173}$$

The exact solution for the above system is $[x_1 \quad x_2]^T = [-1000 \quad 1000]^T$. Suppose we have obtained an approximate solution $[\bar{x}_1 \quad \bar{x}_2]^T = [-1001 \quad 1000]^T$, which is very close to **x** in terms of relative and absolute errors. However, the norm of the residual vector $\mathbf{r}(\bar{\mathbf{x}}) = [1, 1]^T$ is not small at all, if compared with the load vector. On the other hand, for the following system of equations:

$$\begin{bmatrix} 1.000 & 1.001 \\ 1.000 & 1.000 \end{bmatrix} \begin{Bmatrix} x_1 \\ x_2 \end{Bmatrix} = \begin{Bmatrix} 2.001 \\ 2.000 \end{Bmatrix} \tag{3.174}$$

the exact solution can be obtained as $[x_1 \quad x_2]^T = [1 \quad 1]^T$. Suppose we obtain an approximate solution $[\bar{x}_1 \quad \bar{x}_2]^T = [2 \quad 0]^T$, which is not close to **x** at all. However, in this case, the residual vector $\mathbf{r}(\bar{\mathbf{x}}) = [0.001 \quad 0]^T$ has a very small norm. This simple example shows that for an ill-conditioned system, the magnitude of residual vector does not necessarily represent the accuracy of the solution vector. Caution and judgment must be exercised when solving an ill-conditioned system of equations.

3.7.16 Condition number and ill-conditioning in computational mechanics

We can look at the effect of condition number from a different perspective. From the above relations on solution error bounds, we can observe that the error in solution vector (displacement vector) $\|\delta\mathbf{x}\|$ would be small if the condition number $\kappa(\mathbf{A})$ is small. This means that small changes in data would have little effect on the solution vector **x** if the condition number of the system coefficient matrix is small. In other words, the solution vector is not affected seriously if the largest eigenvalue of the system, λ_{max}, is not too large as compared with the smallest eigenvalue λ_{min}. Such a system of equations is said to be "well-conditioned." Otherwise, if the condition number is very large, the system is called "ill-conditioned." In an ill-conditioned system, the resulting error in computation could be amplified even with very small perturbation in the original data. We note that the condition of a system is an inherent property. It does not have anything to do with the numerical method used in the analysis.

In computational mechanics, especially in nonlinear finite element analysis of large structural systems modeled with different types of material models and elements, it is very important to identify the causes of ill-conditioning so that the system of equations obtained in the modeling process can be properly solved. There are several extreme cases that cause the ill-conditioning of a system. From the definition of the condition number, we observe that if the smallest eigenvalue of the system $\lambda_1 \cong 0$, then the system would be extremely ill-conditioned. This case arises from the presence of rigid body modes in the system where the stiffness matrix becomes singular or nearly singular. Since an eigenvalue in a way represents the energy required to generate the deformation mode in the corresponding eigenvector, obviously, strain energy associated with rigid body modes is close to zero. In nonlinear analysis, especially with softening-type structural behavior, around the limit points and bifurcation points, the smallest eigenvalue of the tangent stiffness matrix approaches zero. Thus, the system becomes ill-conditioned in the neighborhood of limit points or bifurcation points.

On the other hand, just from a mathematical observation, we note that if the largest eigenvalue of the system, $\lambda_n \to \infty$, and the smallest eigenvalue is finite, then ill-conditioning also occurs. In finite element analysis, this corresponds to the presence of a rigid subsystem within the structural system, that is, some elements are rigid or nearly rigid. This could also happen with the use of metal plasticity theory to model the material behavior. In the case of incompressible material, as the Poisson's ratio approaches 0.5, material becomes incompressible and some of the deformation modes may require an infinite amount of energy to materialize.

In modeling structures with elements of mixed types of deformation modes, it is possible that $\lambda_n \gg \lambda_1$. This may happen in analyzing plates and shells where the in-plane deformations are coupled with bending modes of deformation. In these structures, usually λ_1 is associated with the bending mode that is flexible, whereas λ_n is associated with the axial and in-plane modes of deformation that tend to be fairly stiff, if not rigid. As a result, the condition number of such a system may become very large, resulting in ill-conditioning. Obviously, ill-conditioning may occur in many other cases where very stiff zones get embedded in soft regions such as in sandwich beams that are strong in bending but weak in shear.

Another case that may potentially result in ill-conditioning is through mesh refinement in adaptive finite element analysis. In structural dynamics, the eigenvalue problem can be written in the following form:

$$\mathbf{K}\Phi_j = \omega_j \mathbf{M}\Phi_j \tag{3.175}$$

where:

- $\mathbf{K}$ and $\mathbf{M}$ are the stiffness and mass matrices, respectively
- ω_j is a natural frequency
- Φ_j is the corresponding mode shape

Assume that the mass matrix is an identity matrix, $\mathbf{M} = \mathbf{I}$, as in the case of uniform elements. Then we have $\lambda = \omega^2$. The stiffness matrix is symmetric and positive definite. Thus, the condition number of a structural system is the square of the ratio of the highest and the lowest natural frequencies of the system:

$$\kappa(\mathbf{A}) = \left(\frac{\omega_{max}}{\omega_{min}}\right)^2 \tag{3.176}$$

With mesh refinement, the lowest natural frequency will be reduced to a certain degree as the structure becomes softer, but will taper off to a constant value. In general, the lowest modes and corresponding eigenvalues are fairly insensitive to further mesh refinement. A reasonably coarse mesh would represent these lowest modes reasonably well. This is because lowest modes are more global in nature and therefore insensitive to element size.

However, the highest modes and corresponding eigenvalues are quite sensitive to element size. The highest frequency is roughly inversely proportional to travel time of a stress wave passing across the smallest elements. For a given problem, with the progressive refinement of mesh, ω_{max} will keep on increasing because with mesh refinement many of the higher modes can be described in the refined finite element mesh. As a consequence, with the progression of mesh refinement, the condition number will keep on increasing. In other words, the condition of the resulting systems is moving toward ill-conditioning with progressive mesh refinement. In the worst scenario, the condition number will increase to such an extent that eventually it will exceed the machine precision. This is the *limit of computability* for any problem.

This poses the question that how far should we carry out mesh refinement in finite element analysis. On the one hand, the mesh refinement procedure provides more accurate results in finite element modeling and analysis. However, on the other hand, the mesh refinement process increases the condition number and the system may become ill-conditioned. With an ill-conditioned system, the solution process will be very sensitive to small perturbations in the initial data. Higher precision in computations is also necessitated. This shows that a good understanding of the fundamentals of numerical methods and their implication in affecting the physical system is essential in making appropriate use of computational methods.

Chapter 4

Iterative methods for the solution of systems of linear equations

4.1 INTRODUCTION

In Chapter 3, we have discussed direct methods for solving a system of linear equations. Using the direct method, the solution vector can be obtained after a finite number of operations. In this chapter, we will discuss iterative solution methods that iteratively approach the solution of a system of linear equations:

$$\mathbf{A}\,\mathbf{x} = \mathbf{b} \tag{4.1}$$

Given an arbitrary initial solution vector $\mathbf{x}_0$, we use the iterative methods to generate a series of solution vectors according to certain rules such that the series of solution vectors converges to the exact solution $\mathbf{x}$:

$$\mathbf{x}_0, \mathbf{x}_1, \dots, \mathbf{x}_k, \dots \tag{4.2}$$

$$\lim_{k \to \infty} \mathbf{x}_k = \mathbf{x} \tag{4.3}$$

The iterative approach for the solution of linear systems is well suited for systems with very large *sparse* coefficient matrices or system stiffness matrices. With sparse matrices, it is not economical to carry out the standard triangular decomposition procedure used in the direct solution methods. It should also be noted that for most systems of nonlinear equations, the solution vector is obtained through iterative methods. However, the solution methods in solving system of nonlinear equations require special treatment and they will be discussed in Chapter 6. The iterative methods discussed in this chapter are iterative solution methods for system of linear equations.

The basic approach in the iterative solution methods is to start with splitting the coefficient matrix **A** into the addition of two matrices and rewrite the original equation in the following form:

$$\mathbf{A} = \mathbf{A}_1 + \mathbf{A}_2 \tag{4.4}$$

$$\mathbf{A}_1\,\mathbf{x} = \mathbf{b} - \mathbf{A}_2\,\mathbf{x} \tag{4.5}$$

We assume that matrix $\mathbf{A}_1$ is nonsingular and always treat the right-hand side as known. This equation is the fundamental equation in all iterative methods.

To carry out the solution process iteratively, we start with an initial solution vector $\mathbf{x}_0$ and go through the following iterative process to generate a sequence of the solution vector:

$$\begin{cases} \mathbf{A}_1\ \mathbf{x}_1 = \mathbf{b} - \mathbf{A}_2\ \mathbf{x}_0 \\ \mathbf{A}_1\ \mathbf{x}_2 = \mathbf{b} - \mathbf{A}_2\ \mathbf{x}_1 \\ \qquad \vdots \\ \mathbf{A}_1\ \mathbf{x}_k = \mathbf{b} - \mathbf{A}_2\ \mathbf{x}_{k-1} \\ \qquad \vdots \end{cases} \tag{4.6}$$

At each iterative step, the right-hand side of the equation is known and the next solution vector is determined by solving the system of equation with $\mathbf{A}_1$.

We can carry out the same procedure successively until a satisfactory solution vector is obtained. Usually, an iterative solution process terminates when certain prescribed convergence tolerances are satisfied.

Obviously, to make this iterative scheme work, we need to impose at least the following two conditions. First, $\mathbf{A}_1$ should be nonsingular and the solution of the system of equations with $\mathbf{A}_1$ as coefficient matrix should not require triangular decomposition. It should be at most a triangular matrix and at least a diagonal matrix so that the solution vector can be obtained through a direct back-substitution process. Second, since we expect that the solution vector obtained from an iterative procedure should eventually approach the exact solution vector, the iterative process must be convergent, as indicated in Equation 4.3.

As with any iterative process, the convergence property of an iterative solution method is the most important aspect to be studied. In addition, it should also be mentioned that the coefficient matrix can be split into the addition of more than two matrices, as will be discussed in later sections in this chapter. However, for the sake of general discussion, we consider the generic case of splitting a coefficient matrix into the addition of two matrices in the convergence analysis.

4.2 CONVERGENCE OF ITERATIVE METHODS

4.2.1 Fundamental results

To study the convergence property of iterative methods, we first need to define the approximation errors in the solution vector. We achieve this by taking the difference between the basic iterative equation and the same equation with the exact solution:

$$\mathbf{A}_1\ \mathbf{x}_j = \mathbf{b} - \mathbf{A}_2\ \mathbf{x}_{j-1} \tag{4.7}$$

$$\mathbf{A}_1\ \mathbf{x} = \mathbf{b} - \mathbf{A}_2\ \mathbf{x} \tag{4.8}$$

$$\mathbf{A}_1\ (\mathbf{x} - \mathbf{x}_j) = -\mathbf{A}_2\ (\mathbf{x} - \mathbf{x}_{j-1}) \tag{4.9}$$

Since the error in solution vector is the difference between the approximate solution vector and the exact solution vector, we define the error in the solution vector at jth iteration as $e_j = \mathbf{x} - \mathbf{x}_j$:

$$\mathbf{A}_1\ e_j = -\mathbf{A}_2\ e_{j-1} \tag{4.10}$$

$$e_j = -\mathbf{A}_1^{-1}\mathbf{A}_2\ e_{j-1} = \mathbf{G}\ e_{j-1} \tag{4.11}$$

This is the standard equation for convergence analysis of iterative methods. The matrix **G** is called the iteration matrix. The iteration matrix **G** is of the following alternative form:

$$\begin{aligned} \mathbf{G} &= -\mathbf{A}_1^{-1}\mathbf{A}_2 \\ &= -\mathbf{A}_1^{-1}(\mathbf{A}-\mathbf{A}_1) \\ &= (\mathbf{I}-\mathbf{A}_1^{-1}\mathbf{A}) \end{aligned} \tag{4.12}$$

The procedure used in convergence analysis of iterative methods is a standard method. Assume that the initial solution error is $e_0 = \mathbf{x} - \mathbf{x}_0$; then we can expand the recursive relation between solution error vectors:

$$\begin{aligned} e_j &= \mathbf{G}\ e_{j-1} \\ &= \mathbf{G}\ \mathbf{G}\ e_{j-2} \\ &= \mathbf{G}\ldots\mathbf{G}\ e_0 \\ &= \mathbf{G}^j\ e_0 \end{aligned} \tag{4.13}$$

For the iterative method to converge, the iteration matrix G must be convergent. Since matrix G is not necessarily symmetric and positive definite, the eigenvalues could be complex. Therefore, the spectral radius of the eigenvalues should be less than 1:

$$\rho(\mathbf{G}) < 1 \tag{4.14}$$

Next, we show that if the method converges, then the solution vector also converges to the exact solution. We start by getting the solution vector by solving the basic iterative equation (Equation 4.7):

$$\begin{aligned} \mathbf{x}_j &= -\mathbf{A}_1^{-1}\mathbf{A}_2\ \mathbf{x}_{j-1} + \mathbf{A}_1^{-1}\mathbf{b} \\ &= \mathbf{G}\ \mathbf{x}_{j-1} + \bar{\mathbf{b}} \end{aligned} \tag{4.15}$$

We can expand the above equation through recursive substitution:

$$\begin{aligned} \mathbf{x}_j &= \mathbf{G}(\mathbf{G}\ \mathbf{x}_{j-2} + \bar{\mathbf{b}}) + \bar{\mathbf{b}} \\ &= \mathbf{G}^2\mathbf{x}_{j-2} + \mathbf{G}\ \bar{\mathbf{b}} + \bar{\mathbf{b}} \\ &= \mathbf{G}^j\mathbf{x}_0 + (\mathbf{G}^{j-1} + \cdots + \mathbf{G} + \mathbf{I})\bar{\mathbf{b}} \end{aligned} \tag{4.16}$$

Now we evaluate the limit of this series using the results obtained from the convergence of a Neumann (geometric) series (discussed in Chapter 3). In this case, the iterative matrix **G** is a convergent matrix:

$$\begin{aligned} \lim_{j\to\infty} \mathbf{x}_j &= \lim_{j\to\infty} \mathbf{G}^j\mathbf{x}_0 + \lim_{j\to\infty} (\mathbf{G}^{j-1} + \cdots + \mathbf{G} + \mathbf{I})\bar{\mathbf{b}} \\ &= 0 + (\mathbf{I}-\mathbf{G})^{-1}\bar{\mathbf{b}} \\ &= [\mathbf{I} + \mathbf{A}_1^{-1}(\mathbf{A}-\mathbf{A}_1)]^{-1}\ \mathbf{A}_1^{-1}\mathbf{b} \\ &= [\mathbf{A}_1^{-1}\mathbf{A}]^{-1}\ \mathbf{A}_1^{-1}\mathbf{b} \\ &= \mathbf{A}^{-1}\ \mathbf{b} = \mathbf{x} \end{aligned} \tag{4.17}$$

This shows that if matrix G is a convergent matrix, the iterative method converges to the right solution. We note that the choice of the initial solution vector is irrelevant to the convergence of an iterative method; it can be selected arbitrarily. This is because the convergence of the method is solely controlled by the eigenvalue properties of the iterative matrix G. These methods are called "globally convergent methods." However, it can be seen intuitively that the choice of initial vector does affect the *rate of convergence.*

To discuss the rate of convergence, we carry out the following analysis, starting with Equation 4.13:

$$\|e_j\| = \|G^j\ e_0\| \leq \|G^j\|\|e_0\| \tag{4.18}$$

$$\frac{\|e_j\|}{\|e_0\|} \leq \|G^j\| = [\rho(G)]^j \tag{4.19}$$

Let $\xi = \|e_j\|/\|e_0\|$ be a factor by which the initial error is reduced after j iterations. Then from the above equation, we obtain the following:

$$j \geq \frac{\log(\xi)}{\log[\rho(G)]} \tag{4.20}$$

This indicates that the magnitude of the denominator in the right-hand side of the equation determines how fast the convergence can be achieved. We thus can define the rate of convergence, R, as follows:

$$R = \log[\rho(G)] \tag{4.21}$$

We observe that larger value of R means faster convergence to the solution vector. Since $\rho(G) < 1$ for convergent iterative methods, the smaller the spectral radius, the faster the convergence rate. On the other hand, because of the presence of ξ in Equation 4.20, this shows that the choice of initial solution vector does have some effect on the rate of convergence of the solution process. However, the eigenvalue property of the iterative matrix plays a major role in determining the rate of convergence of the method.

The necessary and sufficient condition for the iterative method to converge is $\rho(G) < 1$. To verify the necessary part, we observe from the definition of convergent matrix that matrix G being convergent is equivalent to

$$\lim_{k \to \infty} \|G^k\| = 0 \tag{4.22}$$

which implies that

$$\begin{aligned} \lim_{k \to \infty} \|G^k\| &\geq \lim_{k \to \infty} \rho(G^k) \\ &= \lim_{k \to \infty} [\rho(G)]^k = 0 \end{aligned} \tag{4.23}$$

Therefore, there must be $\rho(G) < 1$. On the other hand, to verify the sufficient condition, we can state that if $\rho(G) < 1$, then there must exist a sufficiently small positive number η such that

$$\rho(G) + \eta < 1 \tag{4.24}$$

This is because for any sufficiently small positive number η, there always exists a certain norm such that the following condition is satisfied:

$$\rho(G) \le \|G\|_\alpha \le \rho(G) + \eta \tag{4.25}$$

From our discussion on matrix norms, we know that there must exist a certain matrix norm such that the following relation is satisfied:

$$\|G\| \le \rho(G) + \eta \tag{4.26}$$

Hence, we have

$$\|G^k\| \le \|G\|^k \le [\rho(G) + \eta]^k \tag{4.27}$$

Then, taking the limit of the above equation results in the following relation:

$$\begin{aligned} \lim_{k\to\infty} \|G^k\| &\le \lim_{k\to\infty} \|G\|^k \\ &\le \lim_{k\to\infty} [\rho(G) + \eta]^k \\ &= 0 \end{aligned} \tag{4.28}$$

Since all matrix norms are equivalent, this verifies the sufficient condition.

In practice, especially when the dimensional size of the coefficient matrix becomes large, it is expensive to compute the spectral radius of the iteration matrix. In this case, we can estimate the upper limit of the spectral radius by using the relation that $\rho(G) \le \|G\|_\alpha$, where $\alpha = 1, 2, \infty$, or F.

4.2.2 Transpose of conjugate matrix

For a general matrix **A**, unless it is symmetric and positive definite, its eigenvalues may be complex in most cases. Now let us consider a complex m × n matrix $\mathbf{A} = [a_{ij}]$ with complex elements defined as follows:

$$a_{ij} = b_{ij} + ic_{ij} \tag{4.29}$$

where $i = \sqrt{-1}$. We can then write matrix **A** in the following form:

$$\mathbf{A} = \mathbf{B} + i\,\mathbf{C} \tag{4.30}$$

where $\mathbf{B} = [b_{ij}]$ and $\mathbf{C} = [c_{ij}]$ are real matrices. The conjugate of matrix **A** is defined as follows:

$$\bar{\mathbf{A}} = \mathbf{B} - i\,\mathbf{C} \tag{4.31}$$

which is formed by conjugating each entry in the **A** matrix. The transpose of conjugate of matrix **A** is denoted by $\mathbf{A}^H$:

$$\mathbf{A}^H = \bar{\mathbf{A}}^T \tag{4.32}$$

Obviously, if a matrix has only real elements, then $\bar{\mathbf{A}} = \mathbf{A}$ and $\mathbf{A}^H = \mathbf{A}^T$. We can easily verify that if matrix **A** is both real and symmetric, then $\mathbf{A}^H = \mathbf{A}$. The operator H also has the following properties:

$$\begin{cases} \mathbf{A}^H = \bar{\mathbf{A}}^T = \overline{\mathbf{A}^T} \\ (\mathbf{AB})^H = \mathbf{B}^H \mathbf{A}^H \end{cases} \tag{4.33}$$

These relations can be easily verified.

4.2.3 Convergence with symmetric and positive definite matrices

In finite element analysis, the system coefficient matrix or structural stiffness matrix obtained from finite element spatial discretization is always symmetric and positive definite. Because of the special characteristics of symmetric and positive definite matrices, there are some conditions imposed on iterative solution methods. We recall from previous discussion that a real symmetric matrix **A** is said to be positive definite if $\mathbf{x}^T\mathbf{A}\mathbf{x} > 0$ for all nonzero vectors **x** in $\mathbf{R}^n$. And all the eigenvalues of **A** are positive if it is symmetric and positive definite.

For system of linear equations, $\mathbf{Ax} = \mathbf{b}$, where **A** is symmetric and positive definite, the iterative method $\mathbf{A}_1\ \mathbf{x}_j = \mathbf{b} - \mathbf{A}_2\ \mathbf{x}_{j-1}$, where $\mathbf{A} = \mathbf{A}_1 + \mathbf{A}_2$, is convergent if the following conditions are satisfied:

1. The spectral radius of **G** should be less than 1, $\rho(\mathbf{G}) < 1$, where $\mathbf{G} = (\mathbf{I} - \mathbf{A}_1^{-1}\mathbf{A})$.
2. Matrix $\mathbf{A}_1$ is nonsingular, $\det|\mathbf{A}_1| \neq 0$.
3. The symmetric matrix $\mathbf{Q} = \mathbf{A}_1 + \mathbf{A}_1^T - \mathbf{A}$ is positive definite.

To verify this statement, we start with the standard eigenvalue problem for iteration matrix **G**:

$$\mathbf{G}\ \mathbf{y} = \lambda\ \mathbf{y} \tag{4.34}$$

Since **G** is not symmetric in general, its eigenvalues λ and eigenvectors **y** can be *complex*. After substitution, we have the following:

$$(\mathbf{I} - \mathbf{A}_1^{-1}\mathbf{A})\ \mathbf{y} = \lambda\ \mathbf{y} \tag{4.35}$$

$$\mathbf{A}\ \mathbf{y} = (1-\lambda)\mathbf{A}_1\mathbf{y} \tag{4.36}$$

Since matrix **A** is symmetric and matrix $\mathbf{A}_1$ is real, we have $\mathbf{A}^H = \mathbf{A}$, $\mathbf{A}_1^H = \mathbf{A}_1^T$. Applying the H operator to the above equation yields the following:

$$\mathbf{y}^H\mathbf{A} = (1-\bar{\lambda})\mathbf{y}^H\mathbf{A}_1^T \tag{4.37}$$

We get the following relations from Equations 4.36 and 4.37:

$$\begin{cases} \mathbf{y}^H\mathbf{A}\mathbf{y} = (1-\lambda)\mathbf{y}^H\mathbf{A}_1\mathbf{y} \\ \mathbf{y}^H\mathbf{A}\mathbf{y} = (1-\bar{\lambda})\mathbf{y}^H\mathbf{A}_1^T\mathbf{y} \end{cases} \tag{4.38}$$

$$\begin{cases} \dfrac{1}{(1-\lambda)} = \dfrac{\mathbf{y}^H\mathbf{A}_1\mathbf{y}}{\mathbf{y}^H\mathbf{A}\mathbf{y}} \\[2ex] \dfrac{1}{(1-\bar{\lambda})} = \dfrac{\mathbf{y}^H\mathbf{A}_1^T\mathbf{y}}{\mathbf{y}^H\mathbf{A}\mathbf{y}} \end{cases} \tag{4.39}$$

Therefore, we have

$$\frac{1}{(1-\lambda)}+\frac{1}{(1-\bar{\lambda})}=\frac{\mathbf{y}^H(\mathbf{A}_1+\mathbf{A}_1^T)\mathbf{y}}{\mathbf{y}^H\mathbf{A}\mathbf{y}}$$
$$=1+\frac{\mathbf{y}^H\mathbf{Q}\,\mathbf{y}}{\mathbf{y}^H\mathbf{A}\mathbf{y}} \quad (4.40)$$

In arriving at the above equation, we have used the relation $\mathbf{A}_1+\mathbf{A}_1^T=\mathbf{Q}+\mathbf{A}$. Since matrices **A** and **Q** are symmetric and positive definite, we get

$$\frac{\mathbf{y}^H\mathbf{Q}\,\mathbf{y}}{\mathbf{y}^H\mathbf{A}\mathbf{y}}>0 \quad (4.41)$$

$$\frac{1}{(1-\lambda)}+\frac{1}{(1-\bar{\lambda})}>1 \quad (4.42)$$

Let $\lambda=\alpha+i\beta$ and $\bar{\lambda}=\alpha-i\beta$. Substituting in the above equation, we have

$$\frac{2(1-\alpha)}{(1-\alpha)^2+\beta^2}=\frac{1+(1-2\alpha)}{\alpha^2+\beta^2+(1-2\alpha)}>1 \quad (4.43)$$

This leads to the following relation:

$$\alpha^2+\beta^2=|\lambda|^2<1 \quad (4.44)$$

$$\rho(\mathbf{G})<1 \quad (4.45)$$

This means that the spectral radius of the iteration matrix **G** must be less than 1. Thus, the iterative method given in Equation 4.7 for a system with symmetric and positive definite matrix **A** is convergent and the solution vector converges to the exact solution.

4.3 ITERATIVE METHODS

In the following, we will discuss some of the well-known iterative methods proposed for the solution of sparse system of linear equations. These include the RF (Richardson's) method, Jacobi method, Gauss–Seidel method, the successive over-relaxation (SOR) method, and symmetric SOR method (SSOR). The main difference among them lies in the way the coefficient matrix **A** is split. We will also discuss another type of iterative methods that involve direction search schemes that serve as a preliminary for semi-iterative methods. In Chapter 5, we will see that some basic concepts of these iterative methods are utilized in constructing preconditioning schemes for the conjugate gradient method.

4.3.1 RF method

RF method is the simplest form of iterative solution methods. In the RF method, the coefficient matrix **A** is split into the following two matrices: $\mathbf{A}_1=\mathbf{I}$, $\mathbf{A}_2=\mathbf{A}-\mathbf{I}$. The iterative method is thus defined by the following equation:

$$\mathbf{x}_j=(\mathbf{I}-\mathbf{A})\,\mathbf{x}_{j-1}+\mathbf{b} \quad (4.46)$$

With this method, there is obviously no need to carry out back-substitution operations. Since the iteration matrix $\mathbf{G} = \mathbf{I} - \mathbf{A}$, the convergence condition for the RF method is as follows:

$$\begin{aligned} \rho(\mathbf{G}) &= \rho(\mathbf{I} - \mathbf{A}) \\ &= |1 - \rho(\mathbf{A})| \\ &= |1 - \lambda_{max}(\mathbf{A})| < 1 \end{aligned} \tag{4.47}$$

This convergence condition can also be written as $|\rho(\mathbf{A})| < 2$. It can be easily seen that this method is of little practical value because the eigenvalue properties of most coefficient matrices violate the convergence condition. If this convergence condition is not satisfied, the iterative process will result in divergence. For example, for illustrative purpose, we consider the simple one-dimensional structure modeled with four degrees of freedom as shown in Figure 3.1. The system of equilibrium equation is as follows:

$$\begin{bmatrix} 3 & -2 & -1 & 0 \\ -2 & 5 & -2 & -1 \\ -1 & -2 & 5 & -2 \\ 0 & -1 & -2 & 5 \end{bmatrix} \begin{Bmatrix} x_1 \\ x_2 \\ x_3 \\ x_4 \end{Bmatrix} = \begin{Bmatrix} 1 \\ 0 \\ 0 \\ 0 \end{Bmatrix} \tag{4.48}$$

It can be easily obtained that the spectrum radius for the above coefficient (stiffness) matrix is $|\rho(\mathbf{A})| = \lambda_{max}(\mathbf{A}) = 7.38$, which is larger than 2, the convergence limit. We can expect that the RF method will result in divergence in the iterative solution process for the above system of equations. This phenomenon is illustrated in the following numerical simulation results. We start with an initial trail vector and use a convergence tolerance $\|\mathbf{r}\| \leq 0.001$:

$$\mathbf{x}_0 = \begin{Bmatrix} 1 \\ 1 \\ 1 \\ 1 \end{Bmatrix} \tag{4.49}$$

The solution vectors at the 10th, 30th, and 80th iterations are shown as follows:

$$\mathbf{x}_{10} = \begin{Bmatrix} 7.835 \times 10^5 \\ 4.267 \times 10^6 \\ -1.082 \times 10^7 \\ 8.016 \times 10^6 \end{Bmatrix}, \mathbf{x}_{30} = \begin{Bmatrix} -8.286 \times 10^{21} \\ 8.316 \times 10^{22} \\ -1.287 \times 10^{23} \\ 7.400 \times 10^{22} \end{Bmatrix}, \mathbf{x}_{80} = \begin{Bmatrix} -1.755 \times 10^{62} \\ 1.506 \times 10^{63} \\ -2.244 \times 10^{63} \\ 1.252 \times 10^{63} \end{Bmatrix} \tag{4.50}$$

We recall that the exact solution vector is

$$\mathbf{x} = \begin{Bmatrix} 1.20 \\ 0.90 \\ 0.80 \\ 0.50 \end{Bmatrix} \tag{4.51}$$

It is obvious that, as the iteration progresses, the solution vector becomes larger and larger, resulting in the solution process becoming divergent.

4.3.2 Jacobi method

In the Jacobi method, the original coefficient matrix **A** is divided into the summation of two matrices: $\mathbf{A}_1 = \mathbf{D}_A$, $\mathbf{A}_2 = \mathbf{A} - \mathbf{D}_A$, where matrix $\mathbf{D}_A$ is a diagonal matrix that contains all the diagonal terms of **A**. We assume that these diagonal terms are nonzero. In this case, the iterative solution method is defined by the following basic relation:

$$\mathbf{D}_A\mathbf{x}_j = (\mathbf{D}_A - \mathbf{A})\,\mathbf{x}_{j-1} + \mathbf{b} \tag{4.52}$$

The series of solution vector can be obtained by solving a system with the diagonal coefficient matrix $\mathbf{D}_A$ through direct substitution:

$$\mathbf{x}_j = (\mathbf{I} - \mathbf{D}_A^{-1}\mathbf{A})\,\mathbf{x}_{j-1} + \mathbf{D}_A^{-1}\mathbf{b} \tag{4.53}$$

For the Jacobi method, the iteration matrix is $\mathbf{G} = (\mathbf{I} - \mathbf{D}_A^{-1}\mathbf{A})$. In component form, the Jacobi method can be expressed as follows:

$$x_i^{(k+1)} = \frac{1}{a_{ii}}\left(b_i - \sum_{j=1,\, j\neq i}^{n} a_{ij}x_j^{(k)}\right);\ i = 1,\ldots, n;\ k = 0, 1, 2, \ldots \tag{4.54}$$

The convergence criterion for the Jacobi method is $|\rho(\mathbf{G})| < 1$. It can be verified that the sufficient conditions for convergence of the Jacobi method can be any of the following:

$$\|\mathbf{G}\|_\infty = \max_i \sum_{j=1,\, j\neq i}^{n} \frac{|a_{ij}|}{|a_{ii}|} < 1 \tag{4.55}$$

$$\|\mathbf{G}\|_1 = \max_j \sum_{i=1,\, i\neq j}^{n} \frac{|a_{ij}|}{|a_{ii}|} < 1 \tag{4.56}$$

$$\|(\mathbf{I} - \mathbf{D}_A^{-1}\mathbf{A})\|_\infty = \max_i \sum_{j=1,\, j\neq i}^{n} \frac{|a_{ij}|}{|a_{ii}|} < 1 \tag{4.57}$$

To illustrate the iterative solution process of the Jacobi method, we solve the same system of equilibrium equations as used in the RF method (Equation 4.48). It can be easily verified that in this case, the convergence condition is satisfied. In addition, the above sufficient conditions for convergence are also satisfied. With the use of the same initial trial vector and the same convergence tolerance, the solution vectors obtained are as follows:

$$\mathbf{x}_{10} = \begin{Bmatrix} 1.2754 \\ 0.9713 \\ 0.8665 \\ 0.5443 \end{Bmatrix},\ \mathbf{x}_{20} = \begin{Bmatrix} 1.2336 \\ 0.9316 \\ 0.8295 \\ 0.5197 \end{Bmatrix},\ \mathbf{x}_{40} = \begin{Bmatrix} 1.2066 \\ 0.9062 \\ 0.8058 \\ 0.5039 \end{Bmatrix} \tag{4.58}$$

At the 40th iteration, the solution process has converged. It can be seen that with a convergent solution procedure, the solution vector computed from the progression of iterations approaches the exact solution vector.

4.3.3 Gauss–Seidel method

In the Gauss–Seidel method, first the coefficient matrix **A** is divided into the summation of the following three matrices: $\mathbf{A} = \mathbf{A}_L + \mathbf{A}_D + \mathbf{A}_U$, where $\mathbf{A}_L$ contains the terms below diagonal of matrix **A**, $\mathbf{A}_D$ the diagonal terms of **A**, and $\mathbf{A}_U$ the terms above the diagonal of matrix **A**. We also assume that the diagonal matrix does not have zero element. Then we have $\mathbf{A}_1 = \mathbf{A}_L + \mathbf{A}_D$, $\mathbf{A}_2 = \mathbf{A}_U$. Obviously $\mathbf{A}_1$ is a lower triangular matrix and $\mathbf{A}_2$ is an upper triangular matrix with zero diagonal terms. The iterative method is expressed by the following relation:

$$(\mathbf{A}_L + \mathbf{D}_A)\, \mathbf{x}_j = -\mathbf{A}_U\, \mathbf{x}_{j-1} + \mathbf{b} \tag{4.59}$$

$$\mathbf{x}_j = -(\mathbf{A}_L + \mathbf{D}_A)^{-1} \mathbf{A}_U\, \mathbf{x}_{j-1} + (\mathbf{A}_L + \mathbf{D}_A)^{-1} \mathbf{b} \tag{4.60}$$

The solution vector can thus be obtained by back-substitution with a lower triangular coefficient matrix. Therefore, we have the following iteration matrix:

$$\mathbf{G} = -(\mathbf{A}_L + \mathbf{D}_A)^{-1} \mathbf{A}_U \tag{4.61}$$

It can be shown that the relationship given as follows is the sufficient condition for the convergence of Gauss–Seidel method:

$$\max_i \sum_{j=1,\, j\neq i}^{n} \frac{|a_{ij}|}{|a_{ii}|} < 1 \tag{4.62}$$

This is also the first convergence condition for the Jacobi iterative method. Let $\mathbf{G}_J = (\mathbf{I} - \mathbf{D}_A^{-1}\mathbf{A})$, which is the iteration matrix for the Jacobi method, and $\mathbf{G}_{G-S} = -(\mathbf{A}_L + \mathbf{D}_A)^{-1} \mathbf{A}_U$ is the iteration matrix for Gauss–Seidel method. We can prove the following relation, which means that the Gauss–Seidel method is convergent:

$$\|\mathbf{G}_{G-S}\|_\infty \le \|\mathbf{G}_J\|_\infty < 1 \tag{4.63}$$

If matrix **A** is symmetric and positive definite, we can show that Gauss–Seidel method is always convergent. To prove this statement, we first show that if a real matrix **F** is nonsingular, and **F** + **U** and $\mathbf{F} - \mathbf{U}^T$ are n × n symmetric and positive definite matrices, then the following relation is true:

$$\rho(\mathbf{F}^{-1}\, \mathbf{U}) < 1 \tag{4.64}$$

To start with, we consider the standard eigenvalue problem for $(\mathbf{F}^{-1}\mathbf{U})$:

$$(\mathbf{F}^{-1}\, \mathbf{U})\mathbf{y} = \lambda\, \mathbf{y} \tag{4.65}$$

$$\mathbf{U}\, \mathbf{y} = \lambda\, \mathbf{F}\, \mathbf{y} \tag{4.66}$$

Next, we multiply both sides of the above equation by $\mathbf{y}^H$; we have

$$\mathbf{y}^H \mathbf{U}\, \mathbf{y} = \lambda\, \mathbf{y}^H \mathbf{F}\, \mathbf{y} \tag{4.67}$$

$$\mathbf{y}^H \mathbf{U}^T \mathbf{y} = \bar{\lambda}\, \mathbf{y}^H \mathbf{F}^T \mathbf{y} \tag{4.68}$$

Adding ($\mathbf{y}^H\mathbf{F}\mathbf{y}$) and ($\mathbf{y}^H\mathbf{F}^T\mathbf{y}$) to both sides of Equations 4.67 and 4.68, respectively, yields the following:

$$\begin{cases} \mathbf{y}^H(\mathbf{F}+\mathbf{U})\,\mathbf{y} = (1+\lambda)\,\mathbf{y}^H\mathbf{F}\,\mathbf{y} \\ \mathbf{y}^H(\mathbf{F}+\mathbf{U})^T\,\mathbf{y} = (1+\bar{\lambda})\,\mathbf{y}^H\mathbf{F}^T\mathbf{y} \end{cases} \tag{4.69}$$

Since $\mathbf{F}+\mathbf{U}$ is symmetric, we have

$$(1+\lambda)\,\mathbf{y}^H\mathbf{F}\,\mathbf{y} = (1+\bar{\lambda})\,\mathbf{y}^H\mathbf{F}^T\mathbf{y} \tag{4.70}$$

The above equation can be further expanded in the following way:

$$\begin{aligned} (1+\bar{\lambda})\,\mathbf{y}^H\mathbf{F}^T\mathbf{y} &= (1+\lambda)\,\mathbf{y}^H\mathbf{F}\,\mathbf{y} \\ &= (1+\lambda)\,\mathbf{y}^H(\mathbf{F}-\mathbf{U}^T+\mathbf{U}^T)\,\mathbf{y} \\ &= (1+\lambda)\,[\mathbf{y}^H(\mathbf{F}-\mathbf{U}^T)\mathbf{y}+\mathbf{y}^H\mathbf{U}^T\mathbf{y}] \\ &= (1+\lambda)\,[\mathbf{y}^H(\mathbf{F}-\mathbf{U}^T)\mathbf{y}+\bar{\lambda}\mathbf{y}^H\mathbf{F}^T\mathbf{y}] \end{aligned} \tag{4.71}$$

$$(1-|\lambda|^2)\,\mathbf{y}^H\mathbf{F}^T\mathbf{y} - (1+\lambda)\,\mathbf{y}^H(\mathbf{F}-\mathbf{U}^T)\mathbf{y} \tag{4.72}$$

Substituting Equations 4.69 and 4.70 into the above equation, we can obtain the following relation:

$$(1-|\lambda|^2)\,\mathbf{y}^H(\mathbf{F}+\mathbf{U})\mathbf{y} = |1+\lambda|^2\,\mathbf{y}^H(\mathbf{F}-\mathbf{U}^T)\mathbf{y} \tag{4.73}$$

$$\frac{(1-|\lambda|^2)}{|1+\lambda|^2} = \frac{\mathbf{y}^H(\mathbf{F}-\mathbf{U}^T)\mathbf{y}}{\mathbf{y}^H(\mathbf{F}+\mathbf{U})\mathbf{y}} \tag{4.74}$$

Because matrices ($\mathbf{F}+\mathbf{U}$) and ($\mathbf{F}-\mathbf{U}^T$) are symmetric and positive definite, we have the following:

$$\frac{(1-|\lambda|^2)}{|1+\lambda|^2} > 0 \tag{4.75}$$

$$1-|\lambda|^2 > 0 \tag{4.76}$$

Therefore, we have

$$\rho(\mathbf{F}^{-1}\,\mathbf{U}) < 1 \tag{4.77}$$

Now, if matrix $\mathbf{A}$ is symmetric and positive definite, with the use of Gauss–Seidel method, we have the following splitting of the matrix:

$$\mathbf{A} = \mathbf{A}_L + \mathbf{D}_A + \mathbf{A}_L^T \tag{4.78}$$

$$\begin{cases} \mathbf{F} = \mathbf{A}_L + \mathbf{D}_A \\ \mathbf{U} = \mathbf{A}_L^T \end{cases} \Rightarrow \begin{cases} \mathbf{F}+\mathbf{U} = \mathbf{A} \\ \mathbf{F}-\mathbf{U}^T = \mathbf{D}_A \end{cases} \tag{4.79}$$

Matrix **A** being symmetric and positive definite, we can see that $\mathbf{F} + \mathbf{U}$ and $\mathbf{F} - \mathbf{U}^T$ are also symmetric and positive definite (diagonal terms of **A** are positive):

$$\mathbf{G}_{G\text{-}S} = (\mathbf{A}_L + \mathbf{D}_A)^{-1}\mathbf{A}_L^T = \mathbf{F}^{-1}\mathbf{U} \tag{4.80}$$

$$\rho(\mathbf{G}_{G\text{-}S}) = \rho(\mathbf{F}^{-1}\ \mathbf{U}) < 1 \tag{4.81}$$

Equation 4.81 means that the Gauss–Seidel method for systems with symmetric and positive definite coefficient matrices is always convergent.

Sometimes, the Gauss–Seidel method is considered as an accelerated form of the Jacobi method, which means that the Gauss–Seidel method has a faster convergence rate than the Jacobi method on solving certain problems. It should be noted that this statement cannot be generally applied to all cases. In certain special cases, the Jacobi method converges, whereas the Gauss–Seidel method diverges. It is not difficult to construct such kind of special example cases.

For the same example problem, we described earlier in Equation 4.48, because the structural stiffness matrix is symmetric and positive definite, clearly, the Gauss–Seidel method is always convergent. With the use of the same initial starting vector and the same convergence tolerance, we obtained the converged solution vector in 30 iterations. The results are as follows:

$$\mathbf{x}_{10} = \begin{Bmatrix} 1.2815 \\ 0.9720 \\ 0.8639 \\ 0.5400 \end{Bmatrix},\ \mathbf{x}_{20} = \begin{Bmatrix} 1.2162 \\ 0.9143 \\ 0.8127 \\ 0.5079 \end{Bmatrix},\ \mathbf{x}_{30} = \begin{Bmatrix} 1.2032 \\ 0.9028 \\ 0.8025 \\ 0.5016 \end{Bmatrix} \tag{4.82}$$

Compared with the results obtained using the Jacobi method, we observe that for this specific problem, the Gauss–Seidel method has a faster convergence rate.

4.3.4 Successive over-relaxation methods

SOR methods are acceleration schemes for the basic iterative solution methods. They are primarily used in the solution of sparse systems. To explain the acceleration procedure for standard iterative methods, first we take a close look at the simplest iterative method, the RF method discussed in Section 4.3.1. In the RF method, the iterative equation is as follows:

$$\mathbf{x}_k = (\mathbf{I} - \mathbf{A})\ \mathbf{x}_{k-1} + \mathbf{b} \tag{4.83}$$

which can be written as follows:

$$\mathbf{x}_k = \mathbf{x}_{k-1} + (\mathbf{b} - \mathbf{A}\mathbf{x}_{k-1}) = \mathbf{x}_{k-1} + \mathbf{r}_{k-1} \tag{4.84}$$

The vector $\mathbf{r}_{k-1}$ is the residual vector. This indicates that with the use of iterative method, we are in fact trying to improve upon the solution vector with the residual vector at the previous iteration. If we carry out similar manipulations, we will have the following forms for Jacobi and Gauss–Seidel methods:

$$\mathbf{x}_k = \mathbf{x}_{k-1} + \mathbf{D}_A^{-1}\ \mathbf{r}_{k-1} \tag{4.85}$$

$$\mathbf{x}_k = \mathbf{x}_{k-1} + (\mathbf{A}_L + \mathbf{D}_A)^{-1}\ \mathbf{r}_{k-1} \tag{4.86}$$

In general, these two methods can be considered as having higher rate of convergence than the simplest RF method. Therefore, from our observation, we can introduce an accelerated iterative method based on RF method (also called modified Richardson method) in the following form:

$$\mathbf{x}_k = \mathbf{x}_{k-1} + \alpha\, \mathbf{r}_{k-1} \tag{4.87}$$

where α is called the *relaxation parameter.* Depending on the values of α, an improved rate of convergence as compared with RF method can be expected.

In general, if we split the matrix $\mathbf{A}$ as $\mathbf{A} = \mathbf{A}_L + \mathbf{D}_A + \mathbf{A}_U$, then the simplest form of SOR method can be written in the following form:

$$\mathbf{D}_A\, \mathbf{x}_k = \alpha\,[\mathbf{b} - (\mathbf{A}_L + \mathbf{A}_U)\mathbf{x}_{k-1}] + (1-\alpha)\, \mathbf{D}_A\, \mathbf{x}_{k-1} \tag{4.88}$$

where $0 < \alpha \le 1$. This simplest SOR scheme is based on the Jacobi method, because we can write the above equation as follows:

$$\mathbf{x}_k = \mathbf{x}_{k-1} + \alpha\, \mathbf{D}_A^{-1}\, \mathbf{r}_{k-1} \tag{4.89}$$

By setting $\alpha = 1$, we obtain the Jacobi method.

We can investigate the behavior of this simple SOR method by taking $\alpha = 0.5$ and solving the example problem in Equation 4.48. This results in a convergent solution process. Using the same initial trial vector and convergence tolerance, we obtain the following solutions at the 10th, 30th, and 60th iterations:

$$\mathbf{x}_{10} = \begin{Bmatrix} 1.3134 \\ 1.0007 \\ 0.9010 \\ 0.5679 \end{Bmatrix},\ \mathbf{x}_{30} = \begin{Bmatrix} 1.2516 \\ 0.9487 \\ 0.8454 \\ 0.5302 \end{Bmatrix},\ \mathbf{x}_{60} = \begin{Bmatrix} 1.2156 \\ 0.9148 \\ 0.8138 \\ 0.5092 \end{Bmatrix} \tag{4.90}$$

The convergence tolerance was satisfied at the 63rd iteration:

$$\mathbf{x}_{63} = \begin{Bmatrix} 1.2139 \\ 0.9131 \\ 0.8122 \\ 0.5081 \end{Bmatrix} \tag{4.91}$$

It is important to note from this example that although, the SOR method is theoretically considered an acceleration scheme for the Jacobi method, it does not mean that it always has faster convergence rate than Jacobi method itself.

A more interesting form of the SOR method is a variation of the method expressed in Equation 4.88 that can be written in the following form:

$$\mathbf{D}_A\, \mathbf{x}_k = \alpha\,(\mathbf{b} - \mathbf{A}_L\, \mathbf{x}_k - \mathbf{A}_U\, \mathbf{x}_{k-1}) + (1-\alpha)\, \mathbf{D}_A\, \mathbf{x}_{k-1} \tag{4.92}$$

$$(\mathbf{D}_A + \alpha\, \mathbf{A}_L)\, \mathbf{x}_k = [(1-\alpha)\, \mathbf{D}_A - \alpha\, \mathbf{A}_U)]\, \mathbf{x}_{k-1} + \alpha\, \mathbf{b} \tag{4.93}$$

This is the accelerated scheme for Gauss–Seidel method; for $\alpha = 1$, the above SOR method becomes the Gauss–Seidel method. The iteration matrix is thus obtained as follows:

$$\mathbf{G} = (\mathbf{D}_A + \alpha\, \mathbf{A}_L)^{-1}[(1-\alpha)\, \mathbf{D}_A - \alpha\, \mathbf{A}_U)] \tag{4.94}$$

As with all the other iterative methods, the necessary and sufficient condition for the SOR method to converge is that $\rho(G) < 1$.

With the addition of the relaxation parameter in the SOR method, the convergence of the method is also affected by the values of α. It can be proven that the necessary condition for the SOR method to converge is $0 < \alpha < 2$. This can be demonstrated from the following analysis.

If we assume that the eigenvalues of matrix **G** are $\lambda_1, \ldots, \lambda_n$, then

$$|\det(\mathbf{G})| = |\lambda_1 \ldots \lambda_n| \leq [\rho(\mathbf{G})]^n \tag{4.95}$$

Assuming the SOR method converges, we get

$$|\det(\mathbf{G})|^{1/n} \leq \rho(\mathbf{G}) < 1 \tag{4.96}$$

In the iterative matrix (Equation 4.94), since $\mathbf{A}_L$ is a strictly lower triangular matrix and $\mathbf{A}_U$ a strictly upper triangular matrix, we have

$$\begin{cases} \det\left[(\mathbf{D}_A + \alpha\, \mathbf{A}_L)^{-1}\right] = \det\left(\mathbf{D}_A^{-1}\right) \\ \det\left(\mathbf{D}_A - \mathbf{A}_U\right) = \det\left(\mathbf{D}_A\right) \end{cases} \tag{4.97}$$

We have

$$\begin{aligned} \det(\mathbf{G}) &= \det\left((\mathbf{D}_A + \alpha\, \mathbf{A}_L)^{-1}[(1-\alpha)\, \mathbf{D}_A - \alpha\, \mathbf{A}_U)]\right) \\ &= \det\left((\mathbf{D}_A + \alpha\, \mathbf{A}_L)^{-1}\right)\det\left([(1-\alpha)\, \mathbf{D}_A - \alpha\, \mathbf{A}_U)]\right) \\ &= \det\left(\mathbf{D}_A^{-1}\right)\det\left((1-\alpha)\, \mathbf{D}_A\right) \\ &= \det\left((1-\alpha)\, \mathbf{I}\right) \\ &= (1-\alpha)^n < 1 \end{aligned} \tag{4.98}$$

Therefore, we get

$$|1-\alpha| < 1 \tag{4.99}$$

which results in the following necessary condition for convergence:

$$0 < \alpha < 2 \tag{4.100}$$

This convergence property indicates that only when α is taken in the above range of values can convergence of the SOR method be expected. However, this is not a sufficient convergence condition.

In computational mechanics, the stiffness matrix obtained is usually symmetric and positive definite. We can state that if matrix **A** is symmetric and positive definite, and if $0 < \alpha < 2$, then the SOR method is convergent. To prove this assertion, we start by considering the standard eigenvalue solution for the iteration matrix **G**:

$$\mathbf{G}\, \mathbf{y} = \lambda\, \mathbf{y} \tag{4.101}$$

where λ and $\mathbf{y}$ are an eigenvalue and the corresponding eigenvector, respectively. Substituting the relation for $\mathbf{G}$ in the SOR method into the above equation, we have the following relations:

$$(\mathbf{D_A} + \alpha\,\mathbf{A_L})^{-1}[(1-\alpha)\,\mathbf{D_A} - \alpha\,\mathbf{A_U})]\,\mathbf{y} = \lambda\,\mathbf{y} \tag{4.102}$$

$$[(1-\alpha)\,\mathbf{D_A} - \alpha\,\mathbf{A_U})]\,\mathbf{y} = \lambda\,(\mathbf{D_A} + \alpha\,\mathbf{A_L})\,\mathbf{y} \tag{4.103}$$

$$\mathbf{y}^H[(1-\alpha)\,\mathbf{D_A} - \alpha\,\mathbf{A_U})]\,\mathbf{y} = \lambda\,\mathbf{y}^H(\mathbf{D_A} + \alpha\,\mathbf{A_L})\,\mathbf{y} \tag{4.104}$$

$$\lambda = \frac{\mathbf{y}^H\mathbf{D_A}\mathbf{y} - \alpha\,\mathbf{y}^H\mathbf{D_A}\mathbf{y} - \alpha\,\mathbf{y}^H\mathbf{A_U}\mathbf{y}}{\mathbf{y}^H\mathbf{D_A}\mathbf{y} + \alpha\,\mathbf{y}^H\mathbf{A_L}\mathbf{y}} \tag{4.105}$$

It is obvious through carrying out matrix computations that

$$\mathbf{y}^H\mathbf{D_A}\mathbf{y} = \sum_{i=1}^{n} a_{ii}\,|y_i|^2 = \xi > 0 \tag{4.106}$$

Let $\mathbf{y}^H\mathbf{A_L}\mathbf{y} = \beta + i\,\gamma$, and because matrix $\mathbf{A}$ is symmetric, we have $\mathbf{A_U} = \mathbf{A_L^T}$. Then, we have

$$\mathbf{y}^H\mathbf{A_U}\mathbf{y} = \mathbf{y}^H\mathbf{A_L^T}\mathbf{y} = (\mathbf{y}^H\mathbf{A_L}\mathbf{y})^H = \beta - i\,\gamma \tag{4.107}$$

In addition, since matrix $\mathbf{A}$ is symmetric and positive definite, we have by definition

$$\mathbf{y}^H\mathbf{A}\mathbf{y} = \mathbf{y}^H(\mathbf{A_L} + \mathbf{A_D} + \mathbf{A_U})\mathbf{y} = \xi + 2\beta > 0 \tag{4.108}$$

Substituting all these into the equation for λ, we obtain the following relation:

$$\lambda = \frac{(\xi - \alpha\xi - \alpha\beta) + i\alpha\gamma}{(\xi + \alpha\beta) + i\alpha\gamma} \tag{4.109}$$

Hence, we have

$$|\lambda|^2 = \frac{(\xi - \alpha\xi - \alpha\beta)^2 + \alpha^2\gamma^2}{(\xi + \alpha\beta)^2 + \alpha^2\gamma^2} \tag{4.110}$$

If $0 < \alpha < 2$, then we subtract the denominator from the numerator in the above equation for comparison:

$$(\xi - \alpha\xi - \alpha\beta)^2 - (\xi + \alpha\beta)^2 = \alpha\xi(\xi + 2\beta)(\alpha - 2) < 0 \tag{4.111}$$

This shows that the numerator is smaller than the denominator in the above division when $0 < \alpha < 2$. Therefore, we get

$$|\lambda|^2 < 1 \tag{4.112}$$

which indicates that the SOR method for system with symmetric and positive definite coefficient matrix is convergent.

The performance of the SOR method can be easily illustrated with the simple example problem used in this section (Equation 4.48). Because the stiffness matrix is symmetric and positive definite, we obtain a convergent method by taking a value of α in the interval of (0, 2). For example, taking $\alpha = 1$, we can easily verify numerically that the solution obtained is exactly the same as the Gauss–Seidel method. Using the same initial starting vector, we obtain the converged solution in exactly 30 iterations with values the same as in Equation 4.82.

Taking $\alpha = 0.50$, and using the same initial starting vector, we find that it takes 60 iterations for the solution vector to converge:

$$\mathbf{x}_{10} = \begin{Bmatrix} 1.3409 \\ 1.0310 \\ 0.9209 \\ 0.5729 \end{Bmatrix}, \mathbf{x}_{30} = \begin{Bmatrix} 1.2501 \\ 0.9462 \\ 0.8425 \\ 0.5277 \end{Bmatrix}, \mathbf{x}_{60} = \begin{Bmatrix} 1.2105 \\ 0.9097 \\ 0.8089 \\ 0.5058 \end{Bmatrix} \tag{4.113}$$

Using $\alpha = 1.5$ and the same initial starting vector, we obtain the converged solution vector in 12 iterations, shown as follows:

$$\mathbf{x}_{10} = \begin{Bmatrix} 1.1987 \\ 0.8990 \\ 0.7996 \\ 0.4993 \end{Bmatrix}, \mathbf{x}_{12} = \begin{Bmatrix} 1.1997 \\ 0.8996 \\ 0.7993 \\ 0.5000 \end{Bmatrix} \tag{4.114}$$

However, if we use $\alpha = 2.0$, which borders on the convergence interval, the numerical simulation results show that the iterative solution process is not convergent:

$$\mathbf{x}_{10} = \begin{Bmatrix} 2.1323 \\ 1.6320 \\ 1.5033 \\ 0.6260 \end{Bmatrix}, \mathbf{x}_{30} = \begin{Bmatrix} 0.7151 \\ 0.0190 \\ 0.0755 \\ -0.0552 \end{Bmatrix}, \mathbf{x}_{100} = \begin{Bmatrix} 2.0465 \\ 1.7617 \\ 1.5188 \\ 0.6687 \end{Bmatrix} \tag{4.115}$$

As expected, if we use $\alpha = 2.5$, which violates the convergence condition, the iterative solution process becomes divergent:

$$\mathbf{x}_{10} = \begin{Bmatrix} 44.43 \\ 64.63 \\ 85.80 \\ 70.50 \end{Bmatrix}, \mathbf{x}_{50} = \begin{Bmatrix} -3.507\times 10^{9} \\ -4.833\times 10^{9} \\ -5.639\times 10^{9} \\ -4.592\times 10^{9} \end{Bmatrix}, \mathbf{x}_{100} = \begin{Bmatrix} 2.553\times 10^{19} \\ 2.489\times 10^{19} \\ 2.184\times 10^{19} \\ 1.142\times 10^{19} \end{Bmatrix} \tag{4.116}$$

This simple numerical example shows that the convergence property of the SOR method is sensitive with the choice of values for α. In general, it is difficult to determine the optimum value of α in advance.

4.3.5 Symmetric successive over-relaxation method

The SSOR method is obtained by first using the standard SOR method in calculating an intermediate solution vector $\mathbf{x}_{k-1/2}$ and then using the backward SOR method to compute $\mathbf{x}_k$. The SSOR method can be expressed in the following form:

$$\begin{cases} \mathbf{D}_A\, \mathbf{x}_{k-1/2} = \alpha\,(\mathbf{b} - \mathbf{A}_L\, \mathbf{x}_{k-1/2} - \mathbf{A}_U\, \mathbf{x}_{k-1}) + (1-\alpha)\, \mathbf{D}_A\, \mathbf{x}_{k-1} \\ \mathbf{D}_A\, \mathbf{x}_k = \alpha\,(\mathbf{b} - \mathbf{A}_L\, \mathbf{x}_{k-1/2} - \mathbf{A}_U\, \mathbf{x}_k) + (1-\alpha)\, \mathbf{D}_A\, \mathbf{x}_{k-1/2} \end{cases} \tag{4.117}$$

We write the equations for this two-step iteration in terms of the iteration matrices:

$$\mathbf{x}_{k-1/2} = \mathbf{G}_1\, \mathbf{x}_{k-1} + \mathbf{R}_1 \tag{4.118}$$

$$\begin{cases} \mathbf{G}_1 = (\mathbf{D}_A + \alpha\, \mathbf{A}_L)^{-1}[(1-\alpha)\, \mathbf{D}_A - \alpha\, \mathbf{A}_U)] \\ \mathbf{R}_1 = (\mathbf{D}_A + \alpha\, \mathbf{A}_L)^{-1}\alpha\, \mathbf{b} \end{cases} \tag{4.119}$$

$$\mathbf{x}_k = \mathbf{G}_2\, \mathbf{x}_{k-1/2} + \mathbf{R}_2 \tag{4.120}$$

$$\begin{cases} \mathbf{G}_2 = (\mathbf{D}_A + \alpha\, \mathbf{A}_U)^{-1}[(1-\alpha)\, \mathbf{D}_A - \alpha\, \mathbf{A}_L)] \\ \mathbf{R}_2 = (\mathbf{D}_A + \alpha\, \mathbf{A}_U)^{-1}\alpha\, \mathbf{b} \end{cases} \tag{4.121}$$

We note that the first step is the same as the standard SOR method.

Combining the above two equations together, we have the following:

$$\mathbf{x}_k = \mathbf{G}\, \mathbf{x}_{k-1} + \mathbf{R} \tag{4.122}$$

$$\begin{cases} \mathbf{G} = \mathbf{G}_2\, \mathbf{G}_1 \\ \mathbf{R} = \mathbf{G}_2\, \mathbf{R}_1 + \mathbf{R}_2 \end{cases} \tag{4.123}$$

If the matrix **A** is symmetric and positive definite, it can be shown that the SSOR method converges for any α in the interval of (0, 2). On the other hand, the rate of convergence of this method is relatively insensitive to the choice of values of α. This is one of the advantages of SSOR method over the standard SOR method.

The behavior of SSOR method can also be illustrated with the simple example discussed in this section (Equation 4.48). Using the same initial starting vector and convergence tolerance, we try three cases with the value of $\alpha = 0.5$, 1.0, and 1.5. The iterative solution converges for these three cases in 36, 21, and 26 iterations, respectively. The results in the 10th, 20th, and converged iteration are shown as follows:

$$\alpha = 0.5,\ \mathbf{x}_{10} = \begin{Bmatrix} 1.2667 \\ 0.9636 \\ 0.8595 \\ 0.5396 \end{Bmatrix},\ \mathbf{x}_{20} = \begin{Bmatrix} 1.2246 \\ 0.9234 \\ 0.8219 \\ 0.5146 \end{Bmatrix},\ \mathbf{x}_{36} = \begin{Bmatrix} 1.2050 \\ 0.9047 \\ 0.8044 \\ 0.5029 \end{Bmatrix} \tag{4.124}$$

$$\alpha=1.0,\ \mathbf{x}_{10}=\begin{Bmatrix}1.2650\\0.9255\\0.8242\\0.5161\end{Bmatrix},\ \mathbf{x}_{20}=\begin{Bmatrix}1.2026\\0.9027\\0.8025\\0.5017\end{Bmatrix},\ \mathbf{x}_{21}=\begin{Bmatrix}1.2021\\0.9021\\0.8020\\0.5013\end{Bmatrix} \quad (4.125)$$

$$\alpha=1.5,\ \mathbf{x}_{10}=\begin{Bmatrix}1.2433\\0.9514\\0.8506\\0.5334\end{Bmatrix},\ \mathbf{x}_{20}=\begin{Bmatrix}1.2063\\0.9074\\0.8073\\0.5048\end{Bmatrix},\ \mathbf{x}_{26}=\begin{Bmatrix}1.2020\\0.9023\\0.8023\\0.5048\end{Bmatrix} \quad (4.126)$$

This simple numerical simulation indicates that the SSOR method converges fast and is not very sensitive to the choice of values for α, provided that the convergence condition that $\alpha \in (0, 2)$ is satisfied.

So far, we have discussed some of the classical iterative methods that originate from the splitting of the coefficient (stiffness) matrix. They are first-order iterative methods. The advantage of iterative methods is that they are simple and can be implemented in recursive algorithm. Because only the nonzero elements of the stiffness matrix need to be stored, they are effective tools in solving large and spare system of equations, especially in solving some optimization type of problems. The disadvantage is that the solution process is iterative. This requires that special attention be paid to convergence characteristics of the algorithm and the rate of convergence. With SOR method, there is a question regarding the selection of an optimal relaxation parameter that could be hard to choose, and the behavior of the resulting algorithm is sensitive to the value of α. Because of their approximate nature, in computational mechanics, these iterative methods are not extensively used for the solution of a system of equilibrium equations in the finite element analysis.

One important aspect of these iterative methods is that their convergence is completely independent of the initial solution vector; these methods are globally convergent. Even if we start with a good guess on the initial solution vector, it does not make any difference on the convergence property of the method itself. However, as we will discuss in Chapter 6, for nonlinear-type problems, the Newton–Raphson method, which is an iterative solution method, does not have the property of global convergence. In that case, the choice of initial solution vector has a dominant effect on the convergence of the solution process. It should also be noted that the convergence property of iterative methods is discussed in general term under exact mathematics, whereas the rate of convergence of individual algorithm in finite precision computation is a different subject.

4.4 DIRECTION SEARCH METHODS

In this section, we will discuss another type of iterative method based on the idea of direction search that has its origin in the functional optimization. The fundamental idea behind this family of methods is totally different from the splitting coefficient matrix approach in the classical iterative methods discussed so far. This type of methods includes the steepest descent method, conjugate direction method, and conjugate gradient method. Because of its unique characteristics with respect to direct solution methods, the conjugate gradient method is a semi-iterative method, which will be discussed in Chapter 5. Now we discuss some fundamental concepts with respect to functional optimizations.

4.4.1 Quadratic functional, gradient, and Hessian

In most computational mechanics problems, we usually deal with quadratic functional generally defined in the following form:

$$\Phi(\mathbf{x}) = \frac{1}{2}\,\mathbf{x}^T\mathbf{A}\,\mathbf{x} - \mathbf{x}^T\mathbf{b} + c \tag{4.127}$$

where:

A is an n × n symmetric matrix
x is an n × 1 vector
c is a scalar quantity

Assume that a functional f(**x**) is first order continuously differentiable; then the gradient of f(**x**) at **x** is a vector defined as follows:

$$\nabla f(\mathbf{x}) = \frac{\partial f(\mathbf{x})}{\partial \mathbf{x}} = \begin{Bmatrix} \partial f/\partial x_1 \\ \vdots \\ \partial f/\partial x_n \end{Bmatrix} \tag{4.128}$$

Geometrically, the gradient at a point gives the direction in which the functional has the greatest changes. In the positive direction, the function f(**x**) has the greatest increase, whereas in the negative gradient direction, the function has the greatest decrease. If the gradient at certain point becomes zero, it then means that there is no change of the function at that point, which must be a stationary point. The function value at the stationary point can be either a minimum or a maximum or a saddle point in the neighborhood. We can show that the gradient at any point is perpendicular to the contour line of function f(**x**) and points in the direction of the greatest change of f(**x**). Thus, the gradient of f(**x**) in the neighborhood of **x** forms a vector field called the gradient field.

We note that the symbol ∇ is used to represent the gradient. In the three-dimensional Cartesian coordinate system, the symbol by itself is sometimes expressed as follows:

$$\nabla = i\frac{\partial}{\partial x} + j\frac{\partial}{\partial y} + k\frac{\partial}{\partial z} \tag{4.129}$$

This form of ∇ is called a vector differential operator or Hamilton operator, which is in fact a vector. Therefore, all vector operations apply. We can see that the ∇ used to represent the gradient is an extension on the Hamilton operator.

If the functional f(**x**) is second order continuously differentiable, then the *Hessian* of f(**x**) at **x** is a symmetric matrix of the following form:

$$\mathbf{H}(\mathbf{x}) = \left[\frac{\partial^2 f(\mathbf{x})}{\partial x_i \partial x_j}\right]_{n\times n} \tag{4.130}$$

We can see in geometry that the Hessian or the stiffness matrix in finite element analysis represents the change of gradients or the curvature of the function f(**x**) at **x**. The property of this matrix **H** describes the shape of the function. If **H** is positive definite, then the shape of the function is a paraboloid (a bowl shape). If **H** is negative definite, then the shape of the function is an upside-down paraboloid (a flipped-over bowl). In both cases, there always exists a minimum (lowest point) or a maximum (highest point) at **x** on the surface. However, if H is singular, then there is no unique minimum or maximum on the surface of the function; there may be a line or hyperplane with a uniform functional value. If H is none of the above, then there exists a saddle point **x** on the surface.

The first-order *directional derivative* of the functional f(**x**) at **x** in the direction of **δ** is a *scalar* quantity defined as follows:

$$\begin{aligned} f(\mathbf{x};\delta) &= \left.\frac{df(\mathbf{x}+\xi\delta)}{d\xi}\right|_{\xi=0} \\ &= \left.\frac{df}{d\xi}(x_1+\xi\delta_1,\ldots,x_n+\xi\delta_n)\right|_{\xi=0} \\ &= \nabla^T f(\mathbf{x})\delta \end{aligned} \tag{4.131}$$

Geometrically, from the above equation, we can see that the directional derivative is the projection of the gradient vector at **x** in the direction of **δ**. From this definition, we can observe that the directional derivative is the rate of change of the function f(**x**) at **x** in the direction of **δ**. Therefore, when the directional derivative is positive, it means that the value of function increases in the direction of **δ**; otherwise, the value of function decreases in that direction. When the direction **δ** coincides with the direction of ∇f(**x**), the directional derivative takes the maximum value as ‖∇f(**x**)‖ if **δ** is a unit vector. This shows that the gradient vector at point **x** does give the direction in which the function has the greatest changes.

Through continuous differentiation, we can determine the second-order directional derivative of the functional f(**x**) at **x** in the direction of **δ**, which is a scalar quantity defined as follows:

$$f^{(2)}(\mathbf{x};\delta) = \delta^T \mathbf{H}(\mathbf{x})\delta \tag{4.132}$$

In the same way, depending on whether the functional has continuous first-order or second-order differentials, we can write the Taylor expansion for functional f(**x**) at **x** as follows:

$$f(\mathbf{x}+\delta) = f(\mathbf{x}) + \nabla^T f(\mathbf{x})\delta + o(\|\delta\|) \tag{4.133}$$

$$f(\mathbf{x}+\delta) = f(\mathbf{x}) + \nabla^T f(\mathbf{x})\delta + \frac{1}{2}\delta^T \mathbf{H}(\mathbf{x})\delta + o(\|\delta\|^2) \tag{4.134}$$

where we have $\lim_{\|\delta\|\to\infty} o(\|\delta\|) = 0$.

It can be proven that if $\bar{\mathbf{x}}$ is a local minimal of the functional f(**x**), then $\bar{\mathbf{x}}$ is a stationary point in the vector space such that $\nabla[f(\bar{\mathbf{x}})] = 0$. This is the necessary condition for $\bar{\mathbf{x}}$ to be a local minimum. The sufficient condition can be obtained by considering the characteristics of the Hessian of the functional f(**x**). If the Hessian matrix $\mathbf{H}(\bar{\mathbf{x}})$ is positive definite, then $\bar{\mathbf{x}}$ is a local minimum in some neighborhood of $\bar{\mathbf{x}}$. If $\bar{\mathbf{x}}$ is a maximum point in the vector space, we can make analogous statements that $\mathbf{H}(\bar{\mathbf{x}})$ must be negative definite. We can easily show that even if $\bar{\mathbf{x}}$ is a stationary point in the vector space, it is not an optimal point if there are both negative and positive eigenvalues in the matrix $\mathbf{H}(\bar{\mathbf{x}})$.

With the directional derivatives, we can make a similar statement regarding whether $\bar{\mathbf{x}}$ is a local minimal of the functional f(**x**). The necessary condition is that if $\bar{\mathbf{x}}$ is a local minimum, then $f^{(1)}(\mathbf{x};\delta) = 0$ for all directions **δ**.

4.4.2 Direction search method: basic concept

Assume that the system coefficient (stiffness) matrix **A** obtained from finite element analysis is symmetric and positive definite, and we define a functional of the following quadratic form:

$$\Phi(\mathbf{x}) = \frac{1}{2}\mathbf{x}^T\mathbf{A}\,\mathbf{x} - \mathbf{x}^T\mathbf{b} \tag{4.135}$$

Since matrix **A** is symmetric and positive definite, there exists a single optimal for the quadratic functional, which can be determined by letting the gradient be zero:

$$\nabla\Phi(\mathbf{x}) = \frac{\partial\Phi(\mathbf{x})}{\partial\mathbf{x}} = \mathbf{A}\mathbf{x} - \mathbf{b} = 0 \tag{4.136}$$

This shows that the minimization of the quadratic functional is equivalent to the solution of a corresponding linear system of equations, $\mathbf{Ax} = \mathbf{b}$. The optimal solution vector **x** of the quadratic functional is also the solution of the linear system of equations. This is the basic idea behind all the direction search methods.

If we consider an iterative method for approaching the exact solution **x**, then there should exist a residual vector at the end of each iteration. This residual vector can be defined as follows:

$$\mathbf{r}_k = \mathbf{b} - \mathbf{A}\mathbf{x}_k \tag{4.137}$$

Obviously, this residual vector is equal to the negative gradient of the functional:

$$\mathbf{r}_k = -\nabla\Phi(\mathbf{x}_k) \tag{4.138}$$

From our discussion on the idea behind the SOR method as an acceleration scheme for basic iterative methods in Section 4.3.4, we note that all the standard iterative methods perform modifications on the solution vector based on information on the residual vector:

$$\mathbf{x}_{k+1} = \mathbf{x}_k + \alpha\,\mathbf{r}_k \tag{4.139}$$

In a more general case, this can be considered as updating the next solution vector based on searching in a certain direction. In the above equation, that certain search direction is $\mathbf{r}_k$. In a general case, we assume that there exists a set of linearly independent vectors:

$$\mathbf{p}_0, \mathbf{p}_1, \ldots, \mathbf{p}_{n-1} \tag{4.140}$$

and we have an arbitrarily selected initial solution vector $\mathbf{x}_0$. Then a general (iterative) searching algorithm can be expressed in the following form:

$$\mathbf{x}_{k+1} = \mathbf{x}_k + \lambda_k\,\mathbf{p}_k \tag{4.141}$$

where $\mathbf{p}_k$ is the search direction and λ_k is the step length at the kth iteration. Our objective is to determine the step length and the search direction of the solution vector such that the quadratic functional $\Phi(\mathbf{x})$ is minimized. Once this functional is minimized, the solution vector obtained is also the solution vector for the corresponding system of linear equations.

4.4.3 Steepest descent method

When we use an iterative approach to search for the optimal (minimum) of the quadratic functional $\Phi(\mathbf{x})$ at an intermediate step, supposing the search direction is known, we expect that there exists a step length λ_k such that

$$\Phi(\mathbf{x}_k + \lambda_k \mathbf{p}_k) < \Phi(\mathbf{x}_k) \tag{4.142}$$

If the above condition is satisfied, then the iterative method is globally convergent. Or precisely, we obtain the solution vector that minimizes the functional $\Phi(\mathbf{x})$:

$$\Phi(\mathbf{x}_{k+1}) = \min_{\lambda_k} \Phi(\mathbf{x}_k + \lambda_k \mathbf{p}_k) \tag{4.143}$$

In this case, since $\mathbf{x}_{k+1} = \mathbf{x}_k + \lambda_k \mathbf{p}_k$, we have

$$\begin{aligned} \Phi(\mathbf{x}_{k+1}) &= \frac{1}{2}(\mathbf{x}_k + \lambda_k \mathbf{p}_k)^T \mathbf{A}(\mathbf{x}_k + \lambda_k \mathbf{p}_k) - \mathbf{b}^T(\mathbf{x}_k + \lambda_k \mathbf{p}_k) \\ &= \frac{1}{2}\left[\mathbf{x}_k^T \mathbf{A}\mathbf{x}_k + 2\lambda_k \mathbf{p}_k^T \mathbf{A}\mathbf{x}_k + \lambda_k^2 \mathbf{p}_k^T \mathbf{A}\mathbf{p}_k\right] - \mathbf{x}_k^T \mathbf{b} - \lambda_k \mathbf{p}_k^T \mathbf{b} \end{aligned} \tag{4.144}$$

Taking the derivative of the above equation with respect to λ_k yields

$$\begin{aligned} \frac{\partial \Phi(\mathbf{x}_{k+1})}{\partial \lambda_k} &= \lambda_k \mathbf{p}_k^T \mathbf{A}\mathbf{p}_k + \mathbf{p}_k^T \mathbf{A}\mathbf{x}_k - \mathbf{p}_k^T \mathbf{b} \\ &= \lambda_k \mathbf{p}_k^T \mathbf{A}\mathbf{p}_k + \mathbf{p}_k^T(\mathbf{A}\mathbf{x}_k - \mathbf{b}) \end{aligned} \tag{4.145}$$

The step length can be determined by letting $\partial\Phi(\mathbf{x}_{k+1})/\partial\lambda_k = 0$:

$$\begin{aligned} \lambda_k &= \frac{-\mathbf{p}_k^T(\mathbf{A}\mathbf{x}_k - \mathbf{b})}{\mathbf{p}_k^T \mathbf{A}\mathbf{p}_k} \\ &= \frac{\mathbf{p}_k^T \mathbf{r}_k}{\mathbf{p}_k^T \mathbf{A}\mathbf{p}_k} \end{aligned} \tag{4.146}$$

With the step length determined from the above equation, we can show that the functional is minimized as the iterative process continues:

$$\Phi(\mathbf{x}_{k+1}) < \Phi(\mathbf{x}_k) \tag{4.147}$$

This search can be conducted along any search directions. However, it is intuitively obvious that the best search direction at $\mathbf{x}_k$ would be the one that goes down fastest in the downhill direction, or along the negative direction of the gradient mathematically. That is, at $\mathbf{x}_k$, the best possible search direction is as follows:

$$\begin{aligned} \mathbf{p}_k &= -\nabla\Phi(\mathbf{x}_k) \\ &= \mathbf{b} - \mathbf{A}\mathbf{x}_k = \mathbf{r}_k \end{aligned} \tag{4.148}$$

Such a direction search method is called the *steepest descent method*. In the steepest descent method, the solution vector is updated by the following relation:

$$\lambda_k = \frac{\mathbf{r}_k^T \mathbf{r}_k}{\mathbf{r}_k^T \mathbf{A} \mathbf{r}_k} \tag{4.149}$$

$$\mathbf{x}_{k+1} = \mathbf{x}_k + \lambda_k \mathbf{r}_k \tag{4.150}$$

In this case, the quadratic function is of the following form:

$$\begin{aligned} \Phi(\mathbf{x}_{k+1}) &= \frac{1}{2}\mathbf{x}_k^T \mathbf{A}\mathbf{x}_k + \lambda_k \mathbf{r}_k^T \mathbf{A}\mathbf{x}_k + \frac{1}{2}\lambda_k^2 \mathbf{r}_k^T \mathbf{A}\mathbf{r}_k - \mathbf{x}_k^T \mathbf{b} - \lambda_k \mathbf{r}_k^T \mathbf{b} \\ &= \left(\frac{1}{2}\mathbf{x}_k^T \mathbf{A}\mathbf{x}_k - \mathbf{x}_k^T \mathbf{b}\right) + \lambda_k \mathbf{r}_k^T (\mathbf{A}\mathbf{x}_k - \mathbf{b}) + \frac{1}{2}\lambda_k^2 \mathbf{r}_k^T \mathbf{A}\mathbf{r}_k \end{aligned} \tag{4.151}$$

which can be written as follows:

$$\Phi(\mathbf{x}_{k+1}) = \Phi(\mathbf{x}_k) - \lambda_k \mathbf{r}_k^T \mathbf{r}_k + \frac{1}{2}\lambda_k^2 \mathbf{r}_k^T \mathbf{A}\mathbf{r}_k \tag{4.152}$$

Now if we substitute λ_k in Equation 4.149 into the above equation, we have

$$\begin{aligned} \Phi(\mathbf{x}_{k+1}) &= \Phi(\mathbf{x}_k) - \left(\frac{\mathbf{r}_k^T \mathbf{r}_k}{\mathbf{r}_k^T \mathbf{A}\mathbf{r}_k}\right)\mathbf{r}_k^T \mathbf{r}_k + \frac{1}{2}\left(\frac{\mathbf{r}_k^T \mathbf{r}_k}{\mathbf{r}_k^T \mathbf{A}\mathbf{r}_k}\right)^2 \mathbf{r}_k^T \mathbf{A}\mathbf{r}_k \\ &= \Phi(\mathbf{x}_k) - \frac{1}{2}\frac{(\mathbf{r}_k^T \mathbf{r}_k)^2}{\mathbf{r}_k^T \mathbf{A}\mathbf{r}_k} \end{aligned} \tag{4.153}$$

We note that $\mathbf{r}_k^T \mathbf{r}_k > 0$ and $\mathbf{r}_k^T \mathbf{A}\mathbf{r}_k > 0$ because matrix **A** is symmetric and positive definite; then we have the following relations:

$$\frac{(\mathbf{r}_k^T \mathbf{r}_k)^2}{\mathbf{r}_k^T \mathbf{A}\mathbf{r}_k} > 0 \tag{4.154}$$

$$\Phi(\mathbf{x}_{k+1}) \le \Phi(\mathbf{x}_k) \tag{4.155}$$

This means that the steepest descent method is globally convergent. The global convergence of this method can be shown to satisfy the following relation:

$$\Phi(\mathbf{x}_{k+1}) \le \left(1 - \frac{1}{\kappa_2(\mathbf{A})}\right)\Phi(\mathbf{x}_k) \tag{4.156}$$

$$\kappa_2(\mathbf{A}) = \lambda_1(\mathbf{A})/\lambda_n(\mathbf{A}) = \lambda_{min}/\lambda_{max} \tag{4.157}$$

It can be proven that the rate of convergence of the steepest descent method is described in the following relation:

$$\|\mathbf{x}_k - \mathbf{x}\|_A \leq \left(\frac{\kappa(\mathbf{A})-1}{\kappa(\mathbf{A})+1}\right)^k \|\mathbf{x}_0 - \mathbf{x}\|_A \tag{4.158}$$

where:

$\kappa(\mathbf{A}) = \lambda_{max}/\lambda_{min}$ is the spectral condition number of $\mathbf{A}$
$\mathbf{x}$ is the exact solution vector of the system

We note that this convergence condition is expressed in terms of energy norm. The energy norm corresponding to a positive definite matrix $\mathbf{A}$ is defined as follows:

$$\|\mathbf{x}\|_A = (\mathbf{x}^T\mathbf{A}\mathbf{x})^{1/2} \tag{4.159}$$

If matrix $\mathbf{A}$ is the identity matrix, then the energy norm reduces to the Euclidean norm. In that case, another formula to estimate the rate of convergence can be expressed as follows:

$$\|\mathbf{x}_k - \mathbf{x}\| \leq \left(\frac{\kappa(\mathbf{A})-1}{\kappa(\mathbf{A})+1}\right)^k \|\mathbf{x}_0 - \mathbf{x}\| \tag{4.160}$$

To arrive at this formula, first we consider the eigenvalues and eigenvectors of matrix $\mathbf{A}$:

$$\mathbf{A}\phi_i = \lambda_i \phi_i \tag{4.161}$$

Let $x = \sum_{j=1}^{n} \eta_j \phi_j$; we can easily verify that

$$\lambda_1 \mathbf{x}^T\mathbf{x} \leq \mathbf{x}^T\mathbf{A}\mathbf{x} \leq \lambda_n \mathbf{x}^T\mathbf{x} \tag{4.162}$$

Let $q(\lambda)$ be a polynomial in λ; then we can obtain that

$$\mathbf{x}^T[q(\mathbf{A})]^2\mathbf{x} \leq \max_i q(\lambda_i)^2\, \mathbf{x}^T\mathbf{x} \tag{4.163}$$

This is because $q(\mathbf{A})\mathbf{x} = \sum_{j=1}^{n} q(\lambda_j)\eta_j\phi_j$ such that

$$\begin{aligned}
[q(\mathbf{A})\mathbf{x}]^T[q(\mathbf{A})\mathbf{x}] &= \sum_{j=1}^{n} q(\lambda_j)^2\eta_j^2 \\
&\leq \max_i q(\lambda_i)^2 \sum_{j=1}^{n} \eta_j^2 \\
&= \max_i q(\lambda_i)^2 \mathbf{x}^T\mathbf{x}
\end{aligned} \tag{4.164}$$

From global convergence of the method, we have $\Phi(\mathbf{x}_1) \leq \Phi(\mathbf{x}_0 + \alpha\mathbf{r}_0)$. It follows that

$$\begin{aligned}
(\mathbf{x}_1 - \mathbf{x})^T(\mathbf{x}_1 - \mathbf{x}) &\leq (\mathbf{x}_0 + \alpha\mathbf{r}_0 - \mathbf{x})^T(\mathbf{x}_0 + \alpha\mathbf{r}_0 - \mathbf{x}) \\
&= [(\mathbf{I} + \alpha\mathbf{A})(\mathbf{x}_0 - \mathbf{x})]^T[(\mathbf{I} + \alpha\mathbf{A})(\mathbf{x}_0 - \mathbf{x})]
\end{aligned} \tag{4.165}$$

$$\begin{aligned}\|\mathbf{x}_1-\mathbf{x}\|^2 &= \|(\mathbf{I}+\alpha\mathbf{A})(\mathbf{x}_0-\mathbf{x})\|^2\\ &\le \max_i (1+\alpha\lambda_i)^2\|\mathbf{x}_0-\mathbf{x}\|^2\\ &\le \max_{\lambda_1\le\lambda\le\lambda_n}(1+\alpha\lambda)^2\|\mathbf{x}_0-\mathbf{x}\|^2\end{aligned} \tag{4.166}$$

We know that the solution for the min–max problem $\min_{\alpha}\max_{\lambda_1\le\lambda\le\lambda_n}|1+\alpha\lambda|$ is $\alpha = -2/(\lambda_1+\lambda_n)$. This means that for any real number α, there must be

$$\max_{\lambda_1\le\lambda\le\lambda_n}\left|1-\frac{2}{\lambda_1+\lambda_n}\lambda\right| \le \max_{\lambda_1\le\lambda\le\lambda_n}|1+\alpha\lambda| \tag{4.167}$$

With this result, we then obtain the following equation:

$$\|\mathbf{x}_1-\mathbf{x}\|^2 \le \max_{\lambda_1\le\lambda\le\lambda_n}\left(1-\frac{2\lambda}{\lambda_1+\lambda_n}\right)^2\|\mathbf{x}_0-\mathbf{x}\|^2 \tag{4.168}$$

$$\max_{\lambda_1\le\lambda\le\lambda_n}\left(1-\frac{2\lambda}{\lambda_1+\lambda_n}\right)^2 = \left(\frac{\lambda_n-\lambda_1}{\lambda_n+\lambda_1}\right)^2 = \left(\frac{\kappa(\mathbf{A})-1}{\kappa(\mathbf{A})+1}\right)^2 \tag{4.169}$$

$$\|\mathbf{x}_1-\mathbf{x}\| \le \left(\frac{\kappa(\mathbf{A})-1}{\kappa(\mathbf{A})+1}\right)\|\mathbf{x}_0-\mathbf{x}\| \tag{4.170}$$

Similarly, we can obtain the following:

$$\|\mathbf{x}_k-\mathbf{x}\| \le \left(\frac{\kappa(\mathbf{A})-1}{\kappa(\mathbf{A})+1}\right)^k\|\mathbf{x}_0-\mathbf{x}\| \tag{4.171}$$

It can be seen that the rate of convergence of the steepest descent method mainly depends on the condition number of matrix **A**. In case of ill-conditioned matrix **A**, the convergence is very slow. This is because when $\kappa(\mathbf{A})$ becomes large, we have the following:

$$\lim_{\kappa(\mathbf{A})\to\infty}\frac{\kappa(\mathbf{A})-1}{\kappa(\mathbf{A})+1}\to 1 \tag{4.172}$$

This means that the right-hand side of Equation 4.171 approaches zero in a very slow pace. From a geometric point of view, in this case, the level curves of the functional Φ are very elongated hyperellipsoids. The minimization process corresponds to finding the lowest point on a relatively flat and steep-sided valley. The search process traverses back and forth across the valley and renders the search algorithm inefficient. Although the gradient descent method is not a practical algorithm for practical usage, it does provide the fundamental ideas in deriving the set of semi-iterative methods.

Chapter 5

Conjugate gradient methods

5.1 INTRODUCTION

Conjugate gradient is a method for solving a system of equations with symmetric positive definite matrices. It is most suitable in cases when the matrix is large and sparse. This is often the case in large finite element or finite difference models of structural systems. The direct solution of a system of equations with the methods described in Chapter 3 can be computationally very extensive when the matrix is very large. The conjugate gradient method is very popular in these situations. In exact mathematics, without a round-off error, the conjugate gradient is a direct solution method and it converges to the exact solution in a finite number of steps. Computationally, with the presence of a round-off error, it is an iterative method, and at each iteration, it requires a matrix vector multiplication. In finite element systems, matrix vector multiplication can be performed efficiently by assembling the element contributions, without actually having to form the stiffness matrix. The conjugate gradient method is also very efficient in solving nonlinear systems. As we will see later in this chapter that preconditioning can improve the convergence properties of the conjugate gradient method.

In this chapter, we will first present some preliminaries before an in-depth discussion of the conjugate gradient method and its practical implementation. We will also present various methods of preconditioning. But first, some preliminaries.

As has been discussed in Chapter 4, the steepest descent method has a very slow convergence rate when the system coefficient matrix is ill-conditioned. This is because the gradient directions generated in the solution process become too similar to each other as the iteration progresses. This suggests that we should successively minimize the quadratic functional $\Phi(\mathbf{x})$ by searching along a set of directions $\{\mathbf{p}_1, \mathbf{p}_2, \ldots, \mathbf{p}_n, \ldots\}$ that does not correspond to the set of residual vectors $\{\mathbf{r}_1, \mathbf{r}_2, \ldots, \mathbf{r}_n, \ldots\}$. We can start with a general direction search scheme and substitute it into the quadratic functional.

$$\mathbf{x}_{j+1} = \mathbf{x}_j + \lambda_j\, \mathbf{p}_j \tag{5.1}$$

$$\Phi(\mathbf{x}) = \frac{1}{2}\mathbf{x}^T \mathbf{A}\,\mathbf{x} - \mathbf{x}^T\mathbf{b} + c \tag{5.2}$$

$$\begin{aligned}
\Phi(\mathbf{x}_{j+1}) &= \frac{1}{2}(\mathbf{x}_j + \lambda_j\, \mathbf{p}_j)^T \mathbf{A}\,(\mathbf{x}_j + \lambda_j\, \mathbf{p}_j) - (\mathbf{x}_j + \lambda_j\, \mathbf{p}_j)^T\mathbf{b} \\
&= \Phi(\mathbf{x}_j) + \lambda_j\, \mathbf{p}_j^T(\mathbf{A}\mathbf{x}_j - \mathbf{b}) + \frac{1}{2}\lambda_j^2\, \mathbf{p}_j^T\mathbf{A}\mathbf{p}_j \\
&= \Phi(\mathbf{x}_j) - \lambda_j\, \mathbf{p}_j^T\mathbf{r}_j + \frac{1}{2}\lambda_j^2\, \mathbf{p}_j^T\mathbf{A}\mathbf{p}_j
\end{aligned} \tag{5.3}$$

Obtaining the minimum of functional $\Phi(\mathbf{x}_{j+1})$ requires the solution of a j-dimensional minimization problem. Similarly, minimization of $\Phi(\mathbf{x}_j)$ is a (j−1)-dimensional problem. Using the approach with the steepest descent method, we can determine the step length λ_j by minimizing $\Phi(\mathbf{x}_{j+1})$ with respect to λ_j.

$$\frac{\partial \Phi(\mathbf{x}_{j+1})}{\partial \lambda_j} = -\mathbf{p}_j^T \mathbf{r}_j + \lambda_j \, \mathbf{p}_j^T \mathbf{A} \mathbf{p}_j = 0 \tag{5.4}$$

$$\lambda_j = \frac{\mathbf{p}_j^T \mathbf{r}_j}{\mathbf{p}_j^T \mathbf{A} \mathbf{p}_j} \tag{5.5}$$

This indicates that the search step length is determined by solving a one-dimensional minimization problem. In so doing, the solutions obtained by solving a *one-dimensional* minimization problem are not guaranteed to satisfying a *j-dimensional* minimization problem. Therefore, a natural way to resolve the problem is to uncouple the original multidimensional minimization problem. If the j-dimensional minimization problem can be uncoupled, then the search step can be computed by solving individual one-dimensional minimization problems. Now, the question is how the search direction can be selected such that the multidimensional minimization problem can be uncoupled?

We start by considering the relationship between the search direction and the gradient vector. This is logical because we observe that it is the property of search directions that result in a slow convergence rate with the steepest descent method. With the step length obtained from solving a one-dimensional minimization problem, we substitute it into the original functional.

$$\begin{aligned}
\Phi(\mathbf{x}_{j+1}) &= \Phi(\mathbf{x}_j) - \lambda_j \, \mathbf{p}_j^T \mathbf{r}_j + \frac{1}{2} \lambda_j^2 \, \mathbf{p}_j^T \mathbf{A} \mathbf{p}_j \\
&= \Phi(\mathbf{x}_j) - \frac{\mathbf{p}_j^T \mathbf{r}_j}{\mathbf{p}_j^T \mathbf{A} \mathbf{p}_j} \, \mathbf{p}_j^T \mathbf{r}_j + \frac{1}{2} \left(\frac{\mathbf{p}_j^T \mathbf{r}_j}{\mathbf{p}_j^T \mathbf{A} \mathbf{p}_j} \right)^2 \mathbf{p}_j^T \mathbf{A} \mathbf{p}_j \\
&= \Phi(\mathbf{x}_j) - \frac{1}{2} \frac{(\mathbf{p}_j^T \mathbf{r}_j)^2}{\mathbf{p}_j^T \mathbf{A} \mathbf{p}_j}
\end{aligned} \tag{5.6}$$

Considering the right side of the above equation, we can easily see that if the search direction $\mathbf{p}_j$ is not orthogonal to the residual vector $\mathbf{r}_j$, that is, $\mathbf{p}_j^T \mathbf{r}_j \neq 0$, and if matrix $\mathbf{A}$ is symmetric and positive definite, which is usually the case for stiffness matrix in computational mechanics, then $\Phi(\mathbf{x}_{j+1}) < \Phi(\mathbf{x}_j)$. This means that if those two conditions are satisfied, the method obtained using the step length from the solution of a one-dimensional minimization problem is globally convergent. Of course, it should be noted that the constraint on the search direction and residual vector serves only as a general guideline on the determination of the search directions, which is not good enough. A better method is to determine the search direction in such a way that the convergence can be guaranteed within a *finite number* of iterations. This in essence is a direct solution method under the guise of iterative methods. In the following section, we will discuss a general conjugate direction method that possesses such characteristics.

5.2 A GENERAL CONJUGATE DIRECTION METHOD

Now we consider selecting search directions $\{\mathbf{p}_0, \mathbf{p}_1, \ldots, \mathbf{p}_n, \ldots\}$ in such a way that for any initial solution vector $\mathbf{x}_0$, the direction search method converges to the exact solution vector within a finite number of iterations. To reach this goal, it requires that starting with any initial solution vector $\mathbf{x}_0$, we generate a minimal solution vector series

$$\{\mathbf{x}_0, \mathbf{x}_1, \ldots, \mathbf{x}_k, \ldots\} \tag{5.7}$$

such that the functional $\Phi(\mathbf{x})$ is minimized at each solution vector.

$$\Phi(\mathbf{x}_k) = \min_{\mathbf{z} \in S_k} \Phi(\mathbf{x}_0 + \mathbf{z}) \tag{5.8}$$

$S_k = \{\mathbf{p}_0, \mathbf{p}_1, \ldots, \mathbf{p}_{k-1}\}$ is a subspace spanned by the k linearly independent search direction vectors. This means that the solution vector is the minimum of the functional $\Phi(\mathbf{x}_k)$ on the following set:

$$\{\mathbf{x} \mid \mathbf{x} - \mathbf{x}_0 + \mathbf{z},\ \mathbf{z} \in S_k\} \tag{5.9}$$

Therefore, from our earlier discussion, we know that the *directional derivative* of $\Phi(\mathbf{x}_k)$ at $\mathbf{x}_k$ along any direction $\mathbf{z} \in S_k$ must be zero, namely,

$$\nabla^T \Phi(\mathbf{x}_k)\, \mathbf{z} = 0 \tag{5.10}$$

This indicates that for the functional Φ to be minimized at $\mathbf{x}_k$, the gradient vector $\nabla \Phi(\mathbf{x}_k)$ must be orthogonal to the vector subspace S_k, $\nabla\Phi(\mathbf{x}_k) \perp S_k$. By satisfying this condition, the k-dimensional minimization problem is solved.

To get a clear picture on the relationship between solution vector and search directions in the direction search method, we consider the successive solution vectors generated as iteration progresses up to the jth iteration.

$$\begin{cases} \mathbf{x}_1 = \mathbf{x}_0 + \lambda_0\, \mathbf{p}_0 \\ \quad\vdots \\ \mathbf{x}_j = \mathbf{x}_{j-1} + \lambda_{j-1}\, \mathbf{p}_{j-1} \\ \mathbf{x}_{j+1} = \mathbf{x}_j + \lambda_j\, \mathbf{p}_j \end{cases} \tag{5.11}$$

in which λ_j is determined from solving the corresponding one-dimensional minimization problem. Because this method is globally convergent, we can start with any arbitrary initial solution vector without affecting the convergence of the method. As a special case, let $\mathbf{x}_0 = 0$, then the solution vector at jth iteration can be written as follows:

$$\mathbf{x}_{j+1} = \sum_{k=0}^{j} \lambda_k\, \mathbf{p}_k \tag{5.12}$$

This means that $\mathbf{x}_k$ is a linear combination of the search directions $\{\mathbf{p}_0, \mathbf{p}_1, \ldots, \mathbf{p}_{k-1}\}$. If for each solution vector the functional is minimized,

$$\min_{\mathbf{x}_{k+1} \in \mathrm{span}\{\mathbf{p}_0, \ldots, \mathbf{p}_k\}} \Phi(\mathbf{x}_{k+1}) \tag{5.13}$$

then from Equation 5.10 if we let $\mathbf{z} = \lambda_j \mathbf{p}_j$, we have the following relation:

$$\nabla^T \Phi(\mathbf{x}_{k+1})\, \mathbf{p}_j = 0; \quad \text{for } j = 0, \ldots, k \tag{5.14}$$

This is a very important property because it implies that we will obtain the exact solution vector within n steps for a system of n linear equations. This is because the vectors (search directions) $\{\mathbf{p}_0, \mathbf{p}_1, \ldots, \mathbf{p}_{n-1}\}$ are linearly independent, and

$$\nabla^T \Phi(\mathbf{x}_n)\, \mathbf{p}_j = 0; \quad \text{for } j = 0, \ldots, n-1 \tag{5.15}$$

Consequently, the following relation must be true:

$$\nabla \Phi(\mathbf{x}_n) = 0 \Rightarrow \mathbf{A}\mathbf{x}_n = \mathbf{b} \tag{5.16}$$

The solution vector at the nth iteration, $\mathbf{x}_n$, is the exact solution of the system of linear equations, $\mathbf{Ax} = \mathbf{b}$. We note that the discussion so far only gives a framework on achieving convergence in a finite number of iterations. The more practical question is: what are these direction vectors and how can we determine them?

It can be shown that if the search direction vector subspace $\{\mathbf{p}_k\}$ is **A**-conjugate, which is defined as

$$\mathbf{p}_i^T \mathbf{A} \mathbf{p}_j = 0, \; i \neq j \tag{5.17}$$

then the solution vector obtained through this direction search method minimizes the functional $\Phi(\mathbf{x}_k)$ and the solution vector obtained from step-by-step iterations converges in a finite number (n) of iterations. This can be explained through mathematical induction.

Since the vectors in subspace $\{\mathbf{p}_k\}$ are **A**-conjugate, they are linearly independent. When $k = 0$, the set is an empty set. Hence we can assume that the statement valid. Let $\mathbf{g}_i = \nabla\, \Phi(\mathbf{x}_i)$. Assume that the statement holds for k.

$$\begin{aligned} &\mathbf{g}_i \perp S_i, \; i = 0, 1, \ldots, k \\ &S_i = \{\mathbf{p}_0, \ldots, \mathbf{p}_{i-1}\} \end{aligned} \tag{5.18}$$

Now we want to show that $\mathbf{g}_{k+1} \perp S_{k+1}$. By definition, we have the following:

$$\mathbf{g}_{k+1} = \mathbf{A}\mathbf{x}_{k+1} - \mathbf{b} \tag{5.19}$$

$$\begin{aligned} \mathbf{g}_{k+1} - \mathbf{g}_k &= \mathbf{A}(\mathbf{x}_{k+1} - \mathbf{x}_k) \\ &= \lambda_k \mathbf{A}\mathbf{p}_k \\ \lambda_k &= -\frac{(\mathbf{p}_k^T \mathbf{g}_k)}{(\mathbf{p}_k^T \mathbf{A}\mathbf{p}_k)} \end{aligned} \tag{5.20}$$

Assuming that $g_{k+1} \neq 0$, we have the following:

$$\begin{aligned} p_k^T g_{k+1} &= p_k^T g_k + \lambda_k \, p_k^T A p_k \\ &= p_k^T g_k - \frac{p_k^T g_k}{p_k^T A p_k} \, p_k^T A p_k = 0 \end{aligned} \tag{5.21}$$

Let i < k, and using the definition of A-conjugacy for search direction vectors, we have

$$p_i^T g_{k+1} = p_i^T g_k + \lambda_k \, p_i^T A p_k = p_i^T g_k \tag{5.22}$$

According to the assumption that the statement holds for $i \leq k$, we have

$$p_i^T g_k = 0, \quad \text{for } i = 0, 1, \ldots, k \tag{5.23}$$

Therefore,

$$p_i^T g_{k+1} = 0, \quad \text{for } i = 0, 1, \ldots, k \tag{5.24}$$

which means that the gradient vector g_{k+1} is orthogonal to the vector subspace spanned by $\{p_0, \ldots, p_k\}$, that is,

$$g_{k+1} \perp S_{k+1} \tag{5.25}$$

This also implies that the exact solution would be obtained within k steps. For a system of n linear equations, the solution vector obtained in n steps is the exact solution vector. We note that this is true if all operations are performed in exact mathematics. Next, we discuss the determination of A-conjugate search directions in the conjugate gradient method.

5.3 CONJUGATE GRADIENT METHOD

5.3.1 Fundamental concepts

From the preceding discussions, we know that to achieve convergence in a finite number of steps, the approach is to choose linearly independent search vectors such that each solution vector obtained from $x_{k+1} = x_k + \lambda_k \, p_k$ minimizes $\Phi(x_k)$. In practice, however, we seek a search direction vector such that when we solve the one-dimensional minimization problem, we also equivalently solve the original k-dimensional minimization problem. This requires the decoupling of the multidimensional minimization problem.

We start by observing that the solution vector x_k is a linear combination of the k search directions, $\{p_0, p_1, \ldots, p_{k-1}\}$.

$$x_{j+1} = P_j \, y_j \tag{5.26}$$

$$\begin{cases} \mathbf{P}_j = [\mathbf{p}_0 \mid \mathbf{p}_1 \mid \ldots \mid \mathbf{p}_j]_{nxj} \\ \mathbf{y}_j = \begin{Bmatrix} \lambda_0 \\ \vdots \\ \lambda_j \end{Bmatrix} \end{cases} \tag{5.27}$$

$$\begin{aligned} \mathbf{x}_{j+1} &= \mathbf{P}_j\,\mathbf{y}_j \\ &= \mathbf{P}_{j-1}\,\mathbf{y}_{j-1} + \lambda_j\,\mathbf{p}_j \\ &= \mathbf{x}_j + \lambda_j\,\mathbf{p}_j \end{aligned} \tag{5.28}$$

Substituting into the original quadratic functional (Equation 5.3), we obtain the following relationship:

$$\begin{aligned} \Phi(\mathbf{x}_{j+1}) &= \Phi(\mathbf{x}_j) + \lambda_j\,\mathbf{p}_j^T(\mathbf{A}\mathbf{x}_j - \mathbf{b}) + \frac{1}{2}\lambda_j^2\,\mathbf{p}_j^T\mathbf{A}\mathbf{p}_j \\ &= \Phi(\mathbf{P}_{j-1}\,\mathbf{y}_{j-1}) + \lambda_j\,\mathbf{p}_j^T\mathbf{A}\mathbf{P}_{j-1}\,\mathbf{y}_{j-1} - \lambda_j\,\mathbf{p}_j^T\mathbf{b} + \frac{1}{2}\lambda_j^2\,\mathbf{p}_j^T\mathbf{A}\mathbf{p}_j \\ &= \Phi(\mathbf{P}_{j-1}\,\mathbf{y}_{j-1}) + \lambda_j\,\mathbf{p}_j^T\mathbf{A}\mathbf{P}_{j-1}\,\mathbf{y}_{j-1} + \left[\frac{1}{2}\lambda_j^2\,\mathbf{p}_j^T\mathbf{A}\mathbf{p}_j - \lambda_j\,\mathbf{p}_j^T\mathbf{b}\right] \end{aligned} \tag{5.29}$$

In the above equation, the first term $\Phi(\mathbf{x}_j)$ is a j-dimensional minimization problem; the term $\left(1/2\lambda_j^2\,\mathbf{p}_j^T\mathbf{A}\mathbf{p}_j - \lambda_j\,\mathbf{p}_j^T\mathbf{b}\right)$ is a one-dimensional minimization problem. The only coupling term between λ_j and $\mathbf{x}_j$ is the second term $\lambda_j\,\mathbf{p}_j^T\mathbf{A}\mathbf{P}_{j-1}\,\mathbf{y}_{j-1}$. If we force this term to become zero, then we achieve decoupling of the multidimensional minimization problem. As a consequence, by solving a decoupled one-dimensional minimization problem for λ_j, we also solved a j-dimensional minimization problem at each iteration. Therefore, the condition for decoupling is as follows:

$$\mathbf{p}_j^T\mathbf{A}\mathbf{P}_{j-1} = 0 \tag{5.30}$$

$$\mathbf{p}_j^T\mathbf{A}\mathbf{p}_k = 0, \quad \text{for } k = 0, 1, \ldots, j-1 \tag{5.31}$$

This means that to achieve decoupling of the multidimensional minimization problem, each search direction must be **A**-conjugate to all the previous search directions. We also observe that for an n-dimensional vector, there are only n-independent directions. Therefore, after n steps, we obtain the exact solution vector **x** in exact mathematics.

Now we consider the determination of search step length and the search directions. By enforcing **A**-conjugacy on the search directions, the coupling term in Φ is dropped and the step length can then be determined from the one-dimensional minimization problem.

$$\Phi(\mathbf{x}_{j+1}) = \Phi(\mathbf{x}_j) + \frac{1}{2}\lambda_j^2\,\mathbf{p}_j^T\mathbf{A}\mathbf{p}_j - \lambda_j\,\mathbf{p}_j^T\mathbf{b} \tag{5.32}$$

$$\frac{\partial\Phi(\mathbf{x}_{k+1})}{\partial\lambda_k} = \lambda_k\,\mathbf{p}_k^T\mathbf{A}\mathbf{p}_k - \mathbf{p}_k^T\mathbf{b} = 0 \tag{5.33}$$

$$\lambda_k = \frac{\mathbf{p}_k^T \mathbf{b}}{\mathbf{p}_k^T \mathbf{A} \mathbf{p}_k} = \frac{\mathbf{p}_k^T (\mathbf{r}_{k-1} - \mathbf{A}\mathbf{x}_{k-1})}{\mathbf{p}_k^T \mathbf{A} \mathbf{p}_k} = \frac{\mathbf{p}_k^T (\mathbf{r}_{k-1} - \mathbf{A}\mathbf{P}_{k-2}\, \mathbf{y}_{k-2})}{\mathbf{p}_k^T \mathbf{A} \mathbf{p}_k} = \frac{\mathbf{p}_k^T \mathbf{r}_{k-1}}{\mathbf{p}_k^T \mathbf{A} \mathbf{p}_k} \tag{5.34}$$

To determine the search direction vector, we recall that in the steepest descent method, we follow the best search direction that is in the negative gradient direction.

$$\mathbf{p}_k = \mathbf{r}_k = -\nabla\Phi(\mathbf{x}_k) \tag{5.35}$$

In this case, the search direction $\mathbf{p}_k$ is forced to be **A**-conjugate to all the previous directions. In addition, it should also be closest to $\mathbf{r}_k$, which is considered to be the best search direction. The best choice is that $\mathbf{p}_k$ must be a projection of $\mathbf{r}_k$ onto the span $[\mathbf{AP}_0, \mathbf{AP}_1, \ldots, \mathbf{AP}_{k-1}]$ = span $[\mathbf{AP}_{k-1}]$. Assuming a symmetric **A** matrix, we have the following search direction:

$$\begin{aligned} \mathbf{p}_k &= \left[\mathbf{I} - (\mathbf{AP}_{k-1})(\mathbf{P}_{k-1}^T \mathbf{A})\right] \mathbf{r}_k \\ &= \mathbf{r}_k - \mathbf{AP}_{k-1}\mathbf{z} \end{aligned} \tag{5.36}$$

$$\mathbf{z} = \mathbf{P}_{k-1}^T \mathbf{A} \mathbf{r}_k \tag{5.37}$$

With the search direction determined by this equation, we have obtained a fundamental version of the conjugate gradient method. However, because of the way the search direction is determined, it is evident that this algorithm is not very useful. However, this basic relation on the search direction contains some interesting properties. Next, we can have a close look at the relation between search directions and the residual vector.

5.3.2 Basic relation between search directions and residual vector

From the following solution vector update formulas, we can arrive at the expression for the residual vector $\mathbf{r}_k$:

$$\mathbf{x}_k = \mathbf{x}_{k-1} + \lambda_{k-1}\mathbf{p}_{k-1} \tag{5.38}$$

$$\begin{aligned} \mathbf{r}_k &= \mathbf{b} - \mathbf{A}\mathbf{x}_k \\ &= \mathbf{b} - \mathbf{A}\mathbf{x}_{k-1} - \lambda_{k-1}\mathbf{A}\mathbf{p}_{k-1} \\ &= \mathbf{r}_{k-1} - \lambda_{k-1}\mathbf{A}\mathbf{p}_{k-1} \end{aligned} \tag{5.39}$$

This shows that the residual vector can be directly updated with information computed in the previous step. From Equation 5.36, the original relation on $\mathbf{p}_k$ can be written in the following form:

$$\begin{aligned} \mathbf{p}_k &= \mathbf{r}_k - \left[\mathbf{A}\mathbf{p}_0, \ldots, \mathbf{A}\mathbf{p}_{k-1}\right]_{n\times k} \begin{Bmatrix} z_0 \\ \vdots \\ z_{k-1} \end{Bmatrix} \\ &= \mathbf{r}_k - \sum_{j=0}^{k-1} z_j \mathbf{A}\mathbf{p}_j \end{aligned} \tag{5.40}$$

Substituting from Equation 5.39 into the above equation yields the following:

$$\mathbf{p}_k = \mathbf{r}_k + \sum_{j=0}^{k-1} z_j \left(\mathbf{r}_{j+1} - \mathbf{r}_j\right) / \lambda_j$$
$$= \sum_{j=0}^{k} \xi_{kj}\, \mathbf{r}_j \tag{5.41}$$

This equation can be written in the following matrix form:

$$\mathbf{P}_k = \mathbf{R}_k \mathbf{T}_k \tag{5.42}$$

$$\left[\mathbf{p}_0 \,|\ldots|\, \mathbf{p}_k\right] = \left[\mathbf{r}_0 \,|\ldots|\, \mathbf{r}_k\right] \left[\mathbf{T}\right]_k \tag{5.43}$$

If we expand the relation above, we have the following:

$$\begin{cases} \mathbf{p}_0 = \xi_{00}\mathbf{r}_0 \\ \mathbf{p}_1 = \xi_{10}\mathbf{r}_0 + \xi_{11}\mathbf{r}_1 \\ \vdots \\ \mathbf{p}_k = \xi_{k0}\mathbf{r}_0 + \cdots + \xi_{kk}\mathbf{r}_k \end{cases} \tag{5.44}$$

Then we obtain the matrix $\mathbf{T}_k$, which is an upper triangular matrix as follows:

$$\mathbf{T}_k = \begin{bmatrix} \xi_{00} & \xi_{10} & \cdots & \xi_{k0} \\ & \xi_{11} & \cdots & \xi_{k1} \\ & & \ddots & \vdots \\ & & & \xi_{kk} \end{bmatrix} \tag{5.45}$$

Since $\mathbf{p}_k$ are linearly independent vectors, matrix $\mathbf{T}_k$ is nonsingular. Subsequently, we have the following relation and its implication:

$$\mathbf{R}_k = \mathbf{P}_k \mathbf{T}_k^{-1} \tag{5.46}$$

This implies the following relation:

$$\text{span}\{\mathbf{p}_0, \ldots, \mathbf{p}_k\} = \text{span}\{\mathbf{r}_0, \ldots, \mathbf{r}_k\} \tag{5.47}$$

5.3.3 Associated k-dimensional minimization problem

We have discussed the fact that in the conjugate gradient method, at each iteration, we are solving a one-dimensional minimization problem when we determine λ_k while, at the same time, we are also solving a k-dimensional minimization problem. Here, we are looking at the consequences of the k-dimensional minimization problem. We start with the solution vector being the linear combination of all the previous k direction vectors.

$$\mathbf{x}_k = \mathbf{P}_{k-1}\, \mathbf{y}_{k-1} \tag{5.48}$$

$$\Phi(\mathbf{x}_k) = \frac{1}{2}\left(\mathbf{y}_{k-1}^T \mathbf{P}_{k-1}^T\right)\mathbf{A}\left(\mathbf{P}_{k-1}\mathbf{y}_{k-1}\right) - \left(\mathbf{y}_{k-1}^T \mathbf{P}_{k-1}^T\right)\mathbf{b} \tag{5.49}$$

Next, we optimize the functional with respect to $\mathbf{y}_{k-1}$.

$$\frac{\partial \Phi(\mathbf{x}_k)}{\partial \mathbf{y}_{k-1}} = 0 \tag{5.50}$$

$$\begin{aligned} \left(\mathbf{P}_{k-1}^T \mathbf{A} \mathbf{P}_{k-1}\right)\mathbf{y}_{k-1} - \mathbf{P}_{k-1}^T \mathbf{b} &= 0 \\ \mathbf{P}_{k-1}^T\left(\mathbf{A}\mathbf{P}_{k-1}\mathbf{y}_{k-1} - \mathbf{b}\right) &= 0 \\ \mathbf{P}_{k-1}^T \mathbf{r}_k &= 0 \end{aligned} \tag{5.51}$$

$$\mathbf{r}_k^T \mathbf{p}_j = 0 \quad \text{for } j = 0, 1, \ldots, k-1 \tag{5.52}$$

This indicates that the current residual vector is orthogonal to the vector subspace spanned by all the previous direction vectors.

5.3.4 Orthogonality of residual vectors

The following two relations lead to the orthogonality of residual vectors:

$$\text{span}\{\mathbf{p}_0, \ldots, \mathbf{p}_k\} = \text{span}\{\mathbf{r}_0, \ldots, \mathbf{r}_k\} \tag{5.53}$$

$$\mathbf{P}_{k-1}^T \mathbf{r}_k = 0 \tag{5.54}$$

This shows that the current residual vector is orthogonal to all the previous residual vectors.

$$\mathbf{r}_j^T \mathbf{r}_k = 0 \quad \text{for } j = 0, 1, \ldots, k-1 \tag{5.55}$$

5.3.5 A-conjugacy between residual vector and direction vectors

To verify the A-conjugacy between residual vector and the direction vectors, we start with the relation derived in Equation 5.39 and premultiply it with the transpose of a residual vector.

$$\mathbf{r}_k = \mathbf{r}_{k-1} - \lambda_{k-1}\mathbf{A}\mathbf{p}_{k-1} \tag{5.56}$$

$$\mathbf{r}_j^T \mathbf{r}_k = \mathbf{r}_j^T \mathbf{r}_{k-1} - \lambda_{k-1}\mathbf{r}_j^T \mathbf{A}\mathbf{p}_{k-1} \tag{5.57}$$

In the above equation, we can choose the value of j. For j = k, k–1, and <k–1 we arrive at the following relations by observing that, as shown in Equation 5.55, the current residual vector is orthogonal to all the previous one:

$$\begin{aligned} \mathbf{r}_k^T \mathbf{r}_k &= \mathbf{r}_k^T \mathbf{r}_{k-1} - \lambda_{k-1}\mathbf{r}_k^T \mathbf{A}\mathbf{p}_{k-1} \\ &= -\lambda_{k-1}\mathbf{r}_k^T \mathbf{A}\mathbf{p}_{k-1} \end{aligned} \tag{5.58}$$

$$\mathbf{r}_{k-1}^T \mathbf{r}_k = 0 = \mathbf{r}_{k-1}^T \mathbf{r}_{k-1} - \lambda_{k-1} \mathbf{r}_{k-1}^T \mathbf{A} \mathbf{p}_{k-1}$$
$$\mathbf{r}_{k-1}^T \mathbf{r}_{k-1} = \lambda_{k-1} \mathbf{r}_{k-1}^T \mathbf{A} \mathbf{p}_{k-1} \tag{5.59}$$

$$\mathbf{r}_j^T \mathbf{A} \mathbf{p}_{k-1} = 0 \quad \text{for } j < k-1 \tag{5.60}$$

This means that the residual vectors are **A**-orthogonal or **A**-conjugate to direction vectors. At the same time, we also obtain the following relation:

$$\lambda_{k-1} = -\frac{\mathbf{r}_k^T \mathbf{r}_k}{\mathbf{r}_k^T \mathbf{A} \mathbf{p}_{k-1}} = \frac{\mathbf{r}_{k-1}^T \mathbf{r}_{k-1}}{\mathbf{r}_{k-1}^T \mathbf{A} \mathbf{p}_{k-1}} \tag{5.61}$$

5.3.6 Computation of direction vectors

It can be proven that the conjugate search directions satisfy the following relation (Golub and Van Loan, 1980):

$$\mathbf{p}_k \in \text{span}\{\mathbf{r}_{k-1}, \mathbf{p}_{k-1}\} \tag{5.62}$$

Therefore, the search direction can be determined from the following relation:

$$\mathbf{p}_k = \mathbf{r}_k + \alpha_k \mathbf{p}_{k-1} \tag{5.63}$$

The step length in the search direction can be determined from using the equation above in the definition of **A**-conjugacy for search directions.

$$\mathbf{p}_{k-1}^T \mathbf{A} \mathbf{p}_k = 0 \tag{5.64}$$

$$\mathbf{p}_{k-1}^T \mathbf{A} \mathbf{p}_k = \mathbf{p}_{k-1}^T \mathbf{A} \mathbf{r}_k + \alpha_k \mathbf{p}_{k-1}^T \mathbf{A} \mathbf{p}_{k-1} = 0 \tag{5.65}$$

$$\alpha_k = -\frac{\mathbf{p}_{k-1}^T \mathbf{A} \mathbf{r}_k}{\mathbf{p}_{k-1}^T \mathbf{A} \mathbf{p}_{k-1}} \tag{5.66}$$

At this point, we can see that the search direction at the current step is determined with information on the previous step, as shown in Equation 5.63. There is a more efficient way of computing α_k. To arrive at this, we will start with Equation 5.59 and substitute for the residual vector from Equation 5.63.

$$\begin{aligned} \mathbf{r}_{k-1}^T \mathbf{r}_{k-1} &= \lambda_{k-1} \mathbf{r}_{k-1}^T \mathbf{A} \mathbf{p}_{k-1} \\ &= \lambda_{k-1} \left(\mathbf{p}_{k-1}^T - \alpha_{k-1} \mathbf{p}_{k-2}^T \right) \mathbf{A} \mathbf{p}_{k-1} \\ &= \lambda_{k-1} \mathbf{p}_{k-1}^T \mathbf{A} \mathbf{p}_{k-1} \end{aligned} \tag{5.67}$$

This leads to an expression for the λ_{k-1}.

$$\lambda_{k-1} = \frac{\mathbf{r}_{k-1}^T \mathbf{r}_{k-1}}{\mathbf{p}_{k-1}^T \mathbf{A} \mathbf{p}_{k-1}} \tag{5.68}$$

Next, we use Equation 5.58 and substitute for the residual vector from Equation 5.63 and for λ_{k-1} from the equation above.

$$\begin{aligned}
\mathbf{r}_k^T \mathbf{r}_k &= -\lambda_{k-1} \mathbf{r}_k^T \mathbf{A} \mathbf{p}_{k-1} \\
&= -\lambda_{k-1} \left(\mathbf{p}_k^T - \alpha_k \mathbf{p}_{k-1}^T \right) \mathbf{A} \mathbf{p}_{k-1} \\
&= \lambda_{k-1} \alpha_k \mathbf{p}_{k-1}^T \mathbf{A} \mathbf{p}_{k-1} \\
&= \frac{\mathbf{r}_{k-1}^T \mathbf{r}_{k-1}}{\mathbf{p}_{k-1}^T \mathbf{A} \mathbf{p}_{k-1}} \alpha_k \mathbf{p}_{k-1}^T \mathbf{A} \mathbf{p}_{k-1} \\
&= \alpha_k \mathbf{r}_{k-1}^T \mathbf{r}_{k-1}
\end{aligned} \tag{5.69}$$

This leads to the efficient expression for computing α_k.

$$\alpha_k = \frac{\mathbf{r}_k^T \mathbf{r}_k}{\mathbf{r}_{k-1}^T \mathbf{r}_{k-1}} \tag{5.70}$$

5.3.7 Optimality of conjugate gradient method

The solution obtained from the conjugate gradient method is the optimal solution in terms of minimizing an error functional. Here, we will prove this important property of the conjugate gradient method. We will start by observing that from our discussions so far, we can list some of the properties of the conjugacy of search directions and residual vectors. We can prove the following relations:

$$\text{span}\{\mathbf{r}_0, \mathbf{r}_1, \ldots, \mathbf{r}_k\} = \text{span}\{\mathbf{r}_0, \mathbf{A}\mathbf{r}_0, \ldots, \mathbf{A}^k \mathbf{r}_0\} \tag{5.71}$$

$$\text{span}\{\mathbf{p}_0, \mathbf{p}_1, \ldots, \mathbf{p}_k\} = \text{span}\{\mathbf{r}_0, \mathbf{A}\mathbf{r}_0, \ldots, \mathbf{A}^k \mathbf{r}_0\} \tag{5.72}$$

We start by recalling the following two equations that were developed earlier, Equations 5.56 and 5.63:

$$\begin{cases} \mathbf{r}_k = \mathbf{r}_{k-1} - \lambda_{k-1} \mathbf{A} \mathbf{p}_{k-1} \\ \mathbf{p}_k = \mathbf{r}_k + \alpha_k \ \mathbf{p}_{k-1} \end{cases} \tag{5.73}$$

Next, we successively apply these equations starting from k = 0.

$$
\begin{aligned}
&\mathbf{p}_0 = \mathbf{r}_0 \\
&\begin{cases}
\mathbf{r}_1 = \mathbf{r}_0 - \lambda_0 \mathbf{A}\mathbf{p}_0 = \mathbf{r}_0 - \lambda_0 \mathbf{A}\mathbf{r}_0 \\
\quad = \mathbf{r}_0 + \gamma_{11}\mathbf{A}\mathbf{r}_0 \\
\mathbf{p}_1 = \mathbf{r}_1 + \alpha_1 \mathbf{p}_0 = \mathbf{r}_0 + \gamma_{11}\mathbf{A}\mathbf{r}_0 + \alpha_1 \mathbf{r}_0 \\
\quad = \beta_{10}\mathbf{r}_0 + \beta_{11}\mathbf{A}\mathbf{r}_0
\end{cases} \\
&\begin{cases}
\mathbf{r}_2 = \mathbf{r}_1 - \lambda_1 \mathbf{A}\mathbf{p}_1 = \mathbf{r}_0 + \gamma_{11}\mathbf{A}\mathbf{r}_0 - \lambda_1 \mathbf{A}\left(\beta_{10}\mathbf{r}_0 + \beta_{11}\mathbf{A}\mathbf{r}_0\right) \\
\quad = \mathbf{r}_0 + \gamma_{21}\mathbf{A}\mathbf{r}_0 + \gamma_{22}\mathbf{A}^2\mathbf{r}_0 \\
\mathbf{p}_2 = \mathbf{r}_2 + \alpha_2 \mathbf{p}_1 = \mathbf{r}_0 + \gamma_{21}\mathbf{A}\mathbf{r}_0 + \gamma_{22}\mathbf{A}^2\mathbf{r}_0 + \alpha_2\left(\beta_{10}\mathbf{r}_0 + \beta_{11}\mathbf{A}\mathbf{r}_0\right) \\
\quad = \beta_{20}\mathbf{r}_0 + \beta_{21}\mathbf{A}\mathbf{r}_0 + \beta_{22}\mathbf{A}^2\mathbf{r}_0
\end{cases} \\
&\quad \vdots \\
&\begin{cases}
\mathbf{r}_k = \mathbf{r}_0 + \gamma_{k1}\mathbf{A}\mathbf{r}_0 + \cdots + \gamma_{kk}\mathbf{A}^k\mathbf{r}_0 \\
\mathbf{p}_k = \beta_{k0}\mathbf{r}_0 + \beta_{k1}\mathbf{A}\mathbf{r}_0 + \cdots + \beta_{kk}\mathbf{A}^k\mathbf{r}_0
\end{cases}
\end{aligned}
\tag{5.74}
$$

The last two expressions for $\mathbf{r}_k$ and $\mathbf{p}_k$ prove Equations 5.71 and 5.72. We can express the last equation above in the form of a polynomial $\mathbf{P}_k(\mathbf{A})$.

$$\mathbf{p}_k = \mathbf{P}_k(\mathbf{A})\mathbf{r}_0 \tag{5.75}$$

Next, we assume a general search method as follows and show that in the conjugate gradient method at each step when we determine a direction vector, we are also solving a minimization problem:

$$
\begin{aligned}
\mathbf{x}_{k+1} &= \mathbf{x}_0 + \lambda_k \mathbf{p}_k \\
&= \mathbf{x}_0 + \mathbf{P}_k(\mathbf{A})\mathbf{r}_0
\end{aligned}
\tag{5.76}
$$

We can see from Equation 5.74 that $\mathbf{P}_k(\mathbf{A})$ is determined from the conjugate gradient method. It can also be shown that it also minimizes the following error function:

$$
\begin{aligned}
&E\left(\mathbf{x}_{k+1}\right) = \frac{1}{2}\left(\mathbf{x}_{k+1} - \bar{\mathbf{x}}\right)^T \mathbf{A}\left(\mathbf{x}_{k+1} - \bar{\mathbf{x}}\right) \\
&\qquad \bar{\mathbf{x}} = \text{exact solution}
\end{aligned}
\tag{5.77}
$$

$$\mathbf{x}_{k+1} = \mathbf{x}_0 + \mathbf{P}_k(\mathbf{A})\mathbf{r}_0 = \mathbf{x}_0 + \mathbf{P}_k(\mathbf{A})\left(\bar{\mathbf{x}} - \mathbf{x}_0\right) \tag{5.78}$$

$$\mathbf{x}_{k+1} - \bar{\mathbf{x}} = \left[\mathbf{I} - \mathbf{P}_k(\mathbf{A})\right]\left(\mathbf{x}_0 - \bar{\mathbf{x}}\right) \tag{5.79}$$

$$E\left(\mathbf{x}_{k+1}\right) = \frac{1}{2}\left(\mathbf{x}_0 - \bar{\mathbf{x}}\right)^T \left[\mathbf{I} - \mathbf{P}_k(\mathbf{A})\right]^T \mathbf{A}\left[\mathbf{I} - \mathbf{P}_k(\mathbf{A})\right]\left(\mathbf{x}_0 - \bar{\mathbf{x}}\right) \tag{5.80}$$

Now we consider the selection of the polynomial such that the error functional is minimized. For generality, we assume the following:

$$P_k(\mathbf{A}) = \xi_0\mathbf{I} + \xi_1\mathbf{A} + \cdots + \xi_k\mathbf{A}^k \tag{5.81}$$

Minimization of the functional leads to the following expression:

$$\mathbf{x}_{k+1} = \mathbf{x}_0 + \xi_0\mathbf{r}_0 + \xi_1\mathbf{A}\mathbf{r}_0 + \cdots + \xi_k\mathbf{A}^k\mathbf{r}_0 \tag{5.82}$$

On the other hand, from Equation 5.71, the approximate solution obtained from the conjugate gradient method at (k + 1)th step is as follows:

$$\mathbf{x}_{k+1} = \mathbf{x}_0 + \lambda_1\mathbf{p}_1 + \cdots + \lambda_k\mathbf{p}_k \tag{5.83}$$

We observe that using Equation 5.72 this equation is exactly of the same form as Equation 5.82 obtained from minimizing the error functional. Therefore, the step length parameter determined from the conjugate gradient algorithm minimizes the main functional, which in turn means that the error functional in Equation 5.77 is minimized as well. This shows that the solution obtained from the conjugate gradient method is the optimal solution that minimizes the error functional.

5.3.8 Rate of convergence

We have observed in earlier discussion that the conjugate gradient method in exact mathematics is a direct method that will arrive at the exact solution in n step for a n x n matrix **A**. However, it is an iterative method in computational environment with the presence of a round-off error. Also, it has to converge much faster than n step to be computationally viable.

The rate of convergence can be considered as how fast the error functional in Equation 5.80 approaches zero. Using Chebyshev polynomial, the following expression for the error functional can be proven:

$$E(\mathbf{x}_{k+1}) \le 4\left(\frac{\sqrt{\kappa}-1}{\sqrt{\kappa}+1}\right)^{2(k+1)} E(\mathbf{x}_0) \tag{5.84}$$

$$\kappa(\mathbf{A}) = \frac{\lambda_n}{\lambda_1} \tag{5.85}$$

We can observe that the rate of reduction in the error functional, or the rate of convergence of conjugate gradient algorithm, are dependent on $\kappa(\mathbf{A})$ that is the condition number of matrix. The condition number of a matrix is the ratio of its highest and lowest eigenvalues. The smallest value of the condition number is equal to one for the identity matrix **I**. We can see from the equation above the error functional goes to zero in one step, since the solution with identity matrix is trivial.

When the system is ill-conditioned, or when $\kappa(\mathbf{A})$ becomes very large, then the rate of reduction of the error function becomes slow. This means that the rate of convergence of the conjugate gradient method is slow for ill-conditioned matrices. Because of this property on the rate of convergence of the standard conjugate gradient method, we need to search for other acceleration schemes in the case of ill-conditioned system. One approach is to transform the original system of equations such that the condition number of the coefficient matrix can be reduced. These methods are referred to as preconditioning methods, which will be discussed in Section 5.6. Meanwhile, we discuss the practical algorithms for implementing the conjugate gradient method.

5.4 PRACTICAL ALGORITHMS

5.4.1 Basic algorithm

Having discussed various aspects of the conjugate gradient method, now we can describe some practical algorithms. To recap, conjugate gradient is a semi-iterative method for solving the following system of equations:

$$\mathbf{A}\mathbf{x} = \mathbf{b} \tag{5.86}$$

By semi-iterative method we mean that in exact mathematics (without a round-off error) it is a direct method and it converges to the exact solution in a finite number of steps. In a computational environment, with the presence of a round-off error, it is an iterative method. The solution is obtained by minimizing the following quadratic functional:

$$\Phi(\mathbf{x}) = \frac{1}{2}\mathbf{x}^T\mathbf{A}\mathbf{x} - \mathbf{x}^T\mathbf{b} \tag{5.87}$$

As an iterative method, each iteration consists of updating the latest estimate of the solution $\mathbf{x}_k$, along a direction $\mathbf{p}_k$, by a scalar step length λ_k, to obtain a better estimate of the solution.

$$\mathbf{x}_{k+1} = \mathbf{x}_k + \lambda_k\,\mathbf{p}_k \tag{5.88}$$

The sequence of estimates should converge to the solution of the system of equations within a reasonable number of steps. This is equivalent to saying that the residual vector $\mathbf{r}_k$ should approach zero.

$$\mathbf{r}_k = \mathbf{b} - \mathbf{A}\mathbf{x}_k \tag{5.89}$$

The search directions are defined as follows:

$$\begin{aligned} \mathbf{p}_0 &= \mathbf{r}_0 \\ &\vdots \\ \mathbf{p}_k &= \mathbf{r}_k + \alpha_k\,\mathbf{p}_{k-1} \\ &\vdots \end{aligned} \tag{5.90}$$

The scalar α_k is determined in such a way that the new direction is **A**-conjugate to the previous direction.

$$\mathbf{p}_{k-1}^T\mathbf{A}\mathbf{p}_k = 0 \tag{5.91}$$

This leads to the following equation:

$$\alpha_k = -\frac{\mathbf{p}_{k-1}^T\mathbf{A}\mathbf{r}_k}{\mathbf{p}_{k-1}^T\mathbf{A}\mathbf{p}_{k-1}} \tag{5.92}$$

A computationally more efficient version of this equation is referred to as the *Fletcher–Reeves formula* (Fletcher and Reeves, 1964).

$$\alpha_k = \frac{\mathbf{r}_k^T \mathbf{r}_k}{\mathbf{r}_{k-1}^T \mathbf{r}_{k-1}} \tag{5.93}$$

Another similar relation is called the *Polak–Ribiere formula* (Polak and Ribiere, 1969).

$$\alpha_k = \frac{\mathbf{r}_k^T (\mathbf{r}_k - \mathbf{r}_{k-1})}{\mathbf{r}_{k-1}^T \mathbf{r}_{k-1}} \tag{5.94}$$

Either of these formulae can be used in computation. The step length is determined in such a way that the functional is minimized for the new estimate of the solution, leading to the following two expressions:

$$\frac{\partial}{\partial \lambda_k} \Phi(\mathbf{x}_{k+1}) = 0 \tag{5.95}$$

$$\lambda_k = \frac{\mathbf{p}_k^T \mathbf{r}_k}{\mathbf{p}_k^T \mathbf{A} \mathbf{p}_k} \tag{5.96}$$

$$\lambda_k = \frac{\mathbf{r}_k^T \mathbf{r}_k}{\mathbf{p}_k^T \mathbf{A} \mathbf{p}_k} \tag{5.97}$$

These two steps in finding the direction and the step length can be interpreted in the following way. The actual solution makes the functional zero. Even though we do not expect to find the solution along the direction $\mathbf{p}_k$, we determine the step length to correspond to the minimum of the functional along that direction. This gets us closer to our final goal, because, by seeking to minimize the functional along the next direction $\mathbf{p}_{k+1}$ we will not undo what has already been done, since the next direction is **A**-conjugate to all the directions which have already been calculated. Therefore, the first practical algorithm can be described as follows:

Initialization

$$\mathbf{r}_0 = \mathbf{b}$$

$$\mathbf{p}_0 = \mathbf{r}_0$$

For $k = 0, 1, 2, \cdots$

$$\lambda_k = \frac{\mathbf{r}_k^T \mathbf{r}_k}{\mathbf{p}_k^T \mathbf{A} \mathbf{p}_k}$$

$$\mathbf{x}_{k+1} = \mathbf{x}_k + \lambda_k \mathbf{p}_k$$

$$\mathbf{r}_{k+1} = \mathbf{b} - \mathbf{A}\mathbf{x}_{k+1}$$

$$\alpha_{k+1} = \frac{\mathbf{r}_{k+1}^T \mathbf{r}_{k+1}}{\mathbf{r}_k^T \mathbf{r}_k}$$

$$\mathbf{p}_{k+1} = \mathbf{r}_{k+1} + \alpha_{k+1} \mathbf{p}_k$$

5.4.2 Conjugate gradient acceleration of the Jacobi method

The conjugate gradient method can also be considered as an acceleration scheme for the basic iterative methods. Combining an iterative method with the conjugate gradient method has been shown to be an effective method for solution of very large system of equations, especially with diagonally dominant matrices. Most problems in computational solid mechanics result in diagonally dominant stiffness matrices. We will consider the Jacobi iterative method. As was described in Chapter 4, the iterative equation in the standard Jacobi method is of the following form:

$$\begin{aligned} \mathbf{x}_{k+1} &= \left(\mathbf{I} - \mathbf{D}_A^{-1}\mathbf{A}\right)\mathbf{x}_k + \mathbf{D}_A^{-1}\mathbf{b} \\ \mathbf{D}_A &= \text{diag}(\mathbf{A}) \end{aligned} \tag{5.98}$$

This is equivalent to solving the following corresponding system of equations:

$$\begin{aligned} &\bar{\mathbf{A}}\mathbf{x} = \bar{\mathbf{b}} \\ &\begin{cases} \bar{\mathbf{A}} = \mathbf{D}_A^{-1}\mathbf{A} \\ \bar{\mathbf{b}} = \mathbf{D}_A^{-1}\mathbf{b} \end{cases} \end{aligned} \tag{5.99}$$

It is likely that equivalent matrix in the above equation may not be symmetric so that we need to symmetrize the system of equation in order to use the conjugate gradient method. A symmetric version of the above equation is as follows:

$$\begin{aligned} &\mathbf{D}_A^{1/2}\bar{\mathbf{A}}\mathbf{D}_A^{-1/2}\mathbf{D}_A^{1/2}\mathbf{x} = \mathbf{D}_A^{1/2}\bar{\mathbf{b}} \\ &\hat{\mathbf{A}}\hat{\mathbf{x}} = \hat{\mathbf{b}} \\ &\begin{cases} \hat{\mathbf{A}} = \mathbf{D}_A^{-1/2}\mathbf{A}\mathbf{D}_A^{-1/2} \\ \hat{\mathbf{x}} = \mathbf{D}_A^{1/2}\mathbf{x} \\ \hat{\mathbf{b}} = \mathbf{D}_A^{-1/2}\mathbf{b} \end{cases} \end{aligned} \tag{5.100}$$

Now we perform the conjugate gradient method on the symmetrized equivalent system obtained from the Jacobi method. We will determine the terms in the conjugate gradient method for the equivalent system and the symmetrized equivalent system.

$$\begin{cases} \hat{\mathbf{r}}_k = \hat{\mathbf{b}} - \hat{\mathbf{A}}\hat{\mathbf{x}} = \mathbf{D}_A^{-1/2}\mathbf{r}_k \\ \bar{\mathbf{r}}_k = \bar{\mathbf{b}} - \bar{\mathbf{A}}\mathbf{x}_k = \mathbf{D}_A^{-1}\mathbf{r}_k \end{cases} \tag{5.101}$$

$$\begin{cases} \hat{\mathbf{r}}_k = \mathbf{D}_A^{1/2}\bar{\mathbf{r}}_k \\ \hat{\mathbf{p}}_k = \mathbf{D}_A^{1/2}\bar{\mathbf{p}}_k \end{cases} \tag{5.102}$$

$$\begin{cases} \hat{\mathbf{p}}_k = \hat{\mathbf{r}}_k + \hat{\alpha}_k\hat{\mathbf{p}}_{k-1} \\ \bar{\mathbf{p}}_k = \bar{\mathbf{r}}_k + \hat{\alpha}_k\bar{\mathbf{p}}_{k-1} \end{cases} \tag{5.103}$$

$$\hat{\alpha}_k = \frac{\hat{\mathbf{r}}_k^T \hat{\mathbf{r}}_k}{\hat{\mathbf{r}}_{k-1}^T \hat{\mathbf{r}}_{k-1}} = \frac{\overline{\mathbf{r}}_k^T \mathbf{D}_A \overline{\mathbf{r}}_k}{\overline{\mathbf{r}}_{k-1}^T \mathbf{D}_A \overline{\mathbf{r}}_{k-1}} \tag{5.104}$$

$$\begin{cases} \hat{\mathbf{x}}_{k+1} = \hat{\mathbf{x}}_k + \hat{\lambda}_k \hat{\mathbf{p}}_k \\ \mathbf{x}_{k+1} = \mathbf{x}_k + \hat{\lambda}_k \overline{\mathbf{p}}_k \end{cases} \tag{5.105}$$

$$\hat{\lambda}_k = \frac{\hat{\mathbf{p}}_k^T \hat{\mathbf{r}}_k}{\hat{\mathbf{p}}_k^T \hat{\mathbf{A}} \hat{\mathbf{p}}_k} = \frac{\overline{\mathbf{p}}_k^T \mathbf{D}_A \overline{\mathbf{r}}_k}{\overline{\mathbf{p}}_{k-1}^T \mathbf{A} \overline{\mathbf{p}}_{k-1}} \tag{5.106}$$

This is the conjugate gradient method used in conjunction with the Jacobi method. Let $\mathbf{Q} = \mathbf{D}_A$. The algorithm can be described as follows:

Initialization

$$\mathbf{r}_0 = \mathbf{b}$$

$$\overline{\mathbf{r}}_0 = \mathbf{Q}^{-1}\mathbf{r}_0$$

$$\mathbf{p}_0 = \overline{\mathbf{r}}_0$$

For $k = 0, 1, 2, \ldots$

$$\lambda_k = \frac{\overline{\mathbf{r}}_k^T \mathbf{Q} \overline{\mathbf{r}}_k}{\mathbf{p}_k^T \mathbf{A} \mathbf{p}_k}$$

$$\mathbf{x}_{k+1} = \mathbf{x}_k + \lambda_k \mathbf{p}_k$$

$$\mathbf{r}_{k+1} = \mathbf{b} - \mathbf{A}\mathbf{x}_{k+1}$$

$$\overline{\mathbf{r}}_{k+1} = \mathbf{Q}^{-1}\mathbf{r}_{k+1}$$

$$\alpha_{k+1} = \frac{\overline{\mathbf{r}}_{k+1}^T \mathbf{Q} \overline{\mathbf{r}}_{k+1}}{\overline{\mathbf{r}}_k^T \mathbf{Q} \overline{\mathbf{r}}_k}$$

$$\mathbf{p}_{k+1} = \overline{\mathbf{r}}_{k+1} + \alpha_{k+1}\mathbf{p}_k$$

In this algorithm, $\mathbf{Q}$ is a diagonal matrix which has the diagonal terms of the matrix $\mathbf{A}$. Numerical results have shown that this accelerated method is very effective in speeding up the convergence process for diagonally dominant systems. This is the Jacobi preconditioned conjugate gradient method that is one of the viable methods for solving very large systems.

5.5 CONJUGATE GRADIENT METHOD IN FINITE ELEMENT ANALYSIS

In finite element analysis of structural systems, if the stiffness matrix is available it can, of course, be directly used in this algorithm. However, conjugate gradient algorithm does not require the explicit formation of the stiffness matrix. This makes the conjugate gradient method suitable for very large problems. We can see that the stiffness matrix appears in the form of a matrix–vector multiplication. The resulting vector **I**, the internal resisting force vector, can be computed by direct assembly of element contributions.

$$\mathrm{I}(\mathbf{x}) = \mathbf{A}\mathbf{x} = \sum \int_{v} \mathrm{B}^{T} \sigma \, dv \tag{5.107}$$

For any given vector **x**, which may be considered as a displacement vector, the corresponding element strains and stresses are computed, and the right hand side of the above equation is directly assembled from the element contributions. Even though the stiffness matrix is not formed, the computation of the internal resisting force vector is still computationally the most intensive part of the algorithm. It may appear that this operation is performed twice per iteration. For linear problems, the internal resisting force vectors need to be evaluated only once per iteration.

5.5.1 Conjugate gradient algorithm for linear problems

The conjugate gradient algorithm with Jacobi acceleration or preconditioning for linear problems can be put in the following procedure:

Initialization

$$\mathbf{r} \leftarrow \mathbf{b}$$

$$\overline{\mathbf{r}} \leftarrow \mathbf{Q}^{-1}\mathbf{r}$$

$$\mathbf{p} \leftarrow \overline{\mathbf{r}}$$

$$\beta_0 \leftarrow \overline{\mathbf{r}}^{T}\mathbf{Q}\overline{\mathbf{r}}$$

For k = 0, 1, 2, …

$$\mathbf{I} \leftarrow \mathbf{I}(\mathbf{p})$$

$$\eta_k \leftarrow \mathbf{p}^{T}\mathbf{I}$$

$$\lambda \leftarrow \frac{\beta_k}{\eta_k}$$

$$\mathbf{x} \leftarrow \mathbf{x} + \lambda\mathbf{p}$$

$$\mathbf{r} \leftarrow \mathbf{r} - \lambda\mathbf{I}$$

$$\overline{\mathbf{r}} \leftarrow \mathbf{Q}^{-1}\mathbf{r}$$

$$\beta_{k+1} \leftarrow \overline{\mathbf{r}}^{T}\mathbf{Q}\overline{\mathbf{r}}$$

$$\alpha_{k+1} \leftarrow \frac{\beta_{k+1}}{\beta_k}$$

$$\mathbf{p} \leftarrow \overline{\mathbf{r}} + \alpha_{k+1}\mathbf{p}$$

In most computational solid mechanics problems, **b** is the load vector, **I(p)** is the internal resisting force vector, and the solution vector **x** is the nodal displacement vector. The matrix Q is a diagonal matrix containing the diagonal elements of the stiffness matrix.

5.5.2 Conjugate gradient algorithm for nonlinear problems

In deriving the conjugate gradient method for linear problems, we are searching for the solution vector that minimizes a quadratic functional. However, the use of conjugate gradient method is not limited to minimization of quadratic functional only. In fact, it can be used to minimize any continuous functions for which the gradient can be computed. These include nonlinear structural optimization problems and nonlinear finite element analysis of structural systems. Because of the nonlinearity of the problem, there are several changes to the standard conjugate gradient algorithm in the calculation of the residual force vector, the determination of step length λ_k and choices for α_k.

For nonlinear problems the internal force vector must be calculated twice per iteration. Moreover, here matrix A is the tangent stiffness matrix. The following algorithm does the job without upsetting the internal workings of the nonlinear material models:

Initialization

$$\mathbf{r} \leftarrow \mathbf{b}$$

$$\bar{\mathbf{r}} \leftarrow \mathbf{Q}^{-1}\mathbf{r}$$

$$\mathbf{p} \leftarrow \bar{\mathbf{r}}$$

$$\beta_0 \leftarrow \bar{\mathbf{r}}^T\mathbf{Q}\bar{\mathbf{r}}$$

For k = 0, 1, 2, ...

$$\mathbf{I} \leftarrow \mathbf{I}_t(\mu\mathbf{p})$$

($\mathbf{I}_t$ = increment of internal resisting force vector)

(μ = a small number relative to overall response of the system)

$$\eta_k \leftarrow \mu\mathbf{p}^T\mathbf{I}$$

$$\lambda \leftarrow \frac{\beta_k}{\eta_k}$$

$$\mathbf{x} \leftarrow \mathbf{x} + \lambda\mathbf{p}$$

$$\mathbf{r} \leftarrow \mathbf{r} - \mathbf{I} - \mathbf{I}_t\left[(\lambda - \mu)\mathbf{p}\right];$$

$$\bar{\mathbf{r}} \leftarrow \mathbf{Q}^{-1}\mathbf{r}$$

$$\beta_{k+1} \leftarrow \bar{\mathbf{r}}^T\mathbf{Q}\bar{\mathbf{r}}$$

$$\alpha_{k+1} \leftarrow \frac{\beta_{k+1}}{\beta_k}$$

$$\mathbf{p} \leftarrow \bar{\mathbf{r}} + \alpha_{k+1}\mathbf{p}$$

5.5.3 General comments on computational aspects

In exact arithmetic, it can be shown that the conjugate gradient method yields the exact solution in n iterations, where n is the number of degrees of freedom. So, in a sense, it is a semi-iterative or even a direct method in exact arithmetic. However, in finite precision arithmetic, this is not guaranteed. Even if it were, we cannot afford it. In practice, however, the convergence is much faster, perhaps around $\sqrt{n}$ iterations. Convergence also, to some extent, depends on the physical nature of the problem.

The convergence criterion is usually based on the norm of the residual vector **r**. The norm of the residual vector must be approaching zero and when it falls below a certain prescribed small value, we accept the approximate solution as sufficiently close to the exact solution. In a plot of the norm of the residual vector versus the iteration number, we can distinguish three distinct stages. In stage one the $|r|$ is rapidly increasing. In stage two it remains more or less constant or decreases somewhat and in stage three it decreases exponentially.

In stage one, the signals propagate similar to an explicit solution, one layer of elements per iteration. The directions found by the solution method in stage one are not very good. The method starts finding very good directions in the second and third stages. In the third stage the convergence is very rapid. In the nonlinear problems, the convergence can generally be expected to be slower.

The main advantages of the conjugate gradient method in solving finite element problems are that it involves vector operations, and there is no need to explicitly form the global stiffness matrix. Therefore, the method is well suited for solving very large problems on the vector or parallel computers. On the other hand, the signal propagation property makes it suitable for solving some interesting problems. For example, we consider a one-dimensional structure modeled with bar element. The structure is acted on by a force P at one end of the structure. After the first conjugate gradient iteration, the signal (stress) travels through the first element; at the end of the second iteration, it reaches the end of the second element. Once the signal reaches the far end, no matter what is the support condition, the results are obtained. With symmetric problems, the conjugate gradient method does not require the enforcement of symmetry beforehand.

Of course, the major problem with the conjugate gradient method is that its convergence is sensitive to the condition number of the system coefficient matrix, $\kappa(\mathbf{A})$. Thus, in practice, we need to employ different preconditioning schemes to improve on the condition of the system. Some of the preconditioning schemes will be discussed in Section 5.6.

5.6 PRECONDITIONED CONJUGATE GRADIENT METHODS

5.6.1 Basic concept

From our discussion on the convergence of the conjugate gradient method, it is obvious that the method is very sensitive to the condition number of the stiffness matrix **A**. For ill-conditioned systems, the mutual orthogonality of residual vectors cannot be maintained precisely, which results in a slow convergence rate. Therefore, it seems that a natural approach to rectify the ill-conditioning problem would be to carry out matrix transformation on the original stiffness matrix **A** such that its condition can be improved. Such numerical schemes for improving the condition number of a matrix are called preconditioning procedures. Preconditioning is a numerical technique performed on the originally ill-conditioned system of equations to change it into a well-conditioned system.

We have discussed earlier that a perfectly conditioned matrix is the identity matrix. This is a very important observation. Suppose that we can find matrix **C** which is symmetric and positive definite and is very close to matrix **A** but easy to compute its inverse as well. Then instead of solving the ill-conditioned system of equations, **A x** = **b**, we can obtain the solution vector by solving a transformed equivalent system of equations.

$$\bar{\mathbf{A}}\mathbf{x} = \bar{\mathbf{b}}$$

$$\begin{cases} \bar{\mathbf{A}} = \mathbf{C}^{-1}\mathbf{A} \\ \bar{\mathbf{b}} = \mathbf{C}^{-1}\mathbf{b} \end{cases} \tag{5.108}$$

Because matrix **C** is a good approximation of **A**, we can expect that the condition of the transformed system will be dramatically improved.

$$\kappa(\mathbf{C}^{-1}\mathbf{A}) \ll \kappa(\mathbf{A}) \tag{5.109}$$

The convergence rate of the conjugate gradient method in solving the transformed system of equations would be faster than solving the original system in Equation 5.108. We have seen from the conjugate acceleration of the Jacobi method that we in fact used a preconditioning matrix $\mathbf{C} = \mathbf{D}_A$ in that algorithm. That's why the algorithm is referred to as the Jacobi preconditioned conjugate gradient method.

With this transformation, it is most likely that matrix may not be symmetric. We thus need to symmetrize the above equivalent system in Equation 5.108. Since the preconditioning is symmetric and positive definite, we can decompose it, $\mathbf{C} = \mathbf{E}\mathbf{E}^T$, which can be accomplished through Cholesky decomposition, leading to the symmetric version through the following steps:

$$\begin{aligned} \mathbf{C}^{-1}\mathbf{A}\mathbf{x} &= \mathbf{C}^{-1}\mathbf{b} \\ \left(\mathbf{E}^{-T}\mathbf{E}^{-1}\right)\mathbf{A}\mathbf{x} &= \left(\mathbf{E}^{-T}\mathbf{E}^{-1}\right)\mathbf{b} \\ \left(\mathbf{E}^{-T}\mathbf{E}^{-1}\right)\mathbf{A}\left(\mathbf{E}^{-T}\mathbf{E}^{T}\right)\mathbf{x} &= \left(\mathbf{E}^{-T}\mathbf{E}^{-1}\right)\mathbf{b} \\ \left(\mathbf{E}^{-1}\mathbf{A}\mathbf{E}^{-T}\right)\left(\mathbf{E}^{T}\mathbf{x}\right) &= \left(\mathbf{E}^{-1}\mathbf{b}\right) \end{aligned} \tag{5.110}$$

$$\hat{\mathbf{A}}\hat{\mathbf{x}} = \hat{\mathbf{b}}$$

$$\begin{cases} \hat{\mathbf{A}} = \mathbf{E}^{-1}\mathbf{A}\mathbf{E}^{-T} \\ \hat{\mathbf{x}} = \mathbf{E}^{T}\mathbf{x} \\ \hat{\mathbf{b}} = \mathbf{E}^{-1}\mathbf{b} \end{cases} \tag{5.111}$$

The symmetrized system of equations above can be solved with the conjugate gradient method, because matrix is symmetric and positive definite. We note that matrices $\hat{\mathbf{A}}$ and $\bar{\mathbf{A}}$ have the same eigenvalues.

In theory, we can solve the symmetrized system of equations through direct use of the conjugate gradient method, which results in the *transformed preconditioned conjugate gradient method*. However, this algorithm is not very efficient because we need to compute the decomposition of matrix $\mathbf{C} = \mathbf{E}\mathbf{E}^T$. From our earlier derivation of the Jacobi preconditioned conjugate gradient, by substituting $\mathbf{C} = \mathbf{D}_A$, we obtained the *untransformed preconditioned conjugate gradient method.*

Initialization

$$\mathbf{r}_0 = \mathbf{b}$$

$$\overline{\mathbf{r}}_0 = \mathbf{C}^{-1}\mathbf{r}_0$$

$$\mathbf{p}_0 = \overline{\mathbf{r}}_0$$

For k = 0, 1, 2, ...

$$\lambda_k = \frac{\overline{\mathbf{r}}_k^T \mathbf{C} \overline{\mathbf{r}}_k}{\mathbf{p}_k^T \mathbf{A} \mathbf{p}_k}$$

$$\mathbf{x}_{k+1} = \mathbf{x}_k + \lambda_k \mathbf{p}_k$$

$$\mathbf{r}_{k+1} = \mathbf{b} - \mathbf{A}\mathbf{x}_{k+1}$$

$$\overline{\mathbf{r}}_{k+1} = \mathbf{C}^{-1}\mathbf{r}_{k+1}$$

$$\alpha_{k+1} = \frac{\overline{\mathbf{r}}_{k+1}^T \mathbf{C} \overline{\mathbf{r}}_{k+1}}{\overline{\mathbf{r}}_k^T \mathbf{C} \overline{\mathbf{r}}_k}$$

$$\mathbf{p}_{k+1} = \overline{\mathbf{r}}_{k+1} + \alpha_{k+1}\mathbf{p}_k$$

In that algorithm we can see that preconditioning increases the computational effort of the conjugate gradient method by mainly requiring a solution of system equations with the preconditioning matrix **C**. In the Jacobi untransformed preconditioned conjugate gradient method, matrix **C** is diagonal and easily inverted. We need an efficient method of accomplishing this task when matrix **C** is not diagonal. We can also obtain preconditioners from other standard iterative solution methods. From our discussion on standard iterative methods in Chapter 4, we can use various forms of splitting the preconditioning matrices.

Some preconditioners are not symmetric and positive definite. The resulting equivalent system may not be symmetrizable. It can be expected that the effectiveness of the preconditioning matrices on real problems may vary. Therefore, judgment must be exercised in selecting preconditioners.

5.6.2 SSOR preconditioning

In this method, the preconditioner is the split matrix in the SSOR method described in Chapter 4.

$$\mathbf{C} = \left(\mathbf{D}_A + \alpha\mathbf{A}_L\right)\mathbf{D}_A^{-1}\left(\mathbf{D}_A + \alpha\mathbf{A}_L\right)^T \tag{5.112}$$

The relaxation factor is restricted in the interval $\alpha \in [0, 2]$. This makes the preconditioning matrix always symmetric and positive definite. It can be proven that with the use of this preconditioner, the preconditioned system has the following property:

$$\kappa\left(\mathbf{C}^{-1}\mathbf{A}\right) = \frac{4\alpha(1+\alpha\delta) + (2-\alpha)^2 \mu}{4\alpha(2-\alpha)} \tag{5.113}$$

$$\begin{cases} \mu = \max\limits_{x \neq 0} \dfrac{\mathbf{x}^T \mathbf{D}_A \mathbf{x}}{\mathbf{x}^T \mathbf{A} \mathbf{x}} > 0 \\ \delta = \max\limits_{x \neq 0} \dfrac{\mathbf{x}^T \left(\mathbf{A}_L \mathbf{D}_A^{-1} \mathbf{A}_L^T + \dfrac{1}{4} \mathbf{D}_A \right) \mathbf{x}}{\mathbf{x}^T \mathbf{A} \mathbf{x}} > -\dfrac{1}{4} \end{cases} \tag{5.114}$$

For system of equations in finite element analysis, we usually have $-0.25 \leq \delta \leq 0$. It can be shown that the bound for the condition number of the preconditioned system is as follows:

$$\min_{0<\alpha<2} \kappa\left(\mathbf{C}^{-1}\mathbf{A}\right) \leq \left[\left(\frac{1}{2} + \delta\right)\kappa(\mathbf{A})\right]^{\frac{1}{2}} + \frac{1}{2} \tag{5.115}$$

This bound indicates that the SSOR preconditioned conjugate gradient method usually results in a greater rate of convergence than the standard algorithm. On the other hand, the rate of convergence of this preconditioned algorithm is not sensitive to the estimates of μ and δ.

5.6.3 Preconditioner derived from incomplete decomposition

In 1977, Meijerink and van der Vorst proposed a preconditioner based on incomplete decomposition of matrix **A** (Meijerink and van der Vorst, 1977). The basic idea is to carry out partial Cholesky decomposition of **A** in the following form:

$$\mathbf{A} = \mathbf{L}\mathbf{L}^T + \mathbf{R} \tag{5.116}$$

L is a lower triangular matrix and **R** a sparse error matrix. The objective of this decomposition is to make the matrix $\mathbf{L}\,\mathbf{L}^T$ as close to **A** as possible, and keep a sparse structure or a similar sparse structure as well. In incomplete decomposition, we are performing the exact Cholesky factorization on the perturbed matrix (**A** – **R**). Therefore, because of the presence of error matrix **R**, we can predetermine or fix the zero elements in the matrix **L**. However, prefixing the zero elements in matrix **L** should be carried out with caution such that the partially decomposed matrix is not too far from the original matrix **A**. Of course, if the matrix **A** is diagonally dominant, then intuitively we can zero some of the far off-diagonal terms and keep a band of elements around the diagonal in the **L** matrix.

To illustrate the incomplete decomposition process, we consider a simple 4 × 4 matrix **A** which is symmetric and positive definite. We have $\mathbf{A} = [a_{ij}]$, $\mathbf{R} = [r_{ij}]$ and we pre-fix the structure of the lower triangular matrix **L** as follows:

$$\mathbf{L} = \begin{bmatrix} l_{11} & & & \\ l_{21} & l_{22} & & \\ 0 & l_{32} & l_{33} & \\ 0 & 0 & l_{43} & l_{44} \end{bmatrix} \tag{5.117}$$

We can use the following relation to obtain a series of equations between element in **L**, **A**, and **R**.

$$\mathbf{L}\mathbf{L}^T = \mathbf{A} - \mathbf{R} \tag{5.118}$$

From these equations, we can determine the terms of matrices **L** and **R**. For example, for elements on the first column of **L**, we have

$$\begin{cases} l_{11}^2 = a_{11} - r_{11} \\ l_{11}\, l_{21} = a_{21} - r_{21} \\ l_{11}\, l_{31} = a_{31} - r_{31} \\ l_{11}\, l_{41} = a_{41} - r_{41} \end{cases} \tag{5.119}$$

Now, we need to make some judgment on the choice of elements in the error matrix **R**. In this case, since **A** is positive definite, then $a_{11} > 0$. Hence, as a safe bet, we can set $r_{11} = 0$. We also set that $r_{21} = 0$. With the predetermined zero terms of **L**, we have the following:

$$\begin{cases} r_{11} = 0 \\ l_{11} = \sqrt{a_{11}} \end{cases} \quad \begin{cases} r_{21} = 0 \\ l_{21} = a_{21}/l_{11} \end{cases} \quad \begin{cases} l_{31} = 0 \\ r_{31} = a_{31} \end{cases} \quad \begin{cases} l_{41} = 0 \\ r_{41} = a_{41} \end{cases} \tag{5.120}$$

To determine elements in the second column of **L**, we have the following relations:

$$\begin{cases} l_{21}^2 + l_{22}^2 = a_{22} - r_{22} \\ l_{21}\, l_{31} + l_{22}\, l_{32} = a_{32} - r_{32} \end{cases} \tag{5.121}$$

In complete decomposition of **A** which is positive and definite, we always have $a_{22} - (l_{21})^2 > 0$. However, with incomplete decomposition, this condition cannot always be guaranteed. Therefore, we need to take a small negative value for r_{22} in order to ensure positive definiteness of the resulting **L** matrix. We assume $r_{32} = 0$. These assumptions lead to the following:

$$\begin{cases} l_{22} = \sqrt{a_{22} - l_{21}^2 - r_{22}} \\ l_{32} = a_{32}/l_{22} \end{cases} \tag{5.122}$$

With similar computations, we can determine the rest of the nonzero elements in the **L** matrix as follows:

$$\begin{cases} l_{33} = \sqrt{a_{33} - l_{32}^2 - r_{33}} \\ l_{43} = a_{43}/l_{33} \\ l_{44} = \sqrt{a_{44} - l_{43}^2 - r_{44}} \end{cases} \tag{5.123}$$

r_{33} and r_{44} are small negative numbers to ensure positive definiteness of the resulting **L** matrix. In addition, we have obtained $r_{42} = a_{42}$ and set $r_{43} = 0$. Therefore, the error matrix corresponding to incomplete decomposition as stipulated in the above form is as follows:

$$\mathbf{R} = \begin{bmatrix} 0 & 0 & a_{31} & a_{41} \\ 0 & r_{22} & 0 & a_{42} \\ a_{31} & 0 & r_{33} & 0 \\ a_{41} & a_{42} & 0 & r_{44} \end{bmatrix} \tag{5.124}$$

From this error matrix, it can be seen that if the matrix **A** is diagonally dominant, and if the far off-diagonal terms are of relatively small value, then **R** is a sparse matrix with many zeroes. Maintaining positive definiteness of the partially decomposed matrix plays a crucial role in determining the zero elements in the **L** matrix. Yet this condition may be easily violated. Nevertheless, it has been proven that if **A** is a symmetric positive and definite M-matrix which means that $a_{ij} \leq 0$, $a_{ii} > 0$ and all elements in $\mathbf{A}^{-1}$ are positive, then there are always $r_{ii} = 0$. It is obvious that there are many ways to do incomplete decomposition of matrix **A**.

With this incomplete decomposition of matrix **A**, we can use it as the preconditioning matrix.

$$\mathbf{C} = \mathbf{L}\mathbf{L}^{\mathrm{T}} \tag{5.125}$$

In carrying out the preconditioned conjugate gradient algorithm, it requires the solution of the following system of equations:

$$\left(\mathbf{L}\mathbf{L}^{\mathrm{T}}\right)\mathbf{y} = \bar{\mathbf{y}} \tag{5.126}$$

Because matrix **L** is a sparse and lower triangular matrix, it takes little computational effort in the solution process. This preconditioned conjugate gradient method has been found to be effective in solving some highly ill-conditioned systems in different fields of science and engineering. However, even with incomplete decomposition, the computational efforts involved are still tremendous when dealing with very large structural problems in engineering.

5.6.4 Element-by-element preconditioning

The element-by-element (EBE) preconditioning method was proposed by Hughes et al. (1983) for solving very large systems in finite element analysis on parallel computers. This method is specially designed to perform the preconditioning process at the element level and it conforms to the element format of data storage inherent in the finite element analysis. The preconditioner consists of the following product decomposition:

$$\mathbf{C} = \mathbf{W}^{1/2} \times \prod_{e=1}^{N_{el}} \mathbf{L}_p^e \times \prod_{e=1}^{N_{el}} \mathbf{D}_p^e \times \prod_{e=N_{el}}^{1} \mathbf{U}_p^e \times \mathbf{W}^{1/2} \tag{5.127}$$

In this equation N_{el} = number of elements, $\mathbf{W}$ = diag ($\mathbf{A}$), and $\mathbf{L}_p^e$, $\mathbf{D}_p^e$, $\mathbf{U}_p^e$ are the lower triangular, diagonal, and upper triangular matrices of the LDU or Crout factorization of the corresponding Winget regularized element matrix defined by the following relation:

$$\bar{\mathbf{A}}^e = \mathbf{I} + \mathbf{W}^{-1/2}\left(\mathbf{A}^e - \mathbf{W}^e\right)\mathbf{W}^{-1/2} \tag{5.128}$$

W^e = diag (A^e). The reverse ordering in the upper triangular product in the decomposition of C matrix insures that it results in a C matrix, which is symmetric. The effect of Winget regularization results in the regularized element matrix being positive definite. Since the regularized element matrix is symmetric, then upper and lower triangular matrices are transpose of each other. The upper triangular and diagonal matrix factors for a given element are computed by the following algorithm, for each column k, which goes from one to the number of element degrees of freedom:

$$
\begin{aligned}
&U_{ik}^{e^*} = \bar{A}_{ik}^{e} && \text{for } i = 1, \ldots, k-1 \\
&U_{ik}^{e^*} = \bar{A}_{ik}^{e} - \sum_{j=1}^{i-1} U_{ji}^{e} U_{jk}^{e^*} && \text{for } i = 2, \ldots, k-1 \\
&U_{ik}^{e} = \frac{U_{ik}^{e^*}}{D_{ii}^{e}} && \text{for } i = 1, \ldots, k-1 \\
&D_{kk}^{e} = \bar{A}_{kk}^{e} - \sum_{j=1}^{k-1} U_{jk}^{e^*} U_{jk}^{e}
\end{aligned}
\tag{5.129}
$$

The factorization is performed for all elements each time the matrix **A** is recomputed in the nonlinear iterative process. For linear analysis, this factorization is performed only once. In practice, all element matrices are stored and manipulated in a compact upper triangular vector form. Performance of the element regularization and factorization are accomplished in blocks of similar, nonconflicting elements using the element computation algorithm.

Chapter 6

Solution methods for systems of nonlinear equations

6.1 INTRODUCTION

In this chapter, the solution methods for a system of nonlinear equations are discussed in the context of nonlinear finite element analysis of solids and structural systems. We will only discuss the static cases. In general, the numerical solution of nonlinear structural analysis problems is usually carried out by means of an incremental solution procedure in which the load vector is applied in a series of small incremental steps. At the beginning of each incremental step, the geometry and material properties and incremental response of the structural system are computed as a linear system, often accompanied with corrective iterations. The equilibrium path of the structure is thereby followed in a series of small tangential linear steps. The standard incremental–iterative numerical solution procedure is called the Newton–Raphson method. The modified Newton–Raphson method characterized by infrequent update on the tangent stiffness matrix is more suitable for a moderate degree of structural nonlinearity. If it is desired to calculate structural response past limit points, special methods, such as an arc length method, are used. These methods allow the load increments to be automatically removed from the structure when appropriate.

Under an incremental formulation for the solution of a nonlinear problem, we obtain the system of equilibrium equations in a set of nonlinear equations. The incremental system of equilibrium equations of the nonlinear structural system discretized with standard finite elements can be written as follows:

$$\mathbf{K}_t(\mathbf{u})\,\Delta\mathbf{u} = \Delta\mathbf{p} \tag{6.1}$$

where:

$\mathbf{K}_t$ is the tangent stiffness matrix
$\Delta\mathbf{u}$ the nodal displacement increments
$\Delta\mathbf{p}$ the nodal force increments

Under the total Lagrangian formulation, the tangent stiffness matrix of a structural system is of the following form:

$$\mathbf{K}_t = \mathbf{K}_{(0)} + \mathbf{K}_{(1)} + \mathbf{K}_{(1)}^T + \mathbf{K}_{(2)} + \mathbf{K}_g + \mathbf{K}_L \tag{6.2}$$

If we ignore the stiffness matrix $\mathbf{K}_L$ due to pressure loading, we can write the tangent stiffness matrix as follows:

$$\mathbf{K}_t = \mathbf{K}_{(0)} + \mathbf{K}_d + \mathbf{K}_g \tag{6.3}$$

In which $\mathbf{K}_{(0)}$ is the linear stiffness matrix, $\mathbf{K}_d = \mathbf{K}_{(1)} + \mathbf{K}_{(1)}^T + \mathbf{K}_{(2)}$ the displacement-dependent stiffness matrix, and $\mathbf{K}_g$ the geometric stiffness matrix that depends on the current state of stresses.

For a system of linear equations, as has been discussed earlier, the solution for displacements can be obtained using standard direct equation solvers such as the Gauss-elimination method or any iterative method. It is important to note that the solution obtained is always *unique*. For a nonlinear system, however, the solution may not necessarily be unique. This nonuniqueness of solution is because factors such as geometric instabilities including buckling and bifurcation behaviors, plasticity, and nonconservative loading make the solution path-dependent. Because of these affecting factors, insights into the behavior of a structural system and an understanding of fundamentals of numerical methods are of primary importance in obtaining reliable solutions for a nonlinear system.

The solution procedures for solving a nonlinear system of equations are incremental and iterative in nature. Within the framework of incremental nonlinear finite element analysis, there are many specially designed iterative methods for solving the above nonlinear system of equations in Equation 6.1. In the following sections, we will discuss some of the well-established methods routinely used in nonlinear finite element analysis computer programs.

6.2 PURE INCREMENTAL (EULER) METHOD

With any incremental method, the load is applied in a series of k load steps or increments which may be preselected or which may be adjusted automatically during the solution process.

At each increment we determine the tangent stiffness matrix $\mathbf{K}_{tn}$ from the displacement vector $\mathbf{u}_n$ at the beginning of the increment. From the incremental equilibrium equation, we solve for the displacement increment corresponding to the load increment and update the displacement vector.

$$\mathbf{K}_{t\,n}\,\Delta\mathbf{u}_{n+1} = \Delta\mathbf{p}_{n+1} \tag{6.4}$$

$$\mathbf{u}_{n+1} = \mathbf{u}_n + \Delta\mathbf{u}_{n+1} \tag{6.5}$$

To solve the above system of equations, we can use either a direct solution method such as the Gauss-elimination method by performing triangularization on the stiffness matrix or semi-iterative methods such as the conjugate gradient method and preconditioned conjugate gradient method. The details of these solution methods have been discussed in previous chapters. In this chapter, when we refer to obtaining the solution of a system of linear equations, it means that any appropriate solution algorithm could be used.

As illustrated in Figure 6.1 for a one-dimensional case, it is obvious that because of the accumulation of numerical errors at each load step, the computed nonlinear response will generally diverge from the exact solution path. There exists a drift-off error from the true solution path. This error can be determined by the residual load vector $\mathbf{r}_n$, defined as $\mathbf{r}_n = \mathbf{p}_n - \mathbf{I}_n$ as shown in Figure 6.2. We note that $\mathbf{I}_n$ is the internal resisting force vector, which should not be mistaken as the identity matrix. It is important that the residual load vector not be allowed to accumulate from increment to increment, otherwise solution accuracy will be seriously degraded. It is obvious that because of the phenomenon of error accumulation during each load increment, the use of this method alone for a nonlinear system of equations is of very limited practical value.

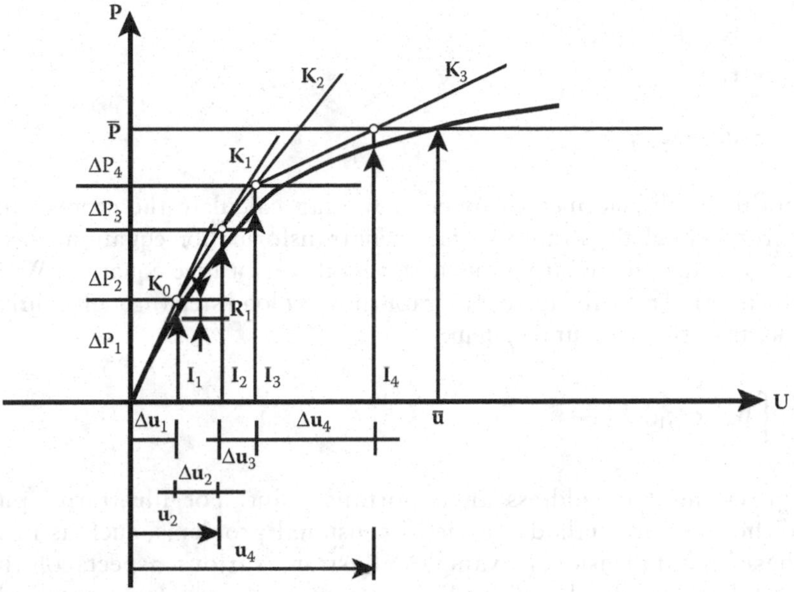

Figure 6.1 Pure incremental method in one-dimensional analysis.

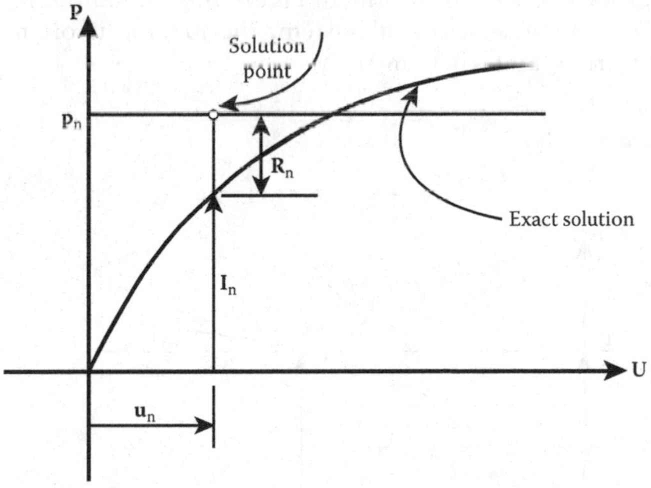

Figure 6.2 Internal resisting force $\mathbf{I}_n$ and residual force $\mathbf{R}_n$ represented in one-dimensional analysis.

6.3 INCREMENTAL METHOD WITH EQUILIBRIUM CORRECTION

From our discussion on the numerical error accumulation problem with a purely incremental method, it is natural that an equilibrium correction scheme is required in order to render the incremental approach viable for practical usage. This rectifying scheme on drift-off involves the application of unbalanced force corrections in addition to the external loads. Therefore, to advance the solution from $\mathbf{u}_n$ to $\mathbf{u}_{n+1}$ a load increment $\mathbf{p}_{n+1} - \mathbf{I}_n$ is applied, rather than $\mathbf{p}_{n+1} - \mathbf{p}_n$. For generality purpose, we consider the equilibrium condition of the system at the nth load step.

$$\begin{aligned} \mathbf{K}_{t\,n}\,\Delta\mathbf{u}_{n+1} &= \mathbf{p}_{n+1} - \mathbf{I}_n \\ &= \mathbf{p}_n - \mathbf{I}_n + \Delta\mathbf{p}_{n+1} \\ &= \mathbf{r}_n + \Delta\mathbf{p}_{n+1} \end{aligned} \tag{6.6}$$

$$\mathbf{u}_{n+1} = \mathbf{u}_n + \Delta\mathbf{u}_{n+1} \tag{6.7}$$

After we obtain the displacement increment, we can calculate the corresponding strain increment vector through the strain–displacement transformation equation. Next, the stress increment vector is determined from the incremental constitutive equation. With these, we can update the internal resisting force vector for the next load step through a direct assembly over the elements of the structural system.

$$\mathbf{I}_{n+1} = \sum \int \mathbf{B}_{n+1}^{T}\,\boldsymbol{\sigma}_{n+1} dv \tag{6.8}$$

At this point we need to address an important point. For illustrative purposes we demonstrate the solution methods in one-dimensional problems, such as in Figure 6.3. Although these one-dimensional examples illustrate various aspects of the solution methods well, they can also be misleading on some aspects. Here we will discuss one of such problems. When we look at Figure 6.3 we see that points represented at the internal resisting force vectors, $\mathbf{I}_k$, are on the actual solution path. This is only true in one-dimensional problems, such as the one in Figure 6.3. In multidimensional problems this is not true. For a multidimensional system, the loading is often represented as a product of load pattern $\mathbf{p}_0$ and a load factor γ.

$$\mathbf{p}_n = \gamma_n\,\mathbf{p}_0 \tag{6.9}$$

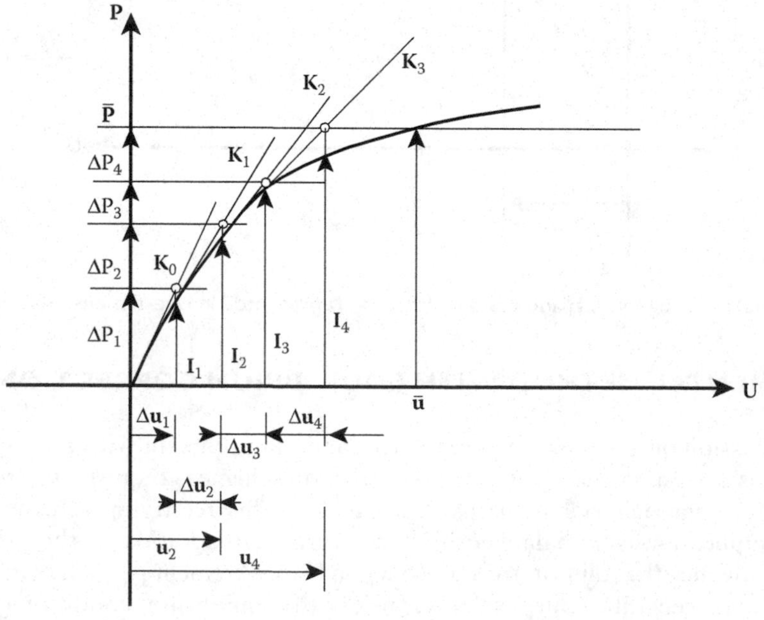

Figure 6.3 Incremental method with equilibrium correction in one-dimensional analysis.

In a multidimensional system, the internal resisting force vector is generally not proportional to the loading pattern. We can illustrate this point by considering a structural system subjected to a single force. In this case, the load pattern $\mathbf{p}_0$ will only have one nonzero term. On the other hand, internal resisting force vector is the direct assembly of element contributions and as such would normally have few, if any, zero terms. Therefore, it is important to remember that points represented by the internal resisting force vector do not lie on the actual solution path in a multidimensional system; they only do on one-dimensional systems.

6.4 ITERATIVE (NEWTON–RAPHSON) METHOD

The Newton–Raphson method is a classical iterative scheme for solving a nonlinear system of equations. For a purely iterative method in finite element analysis, the load is applied in full at the start. Iteratively, the method gradually approaches and eventually converges to the actual solution of the system of equations. Referring to Figure 6.4, at the nth iteration, the internal resistance force vector is computed and used in the incremental system of equilibrium equations.

$$\mathbf{I}_{n-1} = \mathbf{I}(\mathbf{u}_{n-1}) \tag{6.10}$$

$$\mathbf{K}_t(\mathbf{u}_{n-1})\, \Delta\mathbf{u}_n = \bar{\mathbf{p}} - \mathbf{I}_{n-1} \tag{6.11}$$

$$\mathbf{u}_n = \mathbf{u}_{n-1} + \Delta\mathbf{u}_n \tag{6.12}$$

It is expected that as the number of iteration increases, the convergence can eventually be obtained. In fact, mathematical analysis on the convergence of the Newton–Raphson method indicates that if the initial solution point is in the neighborhood of the exact solution

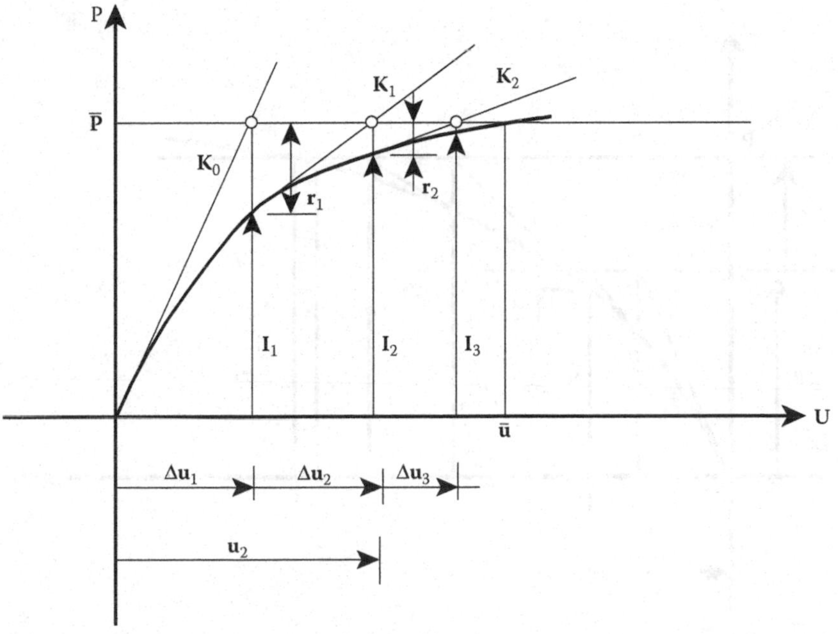

Figure 6.4 Iterative (Newton–Raphson) method in one-dimensional analysis.

point, local convergence can usually be guaranteed. As with a purely incremental method, it is obvious that the direct use of this purely iterative method is not practically efficient in solving a nonlinear system of equations in finite element analysis because of the varying degrees of nonlinearity corresponding to different levels of loading. A natural choice would be to combine the incremental method with the Newton–Raphson iteration (IMNR) scheme, which will be discussed next.

6.5 INCREMENTAL METHOD WITH NEWTON–RAPHSON ITERATIONS

In this method, we apply incremental loading to the structural system, and within each increment, we perform Newton–Raphson iterations until a prespecified convergence criterion is satisfied. This is the most commonly used method for solving nonlinear finite element problems. To clearly describe this method, we need to distinguish the load step number from the iteration number. By convention, we use a subscript to denote the load step number, and a superscript for the iteration number. Referring to Figure 6.5, at the nth load step, the current load vector, internal resisting force vector, and the residual force vector can be computed as follows:

$$\begin{cases} \mathbf{p}_n = \mathbf{p}_{n-1} + \Delta\mathbf{p}_n \\ \mathbf{I}_n^0 = \mathbf{I}_n(\mathbf{u}_{n-1}) \\ \mathbf{r}_n^0 = \mathbf{p}_{n-1} - \mathbf{I}_n^0 \end{cases} \tag{6.13}$$

Now we start the Newton–Raphson iteration process for the nth load step. At the kth iteration, we have the following operations:

$$\mathbf{I}_n^{k-1} = \mathbf{I}_n(\mathbf{u}_n^{k-1}) \tag{6.14}$$

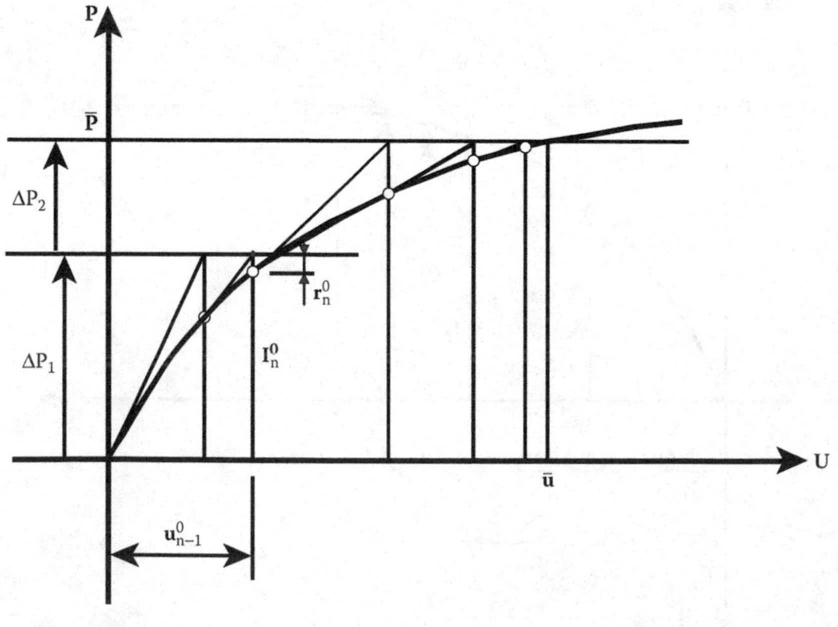

Figure 6.5 Incremental method with Newton–Raphson iterations in one-dimensional analysis.

$$\mathbf{K}_t(\mathbf{u}_n^{k-1})\,\Delta\mathbf{u}_n^k = \mathbf{p}_n - \mathbf{I}_n^{k-1} \tag{6.15}$$

$$\mathbf{u}_n^k = \mathbf{u}_n^{k-1} + \Delta\mathbf{u}_n^k \tag{6.16}$$

The convergence criterion is usually defined as follows, in terms of a prespecified convergence tolerance ε.

$$\frac{\left\|\Delta\mathbf{u}_n^k\right\|_2}{\left\|\mathbf{u}_n^k\right\|_2} \le \varepsilon \tag{6.17}$$

Once the convergence criterion is satisfied for the current load step, the solution process proceeds to the next load step. This process continues until the final solution is obtained.

It should be noted that although, this scheme is reliable in arriving at the solution for an incremental nonlinear finite element system of equations, it is computationally expensive as the current tangent stiffness matrix is computed at every iteration within a load step.

6.6 MODIFIED INCREMENTAL METHOD WITH NEWTON–RAPHSON ITERATIONS

This scheme is a modification of the IMNR method in which the tangent stiffness matrix of the structural system is less frequently updated during each load increment. This is also considered as a quasi-Newton method in which the computation of the tangent stiffness matrix and its triangularization are performed only a few times within each load step. Usually, within a modified method, we compute or update only the initial tangent stiffness matrix and its triangularization at the beginning of each load increment and use this stiffness matrix for Newton–Raphson iterations within the current load increment. This is a numerical scheme derived solely based on considerations for computational efficiency by reducing the computational time spent on updating the tangent stiffness matrix. With this approach, conditions that guarantee the local convergence theorem for Newton–Raphson iteration are violated, which means that local convergence is not guaranteed in this method.

We have the following incremental system of equilibrium equations at the nth load step at the kth iteration in the modified Newton–Raphson iteration:

$$\mathbf{K}_t(\mathbf{u}_n^0)\,\Delta\mathbf{u}_n^k = \mathbf{p}_n - \mathbf{I}_n^{k-1} \tag{6.18}$$

We note the use of the initial tangent stiffness matrix in the above equation. However, the tangent stiffness matrix can also be updated during the iteration process, but not at every iteration. This process is illustrated in Figure 6.6.

In practice, in order to achieve computational efficiency, decisions have to be made on selecting many factors that affect the solution process. For example, very often, we need to determine or preselect the number of load steps or the load step size at the beginning of the solution process. Second, we need to make judgment on the maximum number of iterations within each load step and the frequency of updating the stiffness matrix. With this method, the tangent stiffness update frequency is part of the overall numerical solution strategy; it involves considerations of the number of iterations required for convergence against the computational cost of each iteration, and is highly problem dependent. On the other hand, it can be expected that the modified Newton–Raphson method applies to structural systems with moderate degrees of nonlinearity. Needless to say, information about the structural

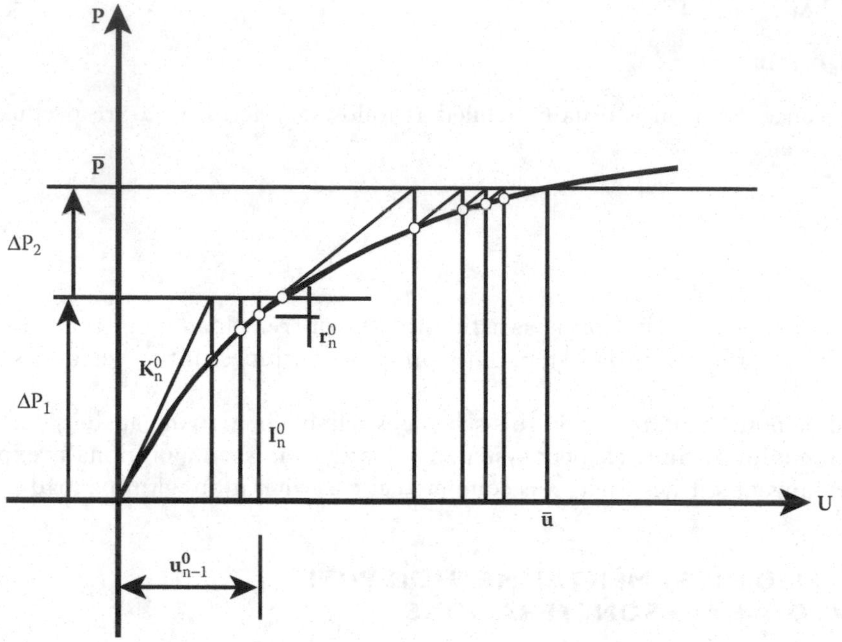

Figure 6.6 Incremental method with modified Newton–Raphson iteration in one-dimensional analysis.

behavior would provide some guidance in determining these parameters for setting up the algorithms according to the characteristic of problems.

6.7 DIAGNOSTIC METHODS AND SPECIAL NUMERICAL TECHNIQUES

The solution methods described so far may not work reliably around the limit points or bifurcation points if the structural system exhibits softening type of behavior. In those special regions, convergence will not be achieved by using the standard incremental method or IMNRs. At the limit point or bifurcation point, at least one of the eigenvalues of the tangent stiffness matrix approaches to zero and it becomes nearly singular. Obviously, we need two types of special numerical schemes. First, we need diagnostic methods that can be used to identify the approaching limit points or bifurcation points. In addition, it is also necessary to use the diagnostic method to adjust load step size when the limit point is approached, and skip the bifurcation point if necessary. Second, once the limit points or bifurcation points are identified, we need special solution techniques to get around or pass the limit point, and to get on to the secondary solution paths at the bifurcation point. In the following sections, we describe some of the numerical techniques commonly used to identify the presence of limit points or bifurcation points in nonlinear finite element analysis.

6.7.1 Current stiffness parameter

This current stiffness parameter (CSP) is based on the observation that the variation in the magnitude of the displacement increment for the same load increment contains information about the degree of nonlinearity of the structural system (Figure 6.7) (Bergan et al., 1978). For linear systems, the displacement increment will not change for the same load increment

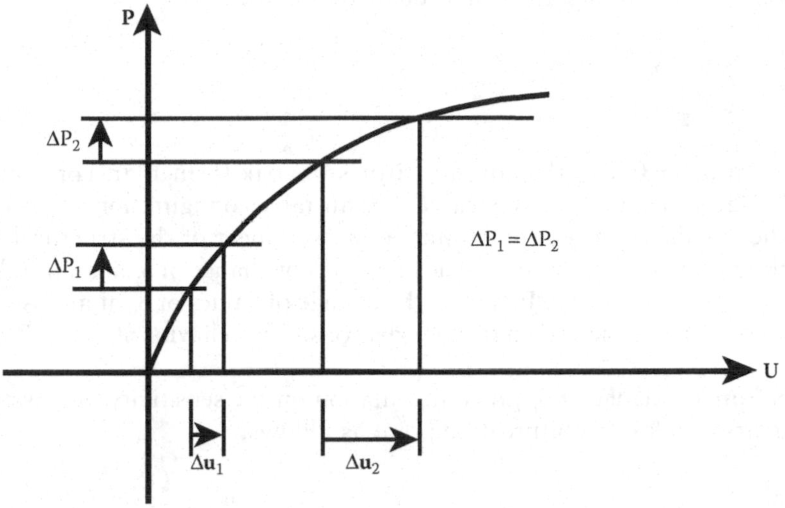

Figure 6.7 Degrees of nonlinearity as measured from strain energy increment under equal load increment in one-dimensional analysis.

applied at different load levels. If it decreases, we have a hardening nonlinear system. If it increases, we have a softening nonlinear system. The rate of increase or decrease indicates the degree of nonlinearity. CSP gives a measure of the degree of nonlinearity.

At any load step n, the incremental load vector $\Delta \mathbf{p}_n$ generates incremental displacement vector $\Delta \mathbf{u}_n$.

$$\mathbf{K}_t \, \Delta \mathbf{u}_n = \Delta \mathbf{p}_n \tag{6.19}$$

$$\Delta \mathbf{u}_n^T \, \mathbf{K}_t \, \Delta \mathbf{u}_n = \Delta \mathbf{u}_n^T \, \Delta \mathbf{p}_n \tag{6.20}$$

The equation above represents the strain energy stored in the system corresponding to the current load increment. It would be an indicator of degree of nonlinearity of the structural response. For a softening type structure, with an equal load increment, the strain energy due to deformation generated in the neighborhood of the limit point would be much higher than in the region with less nonlinearity. We can define a measure of the incremental stiffness with respect to the loading by the inverse of the work associated with the normalized load vector.

$$S_P^n = \frac{\left\| \Delta \mathbf{p}_n \right\|^2}{\Delta \mathbf{u}_n^T \, \Delta \mathbf{p}_n} \tag{6.21}$$

This parameter would provide a measure of the stiffness of the structural system. The larger the value, the stiffer the system. The variation of this parameter will provide an indication of the variation in the degree of nonlinearity.

A normalized version of CSP can be expressed in terms of the load pattern $\mathbf{p}_0$.

$$S_P^n = \frac{\Delta \mathbf{u}_1^T \, \Delta \mathbf{p}_0}{\Delta \mathbf{u}_n^T \, \Delta \mathbf{p}_0} \tag{6.22}$$

The index 1 indicates the first load step when the structural system often behaves almost linearly. Obviously, during the first load step, the CSP always takes a value of 1.0. For softening systems, it decreases and for stiffening systems it increases as the load increases.

For nonproportional loading, the CSP is defined as follows:

$$S_P^n = \frac{\|\Delta p_n\|}{\|\Delta p_1\|} \frac{\Delta u_1^T \Delta p_0}{\Delta u_n^T \Delta p_0} \tag{6.23}$$

At the limit point, $S_p = 0$, and the tangent stiffness matrix is singular. For a stable branch of the load displacement curve, S_p is positive. For unstable configurations, S_p becomes negative. This indicates that S_p gives a quantitative measurement of the structural behavior in the postbuckling range of deformations: the more abrupt change of S_p at an instability point, the more severe type of buckling. In a way, the change of values of S_p at an instability point indicates the sensitivity of the system to perturbations. The behavior of the CSP is illustrated in Figure 6.8.

Because the rate of change of S_p gives information on the sensitivity of a system, we can define a bifurcation index at a bifurcation point as follows:

$$I_P = \frac{S_P^- S_P^+}{S_P^- - S_P^+} \tag{6.24}$$

in which S_P^- = current stiffness parameter prior to bifurcation point, and S_P^+ = current stiffness parameter immediately after bifurcation point. As can be seen from the above equation, this bifurcation index describes the change in direction between the two branches at a bifurcation point. A negative index would indicate a change from an ascending to a descending branch, which signals bifurcation into unstable secondary path. For example, when the secondary path follows a neutral equilibrium, then $S_P^+ = 0$, thus the bifurcation index is zero. If there is no bifurcation at a point on the solution path, then I_p would have a very large value or infinite. Values of I_p for different types of bifurcation are illustrated in Figure 6.9.

Although the concept of CSP is simple, this parameter has been shown to be effective in providing important information about the degree of nonlinearity of the structural behavior and in identifying the approaching of the limit point. However, the CSP gives only information along the primary solution path and does not provide any information on the bifurcation points as well as the secondary solution paths. This appears to be a serious

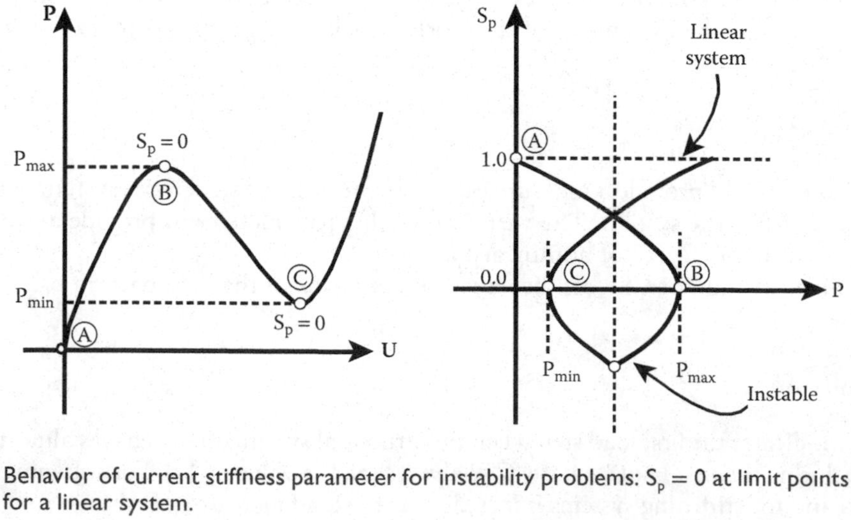

Figure 6.8 Behavior of current stiffness parameter for instability problems: $S_p = 0$ at limit points and $S_p = 1$ for a linear system.

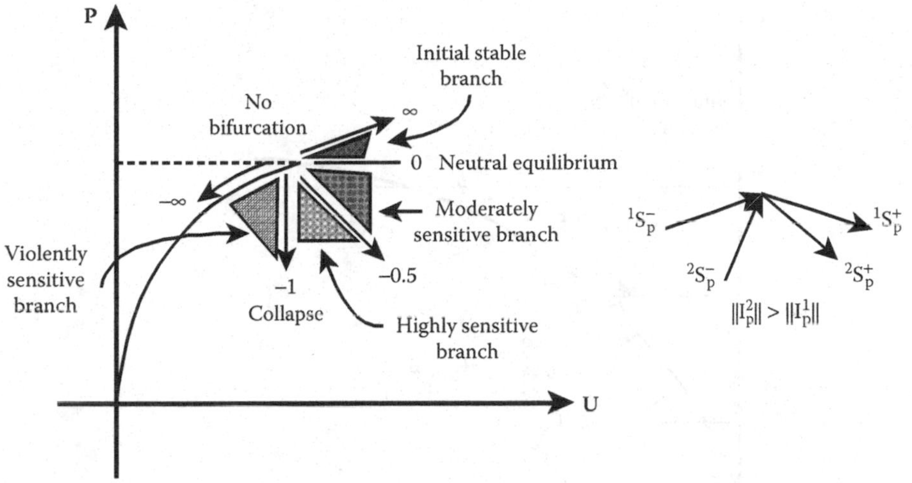

Figure 6.9 Values of bifurcation index for different types of bifurcation.

drawback. Nevertheless, in practice, there are several points to keep in mind. For real structural systems, it is not advisable to use perfect symmetry in modeling the geometry of the structure. If we take advantage of the structural symmetry, for example if we model only half of the structure and apply symmetry condition, we may eliminate some potential buckling and bifurcation points. In modeling perfectly symmetric structural systems we should also introduce some imperfections. In actual structural models it is almost impossible to avoid the introduction of geometric imperfection to the structural system. Without the presence of geometric imperfection, bifurcation may not occur and only the limit point may be reached in the buckling analysis.

6.7.2 Spectral decomposition

Since at the limit point or bifurcation point, the structure becomes unstable, at least one eigenvalue of the tangent stiffness matrix goes to zero. A straightforward approach to identifying the limit point and bifurcation point is to determine the smallest eigenvalue and the corresponding eigenvector of the tangent stiffness matrix at each load step. The eigenvectors associated with zero eigenvalues of the tangent stiffness matrix are often referred to as the buckling modes of the structural system. Computing the eigenvalues and eigenvectors within a reasonable time frame is feasible when the stiffness matrix is reasonably small. Computing the eigenvalues can be accomplished with eigenvalue solution methods to be discussed in Chapter 7. However, when the system becomes large, the computation involved in the eigenvalue solution process may become very time consuming. An alternative approach is based on using the determinant of the tangent stiffness matrix.

The determinant of the tangent stiffness matrix is effective in detecting the limit points and bifurcation points. In addition, the pattern of sign changes of the determinant indicates the presence of bifurcation point even within a load increment. To calculate the determinant, we perform triangular decomposition on the tangent stiffness matrix.

$$\det|\mathbf{K}_t| = \det|\mathbf{L}| \ \det|\mathbf{U}| = \det|\mathbf{U}| \tag{6.25}$$

The sign of determinant of the tangent stiffness matrix is the same as the sign of the determinant of the upper triangular matrix **U**. In general, the number of negative diagonal

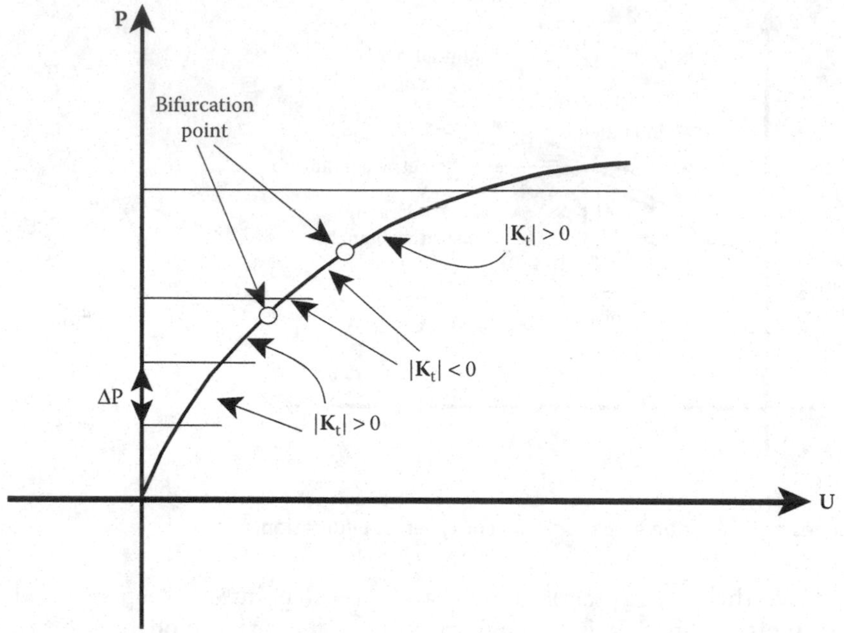

Figure 6.10 Identification of bifurcation point through detecting patterns of sign changes of the determinant of stiffness matrix $|\mathbf{K}_t|$.

terms in the upper triangular matrix is equal to the number of bifurcation points plus the number of limit point that is less than the current load level. In some cases, more than one eigenvalue of the tangent stiffness matrix may go to zero. This may indicate multiple solution paths becoming accessible beyond the bifurcations. The number of negative terms appearing on the diagonal of the upper triangular matrix are the same as the number of zero eigenvalues below the current load. A typical case is schematically illustrated in Figure 6.10. We note that, unlike other similar figures, this is not a one degree-of-freedom system; it represents the response of a multidegree of freedom system (two of its eigenvalues go to zero) along a specific solution path.

6.8 SPECIAL SOLUTION METHODS

As has been discussed before, for snap-through type of behavior, it is very difficult to get around the limit point and to determine the post-buckling behavior. These require special solution methods designed specifically for handling solutions at the limit point. These methods should have the capability to pass over the limit points, accurately locate the critical points, and trace the secondary solution paths if needed. We describe some of the common numerical techniques in these sections.

When a structure is subjected to only a single concentrated load, then we can use displacement increment instead of load increment in approaching the limit point. This can be reasonably carried out because the load-displacement relation at the position where the single load is acted on can be consistently determined. However, when the structure is acted on by more than one loads or a complicated loading pattern, which is usually the case, then it is difficult, if not impossible, to use displacement increments which are consistent with the loading pattern and constraints on the structure.

At the limit point or its vicinity, the incremental stiffness matrix becomes singular and equilibrium iterations fail, resulting in numerical difficulty in the solution process. One option is to avoid Newton–Raphson iteration when the CSP indicates that we are close to a limit point. Around a limit point a pure Eulerian procedure is carried out. The transition zone from a step iterative procedure to a pure incremental procedure is defined using the CSP described in previous section. We can define a threshold value for CSP entering and exiting the transition zone in the neighborhood of the limit point. The equilibrium iterations will be resumed when we exit the transition zone. Usually the threshold values for the CSP is chosen in the range of 0.05–0.10 depending on the nature and nonlinearity of the problem. This method applies only to the primary solution path.

6.8.1 Augmented stiffness method

Another approach is to modify the structure and eliminate the limit point. The response of the original structural system should be recoverable from the response of the modified structure. This is accomplished by augmenting the structural stiffness matrix with a very stiff linear spring. The concept of the augmented stiffness was first proposed by Sharifi and Popov (1971, 1973). This concept is illustrated in Figure 6.11 for a one-dimensional system.

We can use the same approach for a multidegree-of-freedom system by using the load pattern vector $\mathbf{p}_0$ that is used to define the load vector $\mathbf{p} = \lambda\,\mathbf{p}_0$. Rather than solving the original structural problem, we are solving the following augmented structural system:

$$(\mathbf{K}_t + k\,\mathbf{p}_0\,\mathbf{p}_0^T)\,\Delta\mathbf{u}_n = \Delta\lambda_n^*\,\mathbf{p}_0 \tag{6.26}$$

The value of the equivalent linear stiffness of the spring k is chosen to be large enough to eliminate any possibility of a limit point. Obviously, this requires some knowledge of the behavior of the structural system.

The displacement increment $\Delta\mathbf{u}_n$ is determined from the load increment $\Delta\lambda_n^*$ applied to the augmented structural system. Next we need to determine the load increment $\Delta\lambda_n$ acting on the actual structure that would produce the same displacement increment. We can accomplish this by manipulating Equation 6.26.

$$\mathbf{K}_t\,\Delta\mathbf{u}_n = [\Delta\lambda_n^* - k(\mathbf{p}_0^T\,\Delta\mathbf{u}_n)]\,\mathbf{p}_0 \tag{6.27}$$

$$= \Delta\lambda_n\,\mathbf{p}_0$$

$$\Delta\lambda_n = \Delta\lambda_n^* - k(\mathbf{p}_0^T\,\Delta\mathbf{u}_n) \tag{6.28}$$

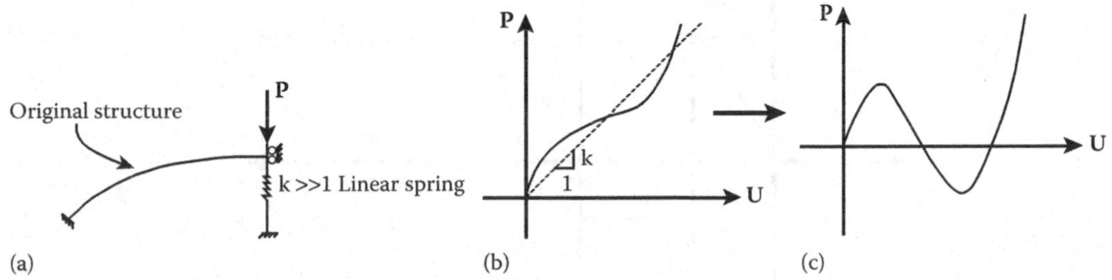

Figure 6.11 Augmented stiffness method illustrated in one-dimensional analysis: (a) augmented structure with linear spring; (b) response of augmented structure; and (c) response of original structure.

In this way, we recover the response of the actual structural system from the response of the augmented structure, where we did not have to deal with limit points. We should point out that this method can only be applied in situations where the load pattern does not change throughout the response of the structure.

A more general method that is widely used in practice is the arc length method that is discussed next.

6.8.2 Arc length method

When we use constant load increments, the solution will encounter difficulty around the limit point. The iterative method may become unstable. A more effective alternative is to use variable load step and determine the load step in order to avoid numerical difficulties near the limit point. In the arc length method (Riks, 1979), as the name implies, we specify the arc length along the solution path, instead of load increment. In that way we can determine the load step to approximately corresponding to the specified arc length. As schematically illustrated in Figure 6.12, the load steps get smaller in the softening structural response.

We recall that in the case of pure incremental Newton–Raphson scheme, we are solving the incremental system of equilibrium equation at the nth increment and kth iteration.

$$
\begin{aligned}
\mathbf{K}_t(\mathbf{u}_n^{k-1})\,\Delta\mathbf{u}_n^k &= \mathbf{p}_n - \mathbf{I}_n^{k-1} \\
&= (\mathbf{p}_n - \mathbf{p}_{n-1}) + (\mathbf{p}_{n-1} - \mathbf{I}_n^{k-1}) \\
&= \Delta\mathbf{p}_n + \mathbf{r}_n^{k-1} \\
&= \Delta\lambda_n \mathbf{p}_0 + \mathbf{r}_n^{k-1}
\end{aligned}
\tag{6.29}
$$

In this equation $\mathbf{r}_n^{k-1}$ is the residual force vector, $\mathbf{p}_0$ is the load pattern, and $\Delta\lambda_n$ is the incremental load parameter. As has been discussed before, in order to get over the limit point, rather than keeping the load increment constant, we need to vary the load increment.

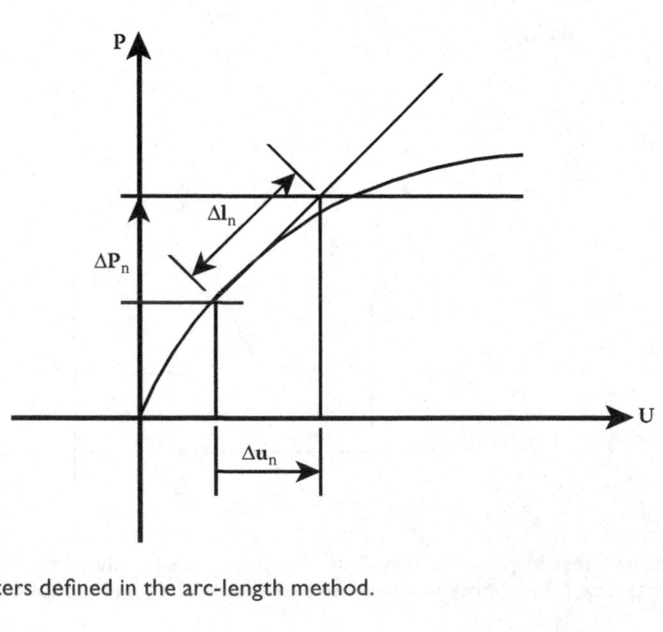

Figure 6.12 Parameters defined in the arc-length method.

We introduce the arc length parameter Δl_n and try to keep it constant. Referring to Figure 6.12, the arc length is defined as follows:

$$\Delta l_n^2 = \Delta\lambda_n^2 \, \mathbf{p}_0^T \, \mathbf{p}_0 + \Delta\mathbf{u}_n^T \Delta\mathbf{u}_n \tag{6.30}$$

At each iteration, we update the increment of load parameter and use it in Equation 6.29 to determine the displacement increment that is used to update the total displacement increment at the end of the iteration.

$$\begin{cases} \Delta\lambda_n \leftarrow \Delta\lambda_n + \Delta\lambda_n^k \\ \Delta\mathbf{u}_n \leftarrow \Delta\mathbf{u}_n + \Delta\mathbf{u}_n^k \end{cases} \tag{6.31}$$

Using these expressions in Equation 6.30 we get the following for the arc length:

$$\Delta l_n^2 = (\Delta\lambda_n + \Delta\lambda_n^k)^2 \, \mathbf{p}_0^T \, \mathbf{p}_0 + (\Delta\mathbf{u}_n + \Delta\mathbf{u}_n^k)^T (\Delta\mathbf{u}_n + \Delta\mathbf{u}_n^k) \tag{6.32}$$

We need to solve the above quadratic equation to determine the load increments at kth iteration $\Delta\lambda_n^k$ in terms of the constant value of the arc length Δl_n. We linearize the above quadratic equation by ignoring all the second-order terms.

$$\Delta l_n^2 = (\Delta\lambda_n{}^2 + 2\,\Delta\lambda_n\Delta\lambda_n^k) \, \mathbf{p}_0^T \, \mathbf{p}_0 + (\Delta\mathbf{u}_n{}^T\Delta\mathbf{u}_n + 2\,\Delta\mathbf{u}_n{}^T\Delta\mathbf{u}_n^k) \tag{6.33}$$

Combining Equations 6.30 and 6.33 we arrive at the following equation:

$$\Delta\lambda_n\Delta\lambda_n^k \, \mathbf{p}_0^T \, \mathbf{p}_0 + \Delta\mathbf{u}_n{}^T\Delta\mathbf{u}_n^k = 0 \tag{6.34}$$

This is the relationship that must be satisfied at each iteration when the arc length is kept constant. The load increments at the kth iteration is then determined.

$$\Delta\lambda_n^k = -\frac{\Delta\mathbf{u}_n{}^T\Delta\mathbf{u}_n^k}{\Delta\lambda_n \, \mathbf{p}_0^T \, \mathbf{p}_0} \tag{6.35}$$

For implementation, we use an approximate version of the above formula.

$$\Delta\lambda_n^k = -\frac{\Delta\mathbf{u}_n{}^T\Delta\mathbf{u}_n^{k-1}}{\Delta\lambda_n \, \mathbf{p}_0^T \, \mathbf{p}_0} \tag{6.36}$$

Now we illustrate the implementation of this algorithm. At the nth increment and kth iteration we are solving the following equation:

$$\mathbf{K}_t(\mathbf{u}_n^{k-1}) \, \Delta\mathbf{u}_n^k = \Delta\lambda_n^k\mathbf{p}_0 + \mathbf{r}_n^{k-1} \tag{6.37}$$

After solving for $\Delta\mathbf{u}_n^k$, we then update the total displacement increment and calculate the next load increment parameter and the residual load vector.

$$\Delta\mathbf{u}_n \leftarrow \Delta\mathbf{u}_n + \Delta\mathbf{u}_n^k \tag{6.38}$$

$$\Delta\lambda_n^{k+1} = -\frac{\Delta\mathbf{u}_n{}^T\Delta\mathbf{u}_n^k}{\Delta\lambda_n \, \mathbf{p}_0^T \, \mathbf{p}_0} \tag{6.39}$$

$$\mathbf{r}_n^k = \mathbf{p}_{n-1} + \Delta\lambda_n^k\mathbf{p}_0 - \mathbf{I}_n^k \tag{6.40}$$

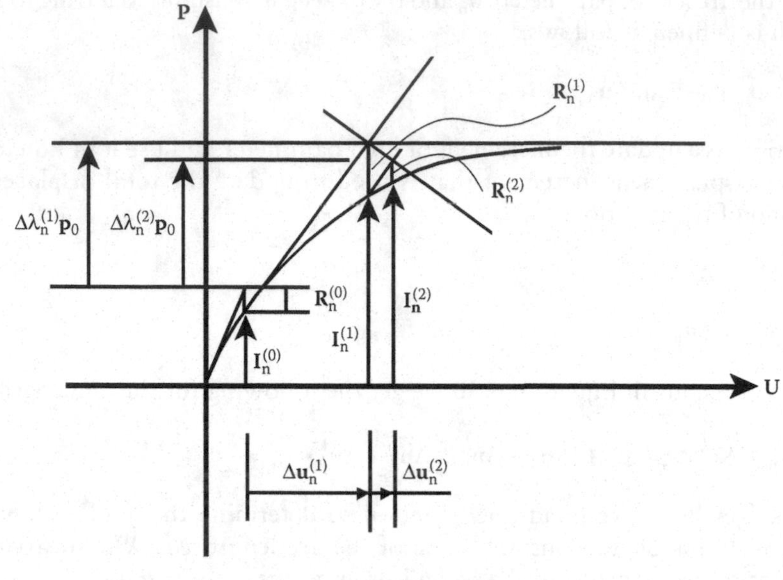

Figure 6.13 The normal plane method in one-dimensional analysis.

We now move to the next iteration.

$$K_t(u_n^k)\ \Delta u_n^{k+1} = \Delta\lambda_n^{k+1} p_0 + r_n^k \tag{6.41}$$

This is called the *normal plane method*, the simplest variation of the arc length method, which is illustrated in Figure 6.13. Another commonly used scheme is called the updated normal plane method, in which the original arc length quadratic equation is solved to determine the load increment and move around a spherical path. Because of this unique feature, it is also called the spherical method. All these methods are derived by keeping some other measure than the load increment constant.

6.9 CONVERGENCE CRITERIA

Since solution methods are used in obtaining the load-displacement relation of a structure, it is natural that the convergence criterion can be defined either in terms of displacement vector or residual force vector.

$$\frac{\left\|\Delta u_n^{(j)}\right\|}{\left\|\Delta u_n\right\|} \le \varepsilon_1 \tag{6.42}$$

$$\frac{\left\|\Delta r_n^{(j)}\right\|}{\left\|\Delta p_n^{(j)}\right\|} \le \varepsilon_2 \tag{6.43}$$

Obviously, we need to use certain norms to evaluate the convergence conditions at each iteration. However, the immediate question is what kind of norm should be used in these calculation. The simplest mathematical norm is the Euclidean norm, which is commonly used in practice. During implementations of solution methods in structural analysis, however, we often use the so-called structural norm, which means that the convergence criterion for different group of degrees of freedom may be different. For example, for bending type problems, such as frames, plates, and shells, there are basically three groups of degrees of freedom: the axial degrees of freedom in frames or in-plane degrees of freedom in plates and shells; out of plane degrees of freedom; and, rotational degrees of freedom. The convergence criterion for these different modes of deformation should be ascribed differently.

Chapter 7

Eigenvalue solution methods

7.1 INTRODUCTION

As we discussed in Chapter 1, any matrix **A** has eigenvalues, λ_j, and corresponding eigenvectors $\mathbf{x}_j$, as expressed in the following standard eigenvalue problem:

$$\mathbf{A}\mathbf{x}_j = \lambda_j \mathbf{x}_j, \quad j = 1, \ldots, n \tag{7.1}$$

The above system of equations can also be written in the following form:

$$(\mathbf{A} - \lambda\mathbf{I})\mathbf{x} = 0 \tag{7.2}$$

For this equation to have a nonzero solution vector **x**, the determinant of its matrix must be zero. This leads to an nth degree polynomial in the variable λ.

$$\det(\mathbf{A} - \lambda\mathbf{I}) = 0 \tag{7.3}$$

$$f(\lambda) = \det(\mathbf{A} - \lambda\mathbf{I}) = \lambda^n + b_1\lambda^{n-1} + \cdots + b_{n-1}\lambda + b_n = 0 \tag{7.4}$$

This is called the characteristic polynomial, and its roots are the eigenvalues of matrix **A**.

We can see from Equation 7.1 that if vector **x** is an eigenvector of matrix **A**, then any multiple of **x** is also an eigenvector of **A**. Another way of looking at this is that multiplication of **A** with the eigenvector **x** does not change the pattern of the vector; it only changes the size and sign (pointing direction) of the vector.

For problems in structural mechanics, matrix **A** is usually symmetric and positive definite. It has n eigenvalues $\lambda_1 \le \lambda_2 \le \ldots \le \lambda_n$ and the corresponding eigenvectors $\{\mathbf{x}_1, \mathbf{x}_2, \ldots, \mathbf{x}_n\}$ are mutually orthogonal and linearly independent. However, there are cases where several eigenvectors correspond to the same eigenvalue; a case of multiple eigenvalues. This happens when there are two or more planes of symmetry of the structural system, or at the bifurcation points where there are several secondary solution paths, or in the case of rigid body modes. For example, the square plate shown in Figure 7.1 has two planes of symmetry. For the deformation modes 1 and 2 as shown in the figure, the eigenvalues corresponding to both modes are the same because the strain energy associated with both deformation configurations is the same. However, the two deformed shapes are two distinct eigenvectors. A three-dimensional structural system without any support has six rigid-body modes, three translations, and three rotations. These are six distinct eigenvectors and their eigenvalues are zero; no strain energy is required for rigid-body motions.

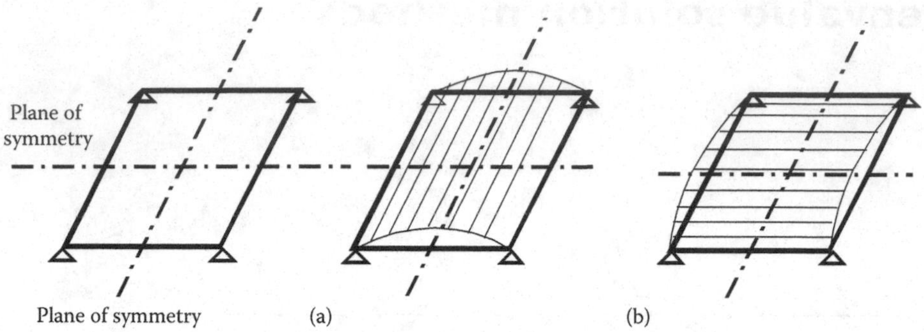

Figure 7.1 Example of multiple eigenvalues in structures with symmetry: (a) mode shape 1 and (b) mode shape 2.

In computational mechanics, we often encounter "generalized eigenvalue" problems of the following form:

$$\mathbf{A}\mathbf{x}_j = \lambda_j \mathbf{B}\mathbf{x}_j, \quad j = 1, \ldots, n \tag{7.5}$$

Matrix **A** is symmetric and positive definite, and matrix **B** is symmetric in general. If we replace the **B** matrix with an identity matrix **I**, then the generalized eigenvalue problem is reduced to the standard eigenvalue problem.

In structural dynamic and vibration problems, we have the following generalized eigenvalue problem in terms of stiffness matrix **K** and mass matrix **M**:

$$\mathbf{K}\phi_j = \omega_j^2 \mathbf{M}\phi_j, \quad j = 1, \ldots, n \tag{7.6}$$

The eigenvalues are the square of the natural frequencies, ω_j, and the corresponding eigenvectors are the mode shapes ϕ_j. This kind of eigenvalue problem is used in modal analysis. We have a similar generalized eigenvalue problem in nonlinear analysis. In linearized buckling analysis, we are interested in the lowest mode response of the structure by solving the following eigenvalue problem:

$$\mathbf{K}_0\phi_j = \lambda_j \mathbf{K}_{g0}\phi_j, \quad j = 1, \ldots, n \tag{7.7}$$

In this equation $\mathbf{K}_0$ = linear stiffness matrix, $\mathbf{K}_{g0}$ = geometric stiffness for a normalized load vector $\mathbf{p}_0$. The linearized buckling load is determined by $\mathbf{p}_{cr} = \lambda \mathbf{p}_0$ and ϕ_1 is the corresponding buckling mode. In general, if the displacement prior to reaching the buckling load is small, then the linearized buckling analysis gives reasonable predictions of the buckling load. However, if the structure has a softening type of behavior and large deformations occur prior to buckling, then the result of linearized buckling analysis may not be accurate.

As will be discussed later, most of the eigenvalue solution methods are developed for the standard eigenvalue problems. For a generalized eigenvalue problem, there are basically two approaches in the solution process: either through transformation to a standard form or through direct solution on the original system.

7.1.1 Eigenvalue tasks in computational mechanics

The solution of eigenvalue problems is essential in numerical analysis of many physical problems. The field of structural mechanics covers a full spectrum of problems that require determination of some of the eigenvalues and eigenvectors of the structural system. Because of the characteristics and need of diverse types of problems, some require solution of a few eigenvalues, whereas others demand information on all eigenvalues. In structural dynamics, we usually are interested in some extreme eigenvalues of the structural system. This is because the dynamic response of a structural system is predominantly determined by the lower modes, often the first few modes. Therefore, in eigenvalue solution for structural dynamic problems, we are looking for the few lowest eigenvalues and eigenvectors. For a structure with a few thousands of degrees of freedom, we may need to determine the lowest 10, or at most 20 eigenvalues and eigenvectors.

With problems such as the linearized buckling analysis, we are interested in obtaining only the lowest eigenvalue and eigenvector. However, sometimes we are interested in the highest eigenvalue, which provides valuable information on the stability of a direct integration method in determining the dynamic response of the structural system. In vibration or structural control problems, we are more interested in obtaining eigenvalues and eigenvectors within a range. In all these cases, the total number of eigenvalues and eigenvectors computed is much smaller than n, the total number of degrees of freedom of the structural system.

However, there are other cases where we need to compute all the eigenvalues and eigenvectors. These cases usually involve small, full matrices, which may arise at an intermediate step in an eigenvalue solution process.

This brief discussion shows that in computational mechanics, we need to have numerical solution methods for obtaining a few eigenvalues and eigenvectors, as well as methods for computing all the eigenvalues and eigenvectors involving small and full matrices. In the following sections, unless stated otherwise, we assume that the matrix **A** is symmetric and positive definite with distinct eigenvalues.

7.2 PRELIMINARIES

We have briefly discussed the basics of eigenvalues and eigenvectors in Chapter 1. In this section, we discuss some important concepts relevant to numerical solution of eigenvalue problems.

7.2.1 Inertia of a matrix

The inertia of a matrix **A** is a three-number set defined as

$$\mathbf{I}(\mathbf{A}) = [\pi, \nu, \zeta] \tag{7.8}$$

where:

π is the number of positive eigenvalues of **A**
υ is the number of negative eigenvalues of **A**
ζ is the number of zero eigenvalues of **A**

We recall that zero eigenvalues correspond to rigid body motion in the structural systems.

7.2.2 Shift of origin

We define the shifting of matrix **A** by μ as follows:

$$\bar{\mathbf{A}} = \mathbf{A} - \mu\mathbf{I} \tag{7.9}$$

Next, we consider the effect of shifting on the eigenvalues and eigenvectors of the original system. We start with the eigenvalue problem for the shifted system and manipulate it to arrive at the original eigenvalue problem.

$$\bar{\mathbf{A}}\,\bar{\mathbf{x}}_j = \bar{\lambda}_j\,\bar{\mathbf{x}}_j \tag{7.10}$$

$$(\mathbf{A} - \mu\,\mathbf{I})\bar{\mathbf{x}}_j = \bar{\lambda}_j\bar{\mathbf{x}}_j \tag{7.11}$$

$$\mathbf{A}\bar{\mathbf{x}}_j = (\bar{\lambda}_j + \mu)\bar{\mathbf{x}}_j \tag{7.12}$$

Comparing this equation with the original eigenvalue problem, Equation 7.1, we arrive at the following relations:

$$\begin{cases} \bar{\mathbf{x}}_j = \mathbf{x}_j \\ \bar{\lambda}_j = \lambda_j - \mu \end{cases} \tag{7.13}$$

This shows that the eigenvectors remain unaffected by shifting. As the name implies, all the eigenvalues of the system are shifted by μ. This indicates that with shifting, we can change the signs of some of eigenvalues in the spectrum, which in turn changes the inertia of the matrix.

Consider a system containing rigid body modes, such as an airplane in air that has six rigid body modes with zero eigenvalues. If we shift the origin in the negative direction, then the zero eigenvalues all become positive. With shifting, we may avoid the numerical difficulties associated with zero eigenvalues.

7.2.3 Similarity transformation

A similarity transformation performed on matrix **A** is defined as follows:

$$\bar{\mathbf{A}} = \mathbf{F}^{-1}\mathbf{A}\mathbf{F} \quad \text{or} \quad \bar{\mathbf{A}} = \mathbf{F}\mathbf{A}\mathbf{F}^{-1} \tag{7.14}$$

If **F** is an orthogonal matrix, then $\mathbf{F}^{-1} = \mathbf{F}^T$, and the above transformation is called *orthogonal similarity transformation*. In order to determine the effect of the similarity transformation on the eigenvalues of eigenvectors of the system, we start with the eigenvalue problem for the transformed system and manipulate it to arrive at the original untransformed eigenvalue problem.

$$\bar{\mathbf{A}}\bar{\mathbf{x}}_j = \bar{\lambda}_j\bar{\mathbf{x}}_j \tag{7.15}$$

$$\mathbf{F}^{-1}\mathbf{A}\mathbf{F}\bar{\mathbf{x}}_j = \bar{\lambda}_j\bar{\mathbf{x}}_j \tag{7.16}$$

$$\mathbf{A}(\mathbf{F}\bar{\mathbf{x}}_j) = \bar{\lambda}_j(\mathbf{F}\bar{\mathbf{x}}_j) \tag{7.17}$$

Comparing this equation with the original eigenvalue problem, Equation 7.1, we arrive at the following relations:

$$\begin{cases} \bar{\mathbf{x}}_j = \mathbf{F}^{-1}\mathbf{x}_j \\ \bar{\lambda}_j = \lambda_j \end{cases} \tag{7.18}$$

This shows that similarity transformation does not change the eigenvalues of the original matrix, but the eigenvectors are modified. From this result, it can be seen that if matrix **A** is similar to a diagonal matrix, then it has n distinct eigenvalues. We can show that if matrix **A** has n linearly independent eigenvectors, then this n × n matrix is similar to a diagonal matrix, which in fact contains all the eigenvalues of **A** as its diagonal terms.

$$\mathbf{X} = \left[\mathbf{x}_1 | \mathbf{x}_2 | \dots | \mathbf{x}_n\right] \tag{7.19}$$

$$\begin{aligned} \mathbf{AX} &= \left[\mathbf{Ax}_1 | \mathbf{Ax}_2 | \dots | \mathbf{Ax}_n\right] \\ &= \left[\lambda_1 \mathbf{x}_1 | \lambda_2 \mathbf{x}_2 | \dots | \lambda_n \mathbf{x}_n\right] \\ &= \left[\mathbf{x}_1 | \mathbf{x}_2 | \dots | \mathbf{x}_n\right] \begin{bmatrix} \lambda_1 & & & \\ & \lambda_2 & & \\ & & \ddots & \\ & & & \lambda_n \end{bmatrix} \\ &= \mathbf{X\Lambda} \end{aligned} \tag{7.20}$$

$$\mathbf{X}^{-1}\mathbf{AX} = \mathbf{\Lambda} \tag{7.21}$$

This result implies that if matrix **A** is similar to a diagonal matrix, then columns in the transformation matrix contains all the eigenvectors and elements in the resulting diagonal matrix are the corresponding eigenvalues of **A**.

However, we should realize that not all matrices can be reduced to a diagonal matrix through similarity transformation. It can be shown that every matrix can be reduced to an upper triangular matrix through similarity transformation.

If matrix **A** is deficient or defective, that is, some of its n eigenvectors are linearly dependent, then it has fewer than n distinct eigenvalues. In other words, there are only k eigenvalues $\lambda_1, \dots, \lambda_k$ with respective multiplicities $m_1, \dots, m_k$, for matrix $\mathbf{A} \in \mathbf{R}^{n \times n}$. As a result, deficient matrices are not similar to diagonal matrices. At best, they are similar to a block-diagonal matrix **J**.

$$\mathbf{J} = \begin{bmatrix} \mathbf{D}_1 & & & \\ & \mathbf{D}_2 & & \\ & & \ddots & \\ & & & \mathbf{D}_k \end{bmatrix} \tag{7.22}$$

$$\mathbf{D}_j = \begin{bmatrix} \lambda_j & 1 & 0 & \cdots & 0 \\ & \lambda_j & 1 & \ddots & \vdots \\ & & \ddots & \ddots & 0 \\ & & & \lambda_j & 1 \\ & & & & \lambda_j \end{bmatrix}_{m_j \times m_j} \tag{7.23}$$

All the terms in a row above the diagonal are 1. The block-diagonal matrix **J** is referred to as the Jordan canonical form of **A**.

The conservation of eigenvalues after similarity transformation has an interesting property. We recall that the eigenvalues are the roots of the corresponding characteristic equation, which is an nth degree polynomial in λ, Equation 7.4. This implies that the coefficients of the polynomial do not change after the similarity transformation. In other words, these coefficients are invariants in similarity transformation.

7.2.4 Congruent transformation

The congruent transformation on matrix **A** is defined as

$$\bar{\mathbf{A}} = \mathbf{F}^T\mathbf{A}\mathbf{F} \tag{7.24}$$

This kind of transformation occurs when the stiffness matrix is transformed from one coordinate system to another; for example, from local to global coordinate system. If matrix **F** is orthogonal, then we have a similarity transformation. In general cases, **F** is not an orthogonal matrix. In mathematics, this kind of transformation occurs in study of quadratic forms $\mathbf{x}^T\mathbf{A}\mathbf{x}$ when the coordinate system is transformed with **F** being the coordinate transformation matrix.

$$\mathbf{x} = \mathbf{F}\mathbf{y} \tag{7.25}$$

$$\begin{aligned} \mathbf{x}^T\mathbf{A}\mathbf{x} &= (\mathbf{F}\mathbf{y})^T\mathbf{A}(\mathbf{F}\mathbf{y}) \\ &= \mathbf{y}^T(\mathbf{F}^T\mathbf{A}\mathbf{F})\mathbf{y} \\ &= \mathbf{y}^T\bar{\mathbf{A}}\mathbf{y} \end{aligned} \tag{7.26}$$

After congruent transformation, the inertia of the original matrix, I(**A**), is maintained. However, it does not maintain the eigenvalues and eigenvectors of the original matrix. This is the result of the Sylvester's inertia theorem.

$$\mathbf{I}(\mathbf{A}) = \mathbf{I}(\mathbf{F}^T\mathbf{A}\mathbf{F}) \tag{7.27}$$

7.2.5 Effect of transposition

From our earlier discussion, we know that the eigenvalues of matrix **A** are the roots of an nth degree polynomial.

$$\det(\mathbf{A} - \lambda\mathbf{I}) = 0 \tag{7.28}$$

We also know that the determinant of transpose of a matrix is the same as the determinant of the matrix itself.

$$\det(\mathbf{A} - \lambda\mathbf{I}) = \det(\mathbf{A}^T - \lambda\mathbf{I}) = 0 \tag{7.29}$$

This indicates that the eigenvalues of **A** and its transpose are the same. However, we will see that the eigenvectors for **A** and its transpose are different. We start with the eigenvalue equation for the matrix and its transpose.

$$\begin{cases} \mathbf{A}\mathbf{x}_j = \lambda_j\mathbf{x}_j \\ \mathbf{A}^T\mathbf{y}_i = \lambda_i\mathbf{y}_i \end{cases} \tag{7.30}$$

Next, we premultiply the first equation with the eigenvector of the transpose of **A** and postmultiply the transpose of the second equation with the eigenvector of **A**, to arrive at the following:

$$\begin{cases} \mathbf{y}_i^T \mathbf{A}\mathbf{x}_j = \lambda_j \mathbf{y}_i^T \mathbf{x}_j \\ \mathbf{y}_i^T \mathbf{A}\mathbf{x}_j = \lambda_i \mathbf{y}_i^T \mathbf{x}_j \end{cases} \tag{7.31}$$

$$(\lambda_j - \lambda_i)\mathbf{y}_i^T \mathbf{x}_j = 0 \Rightarrow \begin{cases} \mathbf{y}_i^T \mathbf{x}_j = 0 & \text{for } i \neq j \\ \mathbf{y}_i^T \mathbf{x}_j \neq 0 & \text{for } i = j \end{cases} \tag{7.32}$$

This means that eigenvectors of matrix **A** and $\mathbf{A}^T$ are orthogonal to each other.

If we represent all the eigenvectors of **A** and $\mathbf{A}^T$ in matrix form as **X** and **Y**, then we have the following relations:

$$\begin{cases} \mathbf{AX} = \mathbf{X\Lambda} \\ \mathbf{A}^T\mathbf{Y} = \mathbf{Y\Lambda} \end{cases} \tag{7.33}$$

$$\begin{cases} \mathbf{A} = \mathbf{X\Lambda X}^{-1} \\ \mathbf{A}^T = \mathbf{Y\Lambda Y}^{-1} \Rightarrow \mathbf{A} = \mathbf{Y}^{-T}\mathbf{\Lambda Y}^T \end{cases} \tag{7.34}$$

$$\mathbf{X}^T\mathbf{Y} = \mathbf{I} \tag{7.35}$$

This relation shows that the eigenvectors of **A** and $\mathbf{A}^T$ are orthonormal.

This leads to the conclusion that for a symmetric matrix, $\mathbf{A} = \mathbf{A}^T$, and $\mathbf{Y} = \mathbf{X}$. As a consequence, $\mathbf{X}^T\mathbf{X} = \mathbf{I}$, which means that eigenvectors of a symmetric matrix are orthonormal.

7.2.6 Sturm sequence

We recall from Chapter 1 that for an $n \times n$ matrix $\mathbf{A} = [a_{ij}]$, we can obtain the following type of principal submatrices:

$$\begin{aligned} &\mathbf{A}_{(1)} = [a_{11}] \\ &\mathbf{A}_{(2)} = \begin{bmatrix} a_{11} & a_{12} \\ a_{21} & a_{22} \end{bmatrix} \\ &\mathbf{A}_{(3)} = \begin{bmatrix} a_{11} & a_{12} & a_{13} \\ a_{21} & a_{22} & a_{23} \\ a_{31} & a_{32} & a_{33} \end{bmatrix} \\ &\vdots \end{aligned} \tag{7.36}$$

These submatrices are called the principal minors of **A**. These principal minors have interesting properties, which show up in the Sturm sequence defined as follows:

$$\begin{aligned} &x_0 = 1 \\ &x_j = \det\left[\mathbf{A}_{(j)}\right], \quad j = 1, \ldots, n \end{aligned} \tag{7.37}$$

The Sturm sequence consists of a first element of 1, followed by the determinants of all the principal minors of matrix **A**.

$$x = \{x_0, x_1, \ldots, x_n\} \tag{7.38}$$

7.2.7 Sturm sequence check

An important Sturm sequence property is that the number of sign changes in the Sturm sequence is equal to the number of negative eigenvalues of matrix **A**. This property is very useful in eigenvalue analysis and can be used in many different ways. One of its usage is to perform the so-called *Sturm sequence check*. This involves the use of Sturm sequence to determine the number of eigenvalues of **A** which are less than a certain value μ. The process can be accomplished through the following steps:

1. Perform a shift μ on **A**
2. Form the Sturm sequence on the shifted matrix
3. Count the number of sign changes in the Sturm sequence

The number of sign changes in the Sturm sequence equals the number of eigenvalues of **A** that are less than the value of shift, μ. Obviously, the most expensive part lies in step 2, in which the determinants of primary minors of matrix **A** are computed. A more practical approach for step 2 is to perform Gaussian triangular decomposition on the shifted matrix. We note that we should not perform Cholesky decomposition because of the presence of negative eigenvalues in the shifted matrix. Then the Sturm sequence can be computed as follows:

$$\begin{aligned}
&\text{Shift: } \bar{\mathbf{A}} = \mathbf{A} - \mu\mathbf{I} \\
&\text{Triangular decomposition: } \bar{\mathbf{A}} = \mathbf{L}\mathbf{D}\mathbf{L}^T \\
&\text{Sturm Sequence:} \\
&x_0 = 1 \\
&x_j = \det(\mathbf{A}_{(j)}) = \det(\mathbf{L}_{(j)}\mathbf{D}_{(j)}\mathbf{L}_{(j)}^T) \\
&\quad = \det(\mathbf{D}_{(j)}) = \prod_j d_{jj}
\end{aligned} \tag{7.39}$$

The number of negative diagonal terms in **D** equals the number of eigenvalues of **A** which are less than the value μ.

With the use of Sturm sequence check, we can count the number of eigenvalues within the interval $[\alpha, \beta]$. We conduct Sturm sequence check once with a shift of α, and repeat with a shift of β. The difference between them is the number of eigenvalues within that interval.

7.2.8 Effect of constraints and Cauchy interlace theorem

First, we state the Cauchy's interlace theorem: eigenvalues of $\mathbf{A}_{(k)}$ interlace the eigenvalues of $\mathbf{A}_{(k+1)}$.

$$\lambda_i^{(k+1)} \le \lambda_i^{(k)} \le \lambda_{i+1}^{(k+1)} \tag{7.40}$$

The eigenvalues of $\mathbf{A}_{(k)}$ and $\mathbf{A}_{(k+1)}$ are $\lambda_i^{(k)}$ and $\lambda_i^{(k+1)}$, respectively. Since eigenvalues are roots of the corresponding characteristic equation, this relationship is illustrated in Figure 7.2.

This is also called the kth constraint problem. We can illustrate the constraint concept by considering matrix $\mathbf{A} = \mathbf{K}$ which is a structural stiffness matrix. For the structure shown in Figure 7.3, there are six lateral displacement degrees of freedom (rotational degrees of freedom are constrained). The six eigenvalues can be determined from the following equation:

$$\det(\mathbf{A}_{(6)} - \lambda \mathbf{I}_{(6)}) = 0 \tag{7.41}$$

If we impose one additional constraint on the system, then the total degrees of freedom of the structure is reduced by one. The corresponding five eigenvalues are computed from the following polynomial:

$$\det(\mathbf{A}_{(5)} - \lambda \mathbf{I}_{(5)}) = 0 \tag{7.42}$$

As shown in Figure 7.3, the five eigenvalues of the constrained system interlace the six eigenvalues of the unconstrained system. The effect of introducing more constraint to a structural system is shown in Figure 7.3.

Physically, if we impose a constraint on a degree of freedom, the resulting structure will become *stiffer* so that the lowest frequency (or eigenvalue) is raised.

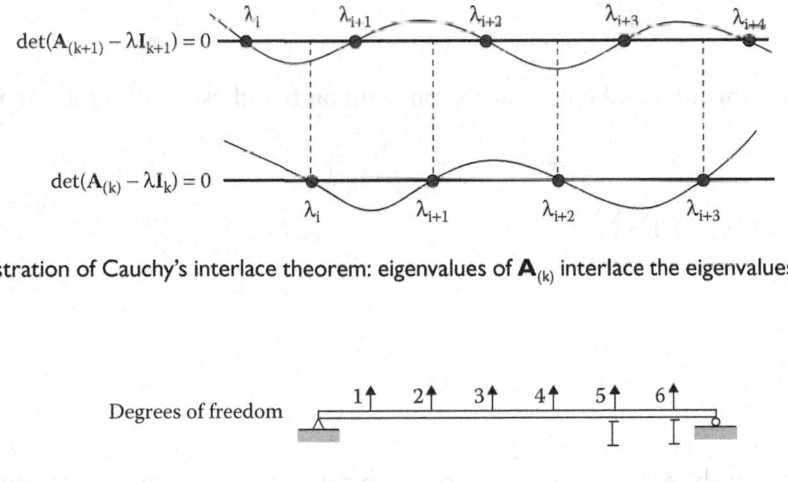

Figure 7.2 Illustration of Cauchy's interlace theorem: eigenvalues of $\mathbf{A}_{(k)}$ interlace the eigenvalues of $\mathbf{A}_{(k+1)}$.

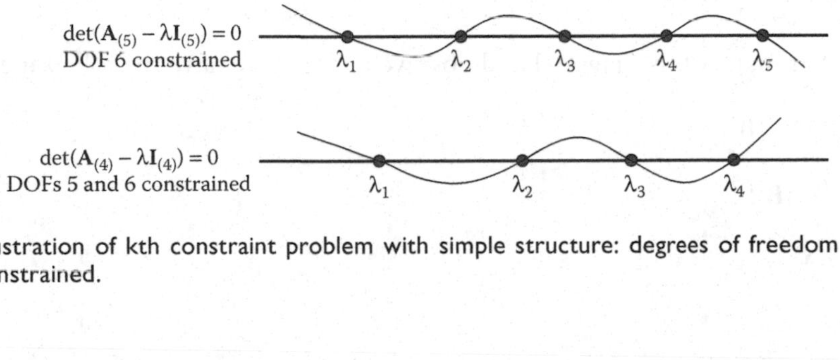

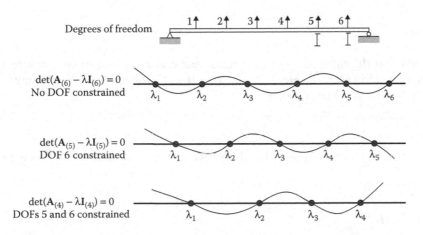

Figure 7.3 Illustration of kth constraint problem with simple structure: degrees of freedom 6 and 5 are constrained.

7.2.9 Hoffman–Wielandt theorem

This theorem (Hoffman and Wielandt, 1953) can be stated as follows: the eigenvalues of symmetric and positive definite matrices **A** and $\bar{\mathbf{A}}$ satisfy the following condition:

$$\max_j \left|\lambda_j(\mathbf{A}) - \lambda_j(\bar{\mathbf{A}})\right| \leq \left\|\mathbf{A} - \bar{\mathbf{A}}\right\| \tag{7.43}$$

This means that eigenvalues of a symmetric and positive definite matrix are continuous function of the matrix itself. We can say that the eigenvalues of symmetric positive definite matrices are well-posed. In general, eigenvectors of symmetric positive definite matrices are also well-posed, except for multiple eigenvalues.

This property of being well-posed means that for any small changes imposed on matrix **A** we can expect small changes on the system's eigenvalues. We can see important implications of this theorem in structural mechanics; it is not possible to cause drastic changes in the eigenvalues of the structural system with minor structural modifications.

7.2.10 Reduction of generalized eigenvalue problems

Most eigenvalue solution methods are for the standard eigenvalue problem. In computational mechanics we often reduce the generalized eigenvalue problem to standard form before applying the eigenvalue solution method. This process is demonstrated in the following equations for a generalized eigenvalue problem where matrix **B** is symmetric and positive definite:

$$\mathbf{A}\mathbf{x} = \lambda\mathbf{B}\mathbf{x} \tag{7.44}$$

We start by performing a Cholesky decomposition on **B** and use that in the equation above.

$$\mathbf{B} = \mathbf{L}\mathbf{L}^{\mathrm{T}} \tag{7.45}$$

$$(\mathbf{L}^{-1}\mathbf{A}\ \mathbf{L}^{-\mathrm{T}})(\mathbf{L}^{\mathrm{T}}\mathbf{x}) = \lambda(\mathbf{L}^{\mathrm{T}}\mathbf{x}) \tag{7.46}$$

$$\bar{\mathbf{A}}\bar{\mathbf{x}} = \lambda\bar{\mathbf{x}}$$

$$\begin{cases} \bar{\mathbf{A}} = \mathbf{L}^{-1}\mathbf{A}\mathbf{L}^{-\mathrm{T}} \\ \bar{\mathbf{x}} = \mathbf{L}^{\mathrm{T}}\mathbf{x} \end{cases} \tag{7.47}$$

It is apparent that the eigenvalues of the original generalized problem are not affected by this transformation. However, the eigenvectors are affected and they can be computed from the eigenvectors of the standard problem.

$$\mathbf{x} = \mathbf{L}^{-\mathrm{T}}\bar{\mathbf{x}} \tag{7.48}$$

In the case that matrix **B** is diagonal and positive definite, we have the following equations:

$$\mathbf{B} = \mathbf{L}\mathbf{L}^{\mathrm{T}} = \mathbf{B}^{1/2}\mathbf{B}^{1/2} \tag{7.49}$$

$$\begin{cases} \bar{\mathbf{A}} = \mathbf{B}^{-1/2}\mathbf{A}\mathbf{B}^{-1/2} \\ \bar{\mathbf{x}} = \mathbf{B}^{1/2}\mathbf{x} \end{cases} \tag{7.50}$$

As a special case, we may need to consider when matrix **B** is diagonal and positive semidefinite. This is the case when there are zero terms in the diagonal of matrix **B**. An example is the case of rotational degrees of freedom of a lumped mass matrix **B** in which the mass is lumped to translational degrees of freedom. To explore this case, we start by rewriting the generalized eigenvalue problem, Equation 7.44, in the following form:

$$\mathbf{Bx} = \frac{1}{\lambda}\mathbf{Ax} \tag{7.51}$$

Assume that the ith element in the diagonal matrix **B** is zero. In the above equation we can use vector $\mathbf{x}_i = [0, \ldots, 0, 1, 0, \ldots, 0]$ in which the nonzero term 1 is at the ith position and obtain the following relation:

$$\begin{Bmatrix} 0 \\ \vdots \\ 0 \\ \vdots \\ 0 \end{Bmatrix} = \frac{1}{\lambda_i}\begin{Bmatrix} a_{1i} \\ \vdots \\ a_{ii} \\ \vdots \\ a_{ni} \end{Bmatrix} \tag{7.52}$$

For the above equation to be true, there must be $\lambda_i \to \infty$. This means that having zero terms in the diagonal of **B** matrix corresponds to the condition that corresponding eigenvalues approach infinite. This kind of problem can be resolved by using the technique of static condensation found in finite element analysis. We can condense out the degrees of freedom with zero terms on the diagonal of the **B**. For example, the rotational degrees of freedom in a lumped mass matrix. The static condensation process proceeds as follows. First, we partition the matrix **B** by moving the zero element to the last position on the diagonal, and partition the matrices accordingly. By manipulating the condensed matrices in the following equation, we arrive at the equation for the condensed system:

$$\mathbf{B} = \text{dia}(\mathbf{B}) = \begin{bmatrix} \mathbf{B}_1 & 0 \\ 0 & 0 \end{bmatrix}; \quad \mathbf{A} = \begin{bmatrix} \mathbf{A}_{11} & \mathbf{A}_{12} \\ \mathbf{A}_{21} & \mathbf{A}_{22} \end{bmatrix}; \quad \mathbf{x} = \begin{Bmatrix} \mathbf{x}_1 \\ \mathbf{x}_2 \end{Bmatrix} \tag{7.53}$$

$$\begin{cases} \mathbf{A}_{11}\mathbf{x}_1 + \mathbf{A}_{12}\mathbf{x}_2 = \lambda\mathbf{B}_1\mathbf{x}_1 \\ \mathbf{A}_{21}\mathbf{x}_1 + \mathbf{A}_{22}\mathbf{x}_2 = 0 \end{cases} \tag{7.54}$$

From the second equation above, we can obtain $\mathbf{x}_2$ and substitute into the first equation to arrive at the eigenvalue problem for $\mathbf{x}_1$.

$$(\mathbf{A}_{11} - \mathbf{A}_{12}\mathbf{A}_{22}^{-1}\mathbf{A}_{21})\mathbf{x}_1 = \lambda\mathbf{B}_1\mathbf{x}_1 \tag{7.55}$$

When matrix **A** is symmetric and positive definite, the equivalent matrix in the above equation is also symmetric and positive definite. This condensed system can be readily transformed to a standard eigenvalue form.

It is possible that matrix **B** may be symmetric but not positive definite. In this case, we can carry out Gauss decomposition on **B**.

$$\mathbf{B} = \mathbf{L}_B\mathbf{D}_B\mathbf{L}_B^T \tag{7.56}$$

$$(\mathbf{L}_B^{-1}\mathbf{A}\mathbf{L}_B^{-T})(\mathbf{L}_B^T\mathbf{x}) = \lambda\ \mathbf{D}_B(\mathbf{L}_B^T\mathbf{x}) \tag{7.57}$$

$$\bar{\mathbf{A}}\bar{\mathbf{x}} = \lambda\mathbf{D}_B\bar{\mathbf{x}}$$

$$\begin{cases} \bar{\mathbf{A}} = \mathbf{L}_B^{-1}\mathbf{A}\mathbf{L}_B^{-T} \\ \bar{\mathbf{x}} = \mathbf{L}_B^T\mathbf{x} \end{cases} \tag{7.58}$$

If the diagonal matrix $\mathbf{D}_B$ is positive semidefinite, then we can carry out static condensation to do further reduction on the above transformed system.

7.2.11 Rayleigh quotient

In Chapter 1 we have introduced the Rayleigh quotient defined as follows:

$$\rho(\mathbf{x}) = \frac{\mathbf{x}^T\mathbf{A}\mathbf{x}}{\mathbf{x}^T\mathbf{x}} \tag{7.59}$$

If the matrix $\mathbf{A}$ represents a structural system, the Rayleigh quotient represents the strain energy stored in the system for a unit displacement vector, $\|\mathbf{x}\| = 1$. Since the capacity of structural systems to sustain deformation or absorb ėnergy is finite, the Rayleigh quotient is thus a bounded number. We have observed that if $\mathbf{A}$ is positive definite, then the Rayleigh quotient is positive. In addition, we have also proven the following relation:

$$\lambda_1 \le \rho(\mathbf{x}) \le \lambda_n \tag{7.60}$$

Another important property is that, if $\mathbf{x}$ is a multiple of an eigenvector, the Rayleigh quotient is equal to the corresponding eigenvalue.

$$\rho(\mathbf{x}) = \lambda_k \text{ if } \mathbf{x} = \alpha\ \mathbf{x}_k \text{ and } \mathbf{A}\mathbf{x}_k = \lambda_k\mathbf{x}_k \tag{7.61}$$

7.2.12 Minimax characterization of eigenvalues

For any positive definite matrix $\mathbf{A}$, we have known from the property of Rayleigh quotient that for any vector $\mathbf{x}$, $\lambda_1 \le \rho(\mathbf{x})$. This implies that if we consider the problem of varying the vector $\mathbf{x}$, we always obtain a value from the Rayleigh quotient which is larger than λ_1, and the minimum value is reached or the equality is satisfied when we have $\mathbf{x} = \mathbf{x}_1$, the eigenvector of matrix $\mathbf{A}$ corresponding to the smallest eigenvalue.

Now, suppose we impose a restriction on the vector $\mathbf{x}$ such that it is orthogonal to a specific vector $\mathbf{y}$, and we consider the problem of minimizing $\rho(\mathbf{x})$ subject to this constraint. After having calculated the minimum of $\rho(\mathbf{x})$ with the condition $\mathbf{x}^T\mathbf{y} = 0$, we could start varying the vector $\mathbf{y}$. For each new vector $\mathbf{y}$, we evaluate a new minimum of $\rho(\mathbf{x})$. If we continue this process, we can expect that the maximum value of all these minimum values obtained would be λ_2. This is intuitively reasonable. This result can be generalized to the principle called the *minimax characterization of eigenvalues*.

$$\lambda_k = \max\left\{\min\frac{\mathbf{x}^T\mathbf{A}\mathbf{x}}{\mathbf{x}^T\mathbf{x}}\right\}, \quad k = 1,\ldots, n \tag{7.62}$$

In this equation the vector **x** satisfies the following orthogonality condition:

$$\mathbf{x}^T\mathbf{y}_j = 0, \; j = 1, \ldots, k-1 \tag{7.63}$$

We choose vectors $\mathbf{y}_j$, j = 1, …, k – 1 and evaluate the minimum of $\rho(\mathbf{x})$ with the vector **x** subject to the constraint that **x** be orthogonal to the vector subspace spanned by $\mathbf{y}_j$, j = 1, …, k – 1. After having computed the minimum of the Rayleigh quotient, we vary the vector $\mathbf{y}_j$ and evaluate another new minimum. The maximum value of all these minima is λ_k.

To obtain this relationship, first we express the vector **x** as a linear combination of eigenvectors and substitute in Equation 7.62 and denote it by R.

$$\mathbf{x} = \sum_{j=1}^{n} \alpha_j \mathbf{x}_j \tag{7.64}$$

$$\begin{aligned} R &= \max\left\{ \min \frac{\sum_{j=1}^{n} \alpha_j^2 \lambda_j}{\sum_{j=1}^{n} \alpha_j^2} \right\} \\ &= \max\left\{ \min\left\{ \lambda_k - \frac{\sum_{j=1}^{k-1} \alpha_j^2(\lambda_k - \lambda_j) + \sum_{j=k+1}^{n} \alpha_j^2(\lambda_k - \lambda_j)}{\sum_{j=1}^{n} \alpha_j^2} \right\} \right\} \end{aligned} \tag{7.65}$$

We can observe the consequence of choosing vector **x** and coefficients α_j to satisfy the following condition.

$$\begin{cases} \mathbf{y}_j \sum_{i=1}^{n} \alpha_i \mathbf{x}_i = 0 \quad \text{for } j = 1, \ldots, k-1 \\ \alpha_{k+1} = \cdots = \alpha_n = 0 \end{cases} \tag{7.66}$$

$$R = \max\left\{ \min\left\{ \lambda_k - \frac{\sum_{j=1}^{k-1} \alpha_j^2(\lambda_k - \lambda_j)}{\sum_{j=1}^{n} \alpha_j^2} \right\} \right\} \le \lambda_k \tag{7.67}$$

On the other hand, if we choose vector **y** in the subspace spanned by the first (k–1) eigenvectors,

$$\mathbf{y} \in \{\mathbf{x}_1, \ldots, \mathbf{x}_{k-1}\} \tag{7.68}$$

then to satisfy the orthogonality condition $\mathbf{x}^T\mathbf{y} = 0$, we must choose **x** to satisfy the following condition, leading to $R = \lambda_k$.

$$\begin{cases} \mathbf{x} \in \{\mathbf{x}_k, \ldots, \mathbf{x}_n\} \\ \alpha_1 = \cdots = \alpha_{k-1} = 0 \end{cases} \tag{7.69}$$

This minimax property of eigenvalues has important practical implications. It basically means that with more constraints introduced to the system, its smallest eigenvalue increases. In other words, the structural system becomes stiffer.

7.3 EIGENVALUE SOLUTION FOR SMALL MATRICES

So far, we have discussed some of the fundamental properties associated with eigenvalues and eigenvectors. Obviously, if the system coefficient matrix or the stiffness matrix is of large dimensions, it would be prohibitively expensive to compute all eigenvalues and eigenvectors or eigenpairs by solving the corresponding characteristic equations which is an nth degree polynomial in λ. In practice, especially in structural analysis, often we do not need to compute all the eigenvalues and eigenvectors. We may be interested in only a few eigenpairs. However, if the system coefficient matrices are of small dimensions, we may compute all the eigenvalues and eigenvectors. In this section, we discuss some numerical methods for computing eigenvalues of small matrices.

Suppose that we have some ways to determine either the eigenvalue or the eigenvector, then we may consider basically two scenarios. The first involves the determination of eigenvector given that the eigenvalue has been determined. In this case, we have the following system of homogeneous equations:

$$(\mathbf{A} - \lambda_j \mathbf{I})\, \mathbf{x}_j = 0 \tag{7.70}$$

For nontrivial solution, we can assume certain value for one element in $\mathbf{x}_j$, and solve the system of equations for the rest of the elements. By so doing, we obtain a nonzero vector which is the eigenvector corresponding to λ_j.

On the other hand, if the eigenvector $\mathbf{x}_j$ is known, then we can normalize the vector such that $\|\mathbf{x}_j\| = 1$. The corresponding eigenvalue can be directly obtained.

$$\|\mathbf{A}\mathbf{x}_j\| = \lambda_j \|\mathbf{x}_j\| = \lambda_j \tag{7.71}$$

This involves only matrix–vector multiplications and is not expensive to compute. Now we consider some basic methods for computing eigenvalues for small matrices.

7.3.1 Bisection method

The fundamental idea in determining eigenvalues is to obtain the roots of the corresponding characteristic equation, which can be expanded as an n-degree polynomial in λ.

$$D(\lambda) = |\mathbf{A} - \lambda \mathbf{I}| = 0 \tag{7.72}$$

We assume that $\lambda_1 < \lambda_2 < \ldots < \lambda_n$. The determination of coefficients for this polynomial requires computation of determinant of a matrix, which is computationally a very time-consuming process. Therefore, this polynomial function should not be solved through direct calculations. A viable procedure involves computing the triangular decomposition of the matrix.

$$\mathbf{A} - \lambda \mathbf{I} = \mathbf{L}_0 \mathbf{D}_0 \mathbf{L}_0^T \tag{7.73}$$

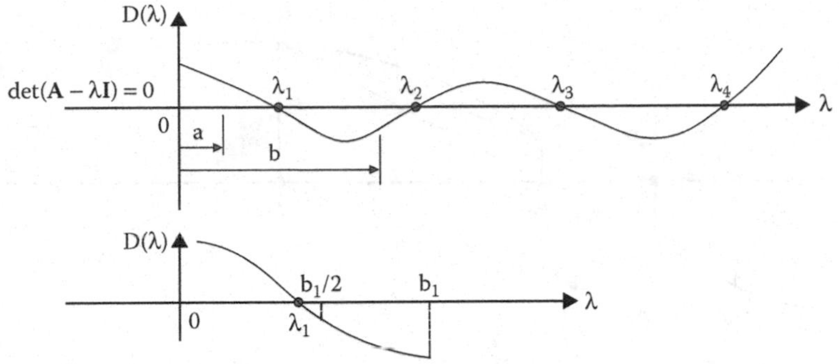

Figure 7.4 Illustration of eigenvalue determination using bisection method.

$$D(\lambda) = \prod_{k=1}^{n} d_{kk} \tag{7.74}$$

Next, we want to determine a specific eigenvalue, say λ_1, we thus need to find an interval [a, b] such that $a < \lambda_1 < b$, as shown in Figure 7.4. In this case, assume that **A** is positive definite, then we need to determine an interval [0, b] such that $0 < \lambda_1 < b$. We can determine this interval using the bisection method, which proceeds in the following way. First, we perform a shift of b_1 to matrix **A** and compute the triangular decomposition on the shifted matrix.

$$\mathbf{A} - b_1\mathbf{I} = \mathbf{L}_1\mathbf{D}_1\mathbf{L}_1^T \tag{7.75}$$

We then perform Sturm sequence check by counting the negative elements on the diagonal of the matrix $\mathbf{D}_1$ to determine the number of eigenvalues in this interval. If there are more than one eigenvalues in this interval, then we carry out another shift to $b_2 = b_1/2$, and perform the similar computations of triangular decomposition and Sturm sequence check. This process continues to the kth iteration until we found from Sturm sequence check that there is only one eigenvalue that is less than b_k. Hence we find the interval $[0, b_k]$ such that $0 < \lambda_1 < b_k$. Next, we divide the interval into two halves and determine which half contains the eigenvalue and continue dividing that interval into halves. We continue this process until the interval containing the eigenvalue is small enough to provide a reasonable approximation of the eigenvalue for practical purposes.

It is apparent that the bisection method is actually a general method to determine the eigenvalues, but it is very inefficient in performing triangular decompositions. A more efficient method would utilize more information from the characteristic equation.

7.3.2 Secant method or determinant search method

The secant search method is a more efficient method than the bisection method. Referring to Figure 7.5, we assume that we have identified the interval $[\lambda_{k-1}, \lambda_k]$ in which there exists an eigenvalue. We have computed the functional values.

$$\begin{cases} D(\lambda_k) = |\mathbf{A} - \lambda_k\mathbf{I}| \\ D(\lambda_{k-1}) = |\mathbf{A} - \lambda_{k-1}\mathbf{I}| \end{cases} \tag{7.76}$$

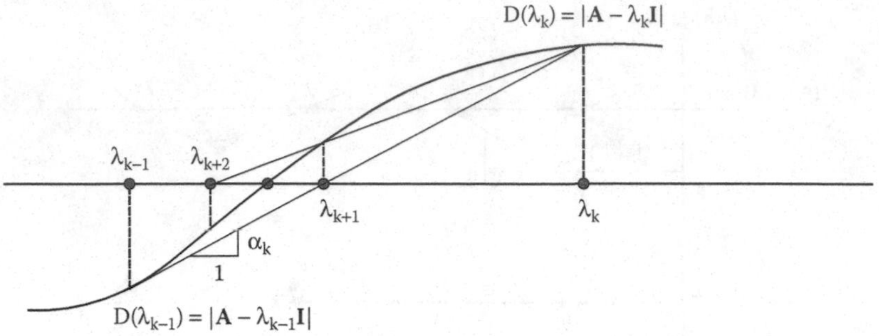

Figure 7.5 Determination of eigenvalue with secant iteration method (determinant search method).

Now we connect the two points $[\lambda_{k-1}, D(\lambda_{k-1})]$ and $[\lambda_k, D(\lambda_k)]$ with a straight line and determine the slope of the line.

$$\alpha_k = \frac{D(\lambda_k) - D(\lambda_{k-1})}{\lambda_k - \lambda_{k-1}} \tag{7.77}$$

Next, we can determine the intersection of this line with the axis λ and denote it as λ_{k+1}.

$$\lambda_{k+1} = \lambda_k - \frac{D(\lambda_k)}{\alpha_k} \tag{7.78}$$

With these equations, the next intersection point can be computed by repeating the above procedures. We note that this method also involves the computation of triangular decomposition of matrices with shift, which is also computationally time-consuming. However, this method is more efficient than the bisection method.

At this juncture, the Newton–Raphson method appears to be readily applicable in the eigenvalue solution process. However, the Newton–Raphson method requires the calculation of the gradient of $D(\lambda_k)$ which is computationally very time-consuming. In practice, therefore, we seldom use Newton–Raphson method in eigenvalue solutions. This shows that the approach of computing the eigenvalues through solving the corresponding characteristic equations is inefficient and not practical, especially when the system matrix is large. However, these methods indicate that the eigenvalue solution process is iterative in nature. We therefore need to resort to other methods that do not take a brute-force approach.

7.3.3 Jacobi method

Jacobi method is an iterative method proposed to find all the eigenvalues and eigenvectors of full symmetric matrices. With a banded matrix, it will become a full matrix after certain number of iterations. As a result, it is mostly applicable to small matrices.

The basic idea behind the Jacobi method is that through a series of plane rotation transformations or orthogonal similarity transformations, a symmetric matrix can be transformed into a diagonal matrix, which contains all the eigenvalues. We have discussed earlier that if $\mathbf{A} \in R^{n \times n}$ is a symmetric matrix, then there exists an orthogonal matrix $\mathbf{P}$ to transform it to a diagonal matrix.

$$\mathbf{P}\mathbf{A}\mathbf{P}^T = \mathbf{D} = \text{diag}\{\lambda_1, \ldots, \lambda_n\} \tag{7.79}$$

The column vectors $\mathbf{v}_i$ in $\mathbf{P}^T$ are the eigenvectors corresponding to eigenvalues λ_i in the diagonal matrix **D**. Therefore, the eigenvalue solution process becomes a process of searching for an orthogonal matrix **P** and performing similarity transformations.

To illustrate the idea, we consider a simple 2 × 2 matrix as follows:

$$\mathbf{A} = \begin{bmatrix} a_{11} & a_{12} \\ a_{21} & a_{22} \end{bmatrix} \tag{7.80}$$

Now we perform a plane rotation transformation $\mathbf{x} = \mathbf{P}\,\mathbf{y}$ with the following orthogonal matrix:

$$\mathbf{P} = \begin{bmatrix} \cos\theta & \sin\theta \\ -\sin\theta & \cos\theta \end{bmatrix} = \begin{bmatrix} c & s \\ -s & c \end{bmatrix} \tag{7.81}$$

With this **P** matrix, we compute the orthogonal similarity transformation on **A** as follows:

$$\mathbf{PAP}^T = \begin{bmatrix} a_{11}c^2 + a_{22}s^2 + a_{21}\sin 2\theta & \frac{1}{2}(a_{22} - a_{11})\sin 2\theta + a_{21}\cos 2\theta \\ \text{Symm.} & a_{11}s^2 + a_{22}c^2 - a_{21}\sin 2\theta \end{bmatrix} \tag{7.82}$$

The value of θ is determined such that the off-diagonal term becomes zero.

$$\tan 2\theta = \frac{2a_{21}}{a_{11} - a_{22}} \tag{7.83}$$

With this value of θ the orthogonal similarity transformation leads to a diagonal matrix with eigenvalues on the diagonal.

$$\mathbf{PAP}^T = \mathrm{diag}\{\lambda_1, \lambda_2\} \tag{7.84}$$

This shows that by carrying out similarity transformation of **A** with a specially defined plane rotation matrix, one of the off-diagonal term can be set to zero in the process. This actually forms the basis of the Jacobi method. Let $\mathbf{R}^{(k)}$ be a plane rotation matrix which is a function of $\theta^{(k)}$. We then perform successively a series of the following operations:

$$\begin{aligned} \mathbf{A}^{(1)} &= \mathbf{A} \\ \mathbf{A}^{(2)} &= \mathbf{R}^{(1)}\mathbf{A}^{(1)}\mathbf{R}^{(1)^T} \\ &\vdots \\ \mathbf{A}^{(k+1)} &= \mathbf{R}^{(k)}\mathbf{A}^{(k)}\mathbf{R}^{(k)^T} \\ &\vdots \end{aligned} \tag{7.85}$$

Our objective is to determine the value of $\theta^{(k)}$ such that a_{ij} and a_{ji} in the symmetric matrix $\mathbf{A}^{(k+1)}$ become zero. Hence, for each iteration, one of the off-diagonal terms is eliminated from matrix **A**. If we carry out this iterative process continuously, eventually, the method will converge.

$$\lim_{k\to\infty} \mathbf{A}^{(k)} \to \Lambda \tag{7.86}$$

We can directly determine the elements of the plane rotation matrix from the Equation 7.85. In the equations $c = \cos\theta^{(k)}$ and $s = \sin\theta^{(k)}$.

$$a_{ij}^{(k+1)} = a_{ij}^{(k)}(c^2 - s^2) + (a_{jj}^{(k)} - a_{ii}^{(k)})sc = 0 \tag{7.87}$$

$$\tan 2\theta^{(k)} = \frac{a_{ij}^{(k)}}{a_{ii}^{(k)} - a_{jj}^{(k)}} \tag{7.88}$$

With the value of $\theta^{(k)}$ computed, the plane rotation matrix is also determined.

To compute the corresponding eigenvectors, when m is sufficiently large, we have the following relations:

$$\mathbf{R}^{(m)}\ldots\mathbf{R}^{(2)}\mathbf{R}^{(1)}\mathbf{A}\,\mathbf{R}^{(1)^T}\mathbf{R}^{(2)^T}\ldots\mathbf{R}^{(m)^T} = \Lambda \tag{7.89}$$

$$\mathbf{P}^{m}\mathbf{A}\mathbf{P}^{m^T} = \Lambda \tag{7.90}$$

$$\mathbf{P}^{m^T} = \mathbf{R}^{(1)^T}\mathbf{R}^{(2)^T}\ldots\mathbf{R}^{(m)^T} \tag{7.91}$$

The column vectors in $\mathbf{P}^{(m)T}$ are the approximate eigenvectors. We can compute the matrix $\mathbf{P}^{(m)T}$ using successive multiplications. We can start with an initial identity matrix $\mathbf{P} \leftarrow \mathbf{I}$. Then, each time when we perform plane rotation similarity transformation on **A**, we update **P**.

$$\mathbf{P} \leftarrow \mathbf{P}\mathbf{R}^{(m)^T} \tag{7.92}$$

Because $\mathbf{R}^{(m)T}$ is a plane rotation matrix, the multiplication of $\mathbf{PR}^{(m)T}$ only requires updating two columns (i and j) of matrix **P**.

It should be noted that in the implementation of Jacobi's method, the sequence of zeroing off-diagonal elements is very important. Usually, we selectively zero the largest off-diagonal term (absolute value) at each iteration instead of performing zero sweeping. If we continue this process, eventually we can reduce the off-diagonal terms to very small values. We then obtain the approximate values of all the eigenvalues of **A**. The convergence criteria can be selected such that the sum of squares of all the off-diagonal terms is smaller than a predetermined tolerance. It is possible that the previously zeroed off-diagonal term may become nonzero after performing subsequent plane rotation transformations. Chances are it may be selected again for elimination. Therefore, the largest off-diagonal term in terms of absolute value must be selected from all the off-diagonal terms of the matrix.

The Jacobi method is a fairly accurate method in computing all the eigenvalues and eigenvectors of a small matrix. For banded matrices, the sparse band would be destroyed after a few plane rotation transformations. Therefore, it requires full storage of the matrix.

7.4 EIGENVALUE SOLUTION FOR LARGE MATRICES

Often, in practice, we are only interested in a few eigenvalues and eigenvectors. On the other hand, for very large problems, it is almost impossible to compute all the eigenvalues and eigenvectors within a reasonable time frame using the most available computational resources. There are many numerical methods to accomplish the task of computing a few eigenvalues and eigenvectors. One of these methods that is simple and of fundamental nature is called the power method.

7.4.1 Power method

The power method is an iterative method for computing the largest eigenvalue and corresponding eigenvector of a matrix. In structural dynamics, this method is usually referred to as Stodola–Vianello method. Assume that matrix **A** has a complete set of eigenpairs with eigenvalues $\lambda_1 < \lambda_2 < \ldots < \lambda_n$ and corresponding eigenvectors $\mathbf{x}_1, \mathbf{x}_2, \ldots, \mathbf{x}_n$. We start with a nonzero initial vector $\mathbf{y}_0$ and construct a vector series in the following procedure:

$$
\begin{aligned}
&\mathbf{y}_0 \\
&\bar{\mathbf{y}}_1 = \mathbf{A}\mathbf{y}_0, \ \mathbf{y}_1 = \frac{1}{\|\bar{\mathbf{y}}_1\|}\bar{\mathbf{y}}_1 \\
&\quad\vdots \\
&\bar{\mathbf{y}}_j = \mathbf{A}\mathbf{y}_{j-1}, \ \mathbf{y}_j = \frac{1}{\|\bar{\mathbf{y}}_j\|}\bar{\mathbf{y}}_j \\
&\quad\vdots
\end{aligned}
\tag{7.93}
$$

This is the power method with normalization. Normalization is not necessary for convergence. Looking at the procedure, it may not be clear about what can be accomplished with this simple algorithm. This aspect will become clear when we discuss the convergence property of the algorithm. First, we observe the above iterations without normalization.

$$
\mathbf{y}_j = \mathbf{A}\mathbf{y}_{j-1} = \mathbf{A}(\mathbf{A}\mathbf{y}_{j-2}) = \cdots = \mathbf{A}^j\mathbf{y}_0 \tag{7.94}
$$

The initial vector can be written as a linear combination of eigenvectors of **A**.

$$
\mathbf{y}_0 = \sum_{k=1}^{n} \alpha_k \mathbf{x}_k \tag{7.95}
$$

$$
\begin{aligned}
\mathbf{y}_j &= \mathbf{A}^j \sum_{k=1}^{n} \alpha_k \mathbf{x}_k = \sum_{k=1}^{n} \alpha_k \mathbf{A}^j \mathbf{x}_k = \sum_{k=1}^{n} \alpha_k \lambda_k^j \mathbf{x}_k \\
&= \lambda_n^j \left[\sum_{k=1}^{n-1} \alpha_k \left(\frac{\lambda_k}{\lambda_n} \right)^j \mathbf{x}_k + \alpha_n \mathbf{x}_n \right]
\end{aligned}
\tag{7.96}
$$

$$
\lim_{j \to \infty}
\begin{cases}
\left(\dfrac{\lambda_k}{\lambda_n} \right)^j \to 0 \quad \text{for } k = 1, \ldots, n-1 \\
\mathbf{y}_j \to \alpha_n \left(\lambda_n^j \right) \mathbf{x}_n \\
\dfrac{\|\mathbf{y}_j\|}{\|\mathbf{y}_{j-1}\|} \to \lambda_n
\end{cases}
\tag{7.97}
$$

This shows that when j becomes sufficiently large, the vector $\mathbf{y}_j$ approximates the eigenvector $\mathbf{x}_n$, corresponding to the largest eigenvalue λ_n.

It is obvious that of all the numbers smaller than 1.0, the ratio $(\lambda_{n-1}/\lambda_n)$ is the largest. Therefore, this ratio $(\lambda_{n-1}/\lambda_n)$ governs the convergence rate in approximating the largest eigenvalue and corresponding eigenvector through the power method. If we have multiple eigenvalues, say $\lambda_{n-1} = \lambda_n$, then the method breaks down.

It can be proven that when we normalize the vector after each iteration, then we have the following convergence:

$$\lim_{j \to \infty} \begin{cases} \mathbf{y}_j \to \mathbf{x}_n \\ \left\|\bar{\mathbf{y}}_j\right\|_2 \to \lambda_n \end{cases} \tag{7.98}$$

Parlett (1980) has given a very interesting proof of this convergence property of the normalized power method. We recall that the basic iteration steps in the power method can be described as follows:

$$\begin{cases} \bar{\mathbf{y}}_j = \mathbf{A}\mathbf{y}_{j-1} \\ \mathbf{y}_j = \dfrac{1}{\left\|\bar{\mathbf{y}}_j\right\|} \bar{\mathbf{y}}_j \end{cases} \tag{7.99}$$

We would like to show that if $\mathbf{y}_0^T \mathbf{x}_n \neq 0$ then we have the convergence condition in Equation 7.98 satisfied. First, we decompose the vector $\mathbf{y}_j$ into sum of its projections on $\mathbf{x}_n$ and $\mathbf{x}_n^\perp$ which is the hyperplane of $\mathbf{x}_n$. Let $\mathbf{u}_j \in \mathbf{x}_n^\perp$ be a unit vector normal to $\mathbf{x}_n$ ($\mathbf{x}_n^T \mathbf{u}_j = 0$). For convenience, we also assume the following are unit vectors.

$$\begin{cases} \left\|\mathbf{x}_n\right\|_2 = 1 \\ \left\|\mathbf{y}_j\right\|_2 = 1 \end{cases} \tag{7.100}$$

Denote the angle between $\mathbf{x}_n$ and $\mathbf{y}_j$ as θ_j, that is, $\theta_j = \cos^{-1}(\mathbf{x}_n^T \mathbf{y}_j)$. Therefore, we want to show the following convergence:

$$\lim_{j \to \infty} \begin{cases} \theta_j \to 0 \\ \mathbf{y}_j \to \mathbf{x}_n \end{cases} \tag{7.101}$$

From the geometric relation, we have the following relation that we can use in the power method:

$$\mathbf{y}_j = \mathbf{x}_n \cos\theta_j + \mathbf{u}_j \sin\theta_j \tag{7.102}$$

$$\begin{aligned} \bar{\mathbf{y}}_{j+1} &= \mathbf{A}\mathbf{y}_j \\ &= \mathbf{A}\mathbf{x}_n \cos\theta_j + \mathbf{A}\mathbf{u}_j \sin\theta_j \\ &= \lambda_n \mathbf{x}_n \cos\theta_j + \frac{\mathbf{A}\mathbf{u}_j}{\left\|\mathbf{A}\mathbf{u}_j\right\|} \left\|\mathbf{A}\mathbf{u}_j\right\| \sin\theta_j \end{aligned} \tag{7.103}$$

$$\begin{aligned} \mathbf{y}_{j+1} &= \frac{1}{\left\|\bar{\mathbf{y}}_{j+1}\right\|} \bar{\mathbf{y}}_{j+1} \\ &= \mathbf{x}_n \frac{\lambda_n \cos\theta_j}{\left\|\bar{\mathbf{y}}_{j+1}\right\|} + \frac{\mathbf{A}\mathbf{u}_j}{\left\|\mathbf{A}\mathbf{u}_j\right\|} \frac{\left\|\mathbf{A}\mathbf{u}_j\right\| \sin\theta_j}{\left\|\bar{\mathbf{y}}_{j+1}\right\|} \end{aligned} \tag{7.104}$$

Comparing the above equation with the following equation, we can arrive at the expression for sin θ and cos θ:

$$\mathbf{y}_{j+1} = \mathbf{x}_n \cos\theta_{j+1} + \mathbf{u}_{j+1} \sin\theta_{j+1} \tag{7.105}$$

$$\begin{cases} \mathbf{u}_{j+1} = \dfrac{\mathbf{Au}_j}{\|\mathbf{Au}_j\|} \\ \cos\theta_{j+1} = \dfrac{\lambda_n \cos\theta_j}{\|\bar{\mathbf{y}}_{j+1}\|} \\ \sin\theta_{j+1} = \dfrac{\|\mathbf{Au}_j\| \sin\theta_j}{\|\bar{\mathbf{y}}_{j+1}\|} \end{cases} \tag{7.106}$$

This leads to the following relation:

$$\frac{\tan\theta_{j+1}}{\tan\theta_j} = \frac{\|\mathbf{Au}_j\|}{\lambda_n} \tag{7.107}$$

We note that $\mathbf{Au}_j$ is also orthogonal to $\mathbf{x}_n$.

$$(\mathbf{Au}_j)^T \mathbf{x}_n = \mathbf{u}_j^T \mathbf{A}\mathbf{x}_n = \lambda_j \mathbf{u}_j^T \mathbf{x}_n = 0 \tag{7.108}$$

Because $\mathbf{u}_j$ is orthogonal to $\mathbf{x}_n$, then the largest value $\mathbf{u}_j$ can take is $\mathbf{x}_{n-1}$. This leads to the following relation:

$$\max_{\mathbf{u}_j} \|\mathbf{Au}_j\| = \|\lambda_{n-1}\mathbf{x}_{n-1}\| = \lambda_{n-1} \tag{7.109}$$

$$\frac{\tan\theta_{j+1}}{\tan\theta_j} = \frac{\|\mathbf{Au}_j\|}{\lambda_n} \le \frac{\lambda_{n-1}}{\lambda_n} < 1 \tag{7.110}$$

This proves that as the iteration number increases, the angle θ keeps on decreasing, leading to the convergence condition stated in Equation 7.101. The above equation also shows that the rate of convergence is governed by the ratio of $(\lambda_{n-1}/\lambda_n)$. We can also show the convergence of the norm of the vector to the largest eigenvalue.

$$\begin{aligned} \lim_{j\to\infty} \|\bar{\mathbf{y}}_{j+1}\|_2 &= \lim_{j\to\infty} \|\lambda_n \mathbf{x}_n \cos\theta_j + \mathbf{Au}_j \sin\theta_j\|_2 \\ &= \lim_{j\to\infty} \left[(\lambda_n \cos\theta_j)^2 + (\|\mathbf{Au}_j\| \sin\theta_j)^2 \right]^{1/2} \\ &= \lambda_n \end{aligned} \tag{7.111}$$

With this relation, we have proven the convergence property of power method with normalization. It should be emphasized that the condition for the power method to converge to the highest eigenvector and eigenvalue is to let the initial trial vector have a component in the direction of the highest eigenvector.

$$\mathbf{y}_0^T \mathbf{x}_n \ne 0 \tag{7.112}$$

This condition can be easily satisfied if we use a vector with all terms being 1.

On the other hand, if we happen to select the initial trial vector in such a manner that it is orthogonal to the highest eigenvector, then from convergence analysis, we can conclude that the power method will converge to the next highest eigenvalue and eigenvector, λ_{n-1} and $\mathbf{x}_{n-1}$. This observation will provide the basic idea (deflation) in computing other eigenvalue and eigenvectors using the power method.

It can be seen that normalization has no effect on the convergence property of the power method. Then one may ask why we want to perform normalization at all? The reason is that as the iteration numbers increase, the nonzero components in the iteration vector may become extremely large values if $|\lambda_n| > 1$, or they can approach zero if $|\lambda_n| < 1$. Without normalization, we may encounter *overflow* problem.

7.4.2 Inverse power method

The power method is an iterative method for determining the largest eigenvalue and eigenvector of a matrix. For many problems in computational mechanics and structural dynamics, we are more interested in the smallest or the smallest few eigenvalues and eigenvectors. We observe from linear algebra that if matrix $\mathbf{A} \in R^{n\times n}$ is symmetric positive definite and nonsingular, and its eigenvalues and corresponding eigenvectors are $\lambda_1 < \lambda_2 < \ldots < \lambda_n$ and $\mathbf{x}_1, \mathbf{x}_2, \ldots, \mathbf{x}_n$, then $\mathbf{A}^{-1}$ has the following eigenvalues:

$$\frac{1}{\lambda_n} < \frac{1}{\lambda_{n-1}} < \cdots < \frac{1}{\lambda_1} \tag{7.113}$$

and the corresponding eigenvectors are $\mathbf{x}_n, \mathbf{x}_{n-1}, \ldots, \mathbf{x}_1$. This shows that we can compute the smallest eigenvalue and eigenvector by using the power method on $\mathbf{A}^{-1}$. This is called the inverse power method, which is the power method but applied on the inverse of the matrix. Starting with an initial vector $\mathbf{y}_0$, we can describe the steps in the method as follows.

$$\begin{aligned}
&\mathbf{y}_0 \\
&\mathbf{A}\bar{\mathbf{y}}_1 = \mathbf{y}_0 \;\Rightarrow\; \text{solve for } \bar{\mathbf{y}}_1 \\
&\mathbf{y}_1 = \frac{1}{\|\bar{\mathbf{y}}_1\|}\bar{\mathbf{y}}_1 \\
&\quad\vdots \\
&\mathbf{A}\bar{\mathbf{y}}_j = \mathbf{y}_{j-1} \;\Rightarrow\; \text{solve for } \bar{\mathbf{y}}_j \\
&\mathbf{y}_j = \frac{1}{\|\bar{\mathbf{y}}_j\|}\bar{\mathbf{y}}_j \\
&\quad\vdots
\end{aligned} \tag{7.114}$$

This procedure involves the solution of system equations using Gauss-elimination or other solution procedures. We can obtain convergence properties directly from the power method. If the initial trial vector has a component along the first eigenvector, then we have the following convergence property:

$$\text{If } \mathbf{y}_0^T \mathbf{x}_1 \neq 0 \text{ then } \lim_{j\to\infty} \begin{cases} \mathbf{y}_j \to \mathbf{x}_1 \\ \|\bar{\mathbf{y}}_j\|_2 \to \dfrac{1}{\lambda_1} \end{cases} \tag{7.115}$$

The convergence rate is governed by the ratio $\lambda_{n-1}(\mathbf{A}^{-1})/\lambda_n(\mathbf{A}^{-1}) = \lambda_1/\lambda_2$. Similarly, this convergence property can be illustrated using the inverse power method without normalization, similar to the convergence property of the power method illustrated in Equations 7.94 through 7.97.

$$\mathbf{y}_j = \mathbf{A}^{-j}\mathbf{y}_0 \tag{7.116}$$

$$\mathbf{y}_0 = \sum_{k=1}^{n} \alpha_k \mathbf{x}_k \tag{7.117}$$

$$\begin{aligned}\mathbf{y}_j &= \mathbf{A}^{-j}\sum_{k=1}^{n} \alpha_k \mathbf{x}_k = \sum_{k=1}^{n} \alpha_k \mathbf{A}^{-j}\mathbf{x}_k = \sum_{k=1}^{n} \alpha_k \left(\frac{1}{\lambda_k}\right)^j \mathbf{x}_k \\ &= \left(\frac{1}{\lambda_1}\right)^j \left[\alpha_1 \mathbf{x}_1 + \sum_{k=2}^{n} \alpha_k \left(\frac{\lambda_1}{\lambda_k}\right)^j \mathbf{x}_k\right]\end{aligned} \tag{7.118}$$

$$\lim_{j \to \infty} \mathbf{y}_j \to \left(\frac{1}{\lambda_1}\right)^j \alpha_1 \mathbf{x}_1 \tag{7.119}$$

This method also breaks down for multiple eigenvalues.

7.4.3 Acceleration of the inverse power method

As has just been discussed, the convergence of the power method depends on the ratio of $(\lambda_{n-1}/\lambda_n)$ and the inverse power methods on (λ_1/λ_2). Both methods break down when encountered with multiple eigenvalues in the system. In reality, it is inevitable that we may have pathologically close eigenvalues or a cluster of close eigenvalues on the round-off error level. If such cases occur, then the convergence slows down dramatically.

Suppose that after j iterations in the inverse power method, we have obtained $\|\mathbf{y}_j\|$, which is very close to $1/\lambda_1$. Now if we perform a shift $\mu = \|\mathbf{y}_j\|$ on $\mathbf{A}$, then all eigenvalues of $\mathbf{A}$ are also shifted by μ.

$$\bar{\mathbf{A}} = \mathbf{A} - \mu\mathbf{I} \tag{7.120}$$

$$\begin{cases}\bar{\lambda}_1 = \lambda_1 - \mu \\ \bar{\lambda}_2 = \lambda_2 - \mu\end{cases} \tag{7.121}$$

We note the first two eigenvalues of the shifted matrix are $\bar{\lambda}_1 \ll \bar{\lambda}_2$, which leads to the following relation:

$$\frac{\bar{\lambda}_1}{\bar{\lambda}_2} \ll \frac{\lambda_1}{\lambda_2} \tag{7.122}$$

This results in faster rate of convergence. This shows that shift improves the convergence rate of the inverse power method. However, we can see that shift may have far less effect on the convergence of the power method.

Now the question is how we select the shift μ such that the convergence rate in the inverse power method can be greatly improved. From our early discussion on Rayleigh quotient, we know that if we compute the Rayleigh quotient at iteration j in the inverse power method,

$$\rho(\mathbf{y}_j) = \frac{\mathbf{y}_j^T \mathbf{A} \mathbf{y}_j}{\mathbf{y}_j^T \mathbf{y}_j} = \mathbf{y}_j^T \mathbf{y}_{j-1} \tag{7.123}$$

where $\mathbf{y}_j = \bar{\mathbf{y}}_j / \|\bar{\mathbf{y}}_j\|$, and if $\mathbf{y}_j$ is close to $\mathbf{x}_j$, then $\rho(\mathbf{y}_j)$ is an order closer to λ_j. Therefore, the best shift we can get is

$$\mu = \rho(\mathbf{y}_j) \tag{7.124}$$

In practice, the Rayleigh quotient shift usually results in very fast convergence in the inverse power method.

7.4.4 Deflation and the inverse power method

The standard procedure in the inverse power method computes only the smallest eigenvalue and eigenvector. Next, we want to compute the second lowest eigenvalue and eigenvector. We observe that if the initial vector $\mathbf{y}_0$ is orthogonal to $\mathbf{x}_1$, $\mathbf{y}_0^T \mathbf{x}_1 = 0$, then we in fact can obtain λ_2 and $\mathbf{x}_2$ from the inverse power method. This indicates that initial vector should be on the hyperplane of the first eigenvector. To accomplish this, we carry out the following replacement operation:

$$\mathbf{y}_0 \leftarrow (\mathbf{I} - \mathbf{x}_1 \mathbf{x}_1^T) \mathbf{y}_0 \tag{7.125}$$

The matrix in the equation is the projection matrix on the hyperplane of $\mathbf{x}_1$. This process is called deflation.

With the deflation process incorporated, we can describe the procedure for computing the eigenvalue and eigenvector, λ_2 and $\mathbf{x}_2$, in the inverse power method as follows:

$$\begin{aligned}
&\mathbf{y}_0 \leftarrow (\mathbf{I} - \mathbf{x}_1 \mathbf{x}_1^T) \mathbf{y}_0 \\
&\mathbf{A} \bar{\mathbf{y}}_1 = \mathbf{y}_0 \Rightarrow \text{solve for } \bar{\mathbf{y}}_1 \\
&\mathbf{y}_1 = \frac{1}{\|\bar{\mathbf{y}}_1\|} \bar{\mathbf{y}}_1 \\
&\vdots \\
&\mathbf{A} \bar{\mathbf{y}}_j = \mathbf{y}_{j-1} \Rightarrow \text{solve for } \bar{\mathbf{y}}_j \\
&\bar{\mathbf{y}}_j \leftarrow (\mathbf{I} - \mathbf{x}_1 \mathbf{x}_1^T) \bar{\mathbf{y}}_j \\
&\mathbf{y}_j = \frac{1}{\|\bar{\mathbf{y}}_j\|} \bar{\mathbf{y}}_j \\
&\vdots
\end{aligned} \tag{7.126}$$

In this procedure, the deflation process needs not be enforced for all the iterations. It should be activated at certain stages to eliminate the round-off error with $\mathbf{x}_1$ components crept into the vector $\mathbf{y}_j$.

The process will converge to $\mathbf{x}_2$ if the initial vector has no component of $\mathbf{x}_1$ and some component of $\mathbf{x}_2$.

$$\text{If } \mathbf{y}_0^T \mathbf{x}_2 \neq 0 \text{ then } \lim_{j \to \infty} \begin{cases} \mathbf{y}_j \to \mathbf{x}_2 \\ \|\bar{\mathbf{y}}_j\|_2 \to \dfrac{1}{\lambda_2} \end{cases} \tag{7.127}$$

The rate of convergence is governed by the ratio (λ_2/λ_3).

Similarly, after we find the first two eigenpairs $(\lambda_1, \mathbf{x}_1)$ and $(\lambda_2, \mathbf{x}_2)$, we can compute the next eigenpair $(\lambda_3, \mathbf{x}_3)$ using a deflation procedure. We apply the following deflation to the initial vector and within the iteration.

In a general case, if we have obtained the first m eigenvalues $\lambda_1, \lambda_2, \ldots, \lambda_m$ and their corresponding eigenvectors $\mathbf{x}_1, \mathbf{x}_2, \ldots, \mathbf{x}_m$, we can compute the next eigenpair $(\lambda_{m+1}, \mathbf{x}_{m+1})$ through a deflation process. We first compute a projection $\mathbf{P}_m$ matrix that projects to the hyperplane of subspace spanned by $\{\mathbf{x}_1, \mathbf{x}_2, \ldots, \mathbf{x}_m\}$.

$$\begin{aligned} \mathbf{P}_m &= (\mathbf{I} - \mathbf{X}_m \mathbf{X}_m^T) \\ \mathbf{X}_m &= \left[\mathbf{x}_1 \middle| \mathbf{x}_2 \middle| \cdots \middle| \mathbf{x}_m\right] \end{aligned} \tag{7.128}$$

The initial trial vector can then be determined as follows:

$$\mathbf{y}_0 \leftarrow \mathbf{P}_m \mathbf{y}_0 \tag{7.129}$$

Within the iteration process of the inverse power method, we perform selectively the deflation process.

$$\bar{\mathbf{y}}_j \leftarrow \mathbf{P}_m \bar{\mathbf{y}}_j \tag{7.130}$$

This process converges to eigenpair $(\lambda_{m+1}, \mathbf{x}_{m+1})$.

$$\lim_{j \to \infty} \begin{cases} \mathbf{y}_j \to \mathbf{x}_{m+1} \\ \left\| \bar{\mathbf{y}}_j \right\|_2 \to \dfrac{1}{\lambda_{m+1}} \end{cases} \tag{7.131}$$

The deflation scheme described so far is sometimes called *the vector deflation* process, in which the size of the original matrix **A** after deflation does not change. After finding the first eigenpair $(\lambda_1, \mathbf{x}_1)$ from the original matrix, we may perform a deflation scheme that reduces the size of the matrix by one, while retaining all its remaining eigenvalues. This process is referred to as *matrix deflation*.

7.4.5 Matrix deflation

The idea behind matrix deflation is to find a n × n nonsingular matrix **H** such that the similarity transformation on the n × n matrix **A**, is of the following form:

$$\mathbf{HAH}^{-1} = \mathbf{HA}_{(1)}\mathbf{H}^{-1} = \begin{bmatrix} \lambda_1 & 0 \\ 0 & \mathbf{A}_{(2)} \end{bmatrix} \tag{7.132}$$

where $\mathbf{A}_{(2)}$ is an (n−1) × (n−1) matrix. Since the similarity transformation does not change the eigenvalues of the original matrix, we have

$$\left|\mathbf{A} - \lambda\mathbf{I}\right| = \left|\mathbf{HAH}^{-1} - \lambda\mathbf{I}\right| = (\lambda_1 - \lambda)\left|\mathbf{A}_{(2)} - \lambda\mathbf{I}\right| \tag{7.133}$$

It thus follows that the rest of the (n−1) eigenvalues $\lambda_2, \ldots, \lambda_n$ are roots of the characteristic polynomial $\left|\mathbf{A}_{(2)} - \lambda\mathbf{I}\right|$ and the solution of the following eigenvalue problem:

$$\mathbf{A}_{(2)}\mathbf{x} = \lambda\mathbf{x} \tag{7.134}$$

Next, we discuss how to find such a transformation matrix that has the above desired property. We note that the first column of $\mathbf{HAH}^{-1}$ is $\lambda_1\mathbf{e}_1$, with $\mathbf{e}_1$ being the first column of the identity matrix, a unit directional vector.

$$\mathbf{HAH}^{-1}\mathbf{e}_1 = \lambda_1\mathbf{e}_1 \tag{7.135}$$

If we consider the application of similarity transformation as a successive operation, we have the first transformation as follows:

$$\mathbf{H}_1\mathbf{AH}_1^{-1}\mathbf{e}_1 = \lambda_1\mathbf{e}_1 \tag{7.136}$$

$$\mathbf{A}(\mathbf{H}_1^{-1}\mathbf{e}_1) = \lambda_1(\mathbf{H}_1^{-1}\mathbf{e}_1) \tag{7.137}$$

Comparing the above relation with the standard eigenvalue equation

$$\mathbf{Ax}_1 = \lambda_1\mathbf{x}_1 \tag{7.138}$$

we obtain the following relation:

$$\mathbf{H}_1^{-1}\mathbf{e}_1 = \mathbf{x}_1 \tag{7.139}$$

$$\mathbf{H}_1\mathbf{x}_1 = \mathbf{e}_1 = \begin{Bmatrix} 1 \\ 0 \\ \vdots \\ 0 \end{Bmatrix} \tag{7.140}$$

This is an interesting observation on the kind of matrix $\mathbf{H}_1$ should be. On the other hand, we observe from the similarity transformation that symmetry in the resulting reduced matrix is destroyed. However, if we impose the condition of orthogonality on the transformation matrix $\mathbf{H}_1$, then we have the following:

$$\mathbf{H}_1^{-1} = \mathbf{H}_1^{T} \tag{7.141}$$

$$\mathbf{H}_1\mathbf{AH}_1^{-1} = \mathbf{H}_1\mathbf{AH}_1^{T} \tag{7.142}$$

The symmetry in the resulting reduced matrix is now conserved. This is evidently very important as otherwise the loss of symmetry may result in programming complexity and extra computational time.

By imposing restriction on orthogonality of the transformation matrix, we can use either Givens rotation matrix or Householder reflection matrix as $\mathbf{H}_1$. Generally speaking, the Householder reflection transformation is more stable than Givens rotation transformation. If we take as a reflection matrix, which is both symmetric and orthogonal, we have

$$\mathbf{H}_1 = \left(\mathbf{I} - \frac{2}{\bar{\mathbf{u}}^T\bar{\mathbf{u}}}\bar{\mathbf{u}}\bar{\mathbf{u}}^T\right) \tag{7.143}$$

We can use the following numerically stable scheme for determining the vector $\bar{\mathbf{u}}$, in which $x_{1,1}$ is the first term in the vector $\mathbf{x}_1$:

$$\bar{\mathbf{u}} = \mathbf{x}_1 + \text{sign}(x_{1,1})\|\mathbf{x}_1\|\mathbf{e}_1 \tag{7.144}$$

It appears that matrix deflation is a viable process in computing the eigenvalues other than the lowest eigenvalue and eigenvector. For sparse or full matrix, it will not encounter any numerical or programming problems. However, with banded matrix, one needs to keep track of the banded structure during the transformation process. This apparently would be a drawback.

7.4.6 LR method

The LR method or Rutishauser's algorithm (Rutishauser, 1958) is based on a simple observation with triangular decomposition of matrix **A** in the Gauss-elimination method. If we perform triangular decomposition on **A**, we have

$$\mathbf{A} = \mathbf{LR} \tag{7.145}$$

where:
- **L** is a unit lower triangular matrix
- **R** is an upper triangular matrix

If we simply reverse the order of the matrices, we have the following:

$$\begin{aligned} \mathbf{RL} &= \left(\mathbf{L}^{-1}\mathbf{L}\right)\mathbf{RL} \\ &= \mathbf{L}^{-1}\left(\mathbf{LR}\right)\mathbf{L} \\ &= \mathbf{L}^{-1}\mathbf{AL} \end{aligned} \tag{7.146}$$

This shows that matrix (**RL**) is similar to matrix **A** so that it will have the same eigenvalues of **A**. As a consequence, this observation serves as the basis of the LR method. Since $\mathbf{R} = \mathbf{L}^{-1}\mathbf{A}$, the LR method can be described in the following procedure:

$$\begin{aligned} &\begin{cases} \mathbf{A}_1 = \mathbf{A} \\ \mathbf{A}_1 = \mathbf{L}_1\mathbf{R}_1, \text{ triangular decomposition} \end{cases} \\ &\begin{cases} \mathbf{A}_2 = \mathbf{R}_1\mathbf{L}_1, \text{ similarity transformation} \\ \mathbf{A}_2 = \mathbf{L}_2\mathbf{R}_2 \end{cases} \\ &\quad\vdots \\ &\begin{cases} \mathbf{A}_k = \mathbf{R}_{k-1}\mathbf{L}_{k-1} \\ \mathbf{A}_k = \mathbf{L}_k\mathbf{R}_k \end{cases} \\ &\quad\vdots \end{aligned} \tag{7.147}$$

We note that this method is computationally time consuming because it requires performing triangular decomposition at each iteration. Also, for symmetric matrix **A**, similarity transformation destroys the symmetry. It can be proven that the convergence property of this method is as follows, where x denotes nonzero terms.

$$\lim_{k\to\infty}\begin{cases}\mathbf{L}_k \to \mathbf{I} \\ \mathbf{R}_k \to \begin{bmatrix} \lambda_n & x & x & \cdots & x \\ & \lambda_{n-1} & x & \cdots & \vdots \\ & & \ddots & \ddots & x \\ & & & \lambda_2 & x \\ & & & & \lambda_1 \end{bmatrix}\end{cases} \tag{7.148}$$

This shows that the eigenvalues emerge on the diagonal of $\mathbf{A}_k$ or $\mathbf{R}_k$ matrix. One unique feature associated with the convergence property of this method is that it converges from both ends and progresses toward the eigenvalues in the middle of the diagonal terms. The convergence to the lowest eigenvalues is faster than to the largest eigenvalues.

To study the convergence of the LR method, we note that it can be expanded in the following form.

$$\mathbf{A}_{k+1} = \mathbf{R}_k\mathbf{L}_k = (\mathbf{L}_k^{-1}\mathbf{L}_k)\mathbf{R}_k\mathbf{L}_k = \mathbf{L}_k^{-1}\mathbf{A}_k\mathbf{L}_k \tag{7.149}$$

$$\begin{aligned}\mathbf{A}_{k+1} &= \mathbf{L}_k^{-1}\mathbf{A}_k\mathbf{L}_k \\ &= \mathbf{L}_k^{-1}(\mathbf{L}_{k-1}^{-1}\mathbf{A}_{k-1}\mathbf{L}_{k-1})\mathbf{L}_k \\ &\ \ \vdots \\ &= \mathbf{L}_k^{-1}\mathbf{L}_{k-1}^{-1}\ldots\mathbf{L}_1^{-1}\mathbf{A}\mathbf{L}_1\ldots\mathbf{L}_{k-1}\mathbf{L}_k\end{aligned} \tag{7.150}$$

$$\mathbf{L}_1\ldots\mathbf{L}_{k-1}\mathbf{L}_k\mathbf{A}_{k+1} = \mathbf{A}\mathbf{L}_1\ldots\mathbf{L}_{k-1}\mathbf{L}_k \tag{7.151}$$

Now, we define a lower triangular matrix $\mathbf{T}_k$ and an upper triangular matrix $\mathbf{U}_k$.

$$\begin{cases}\mathbf{T}_k = \mathbf{L}_1\ldots\mathbf{L}_{k-1}\mathbf{L}_k \\ \mathbf{U}_k = \mathbf{R}_k\mathbf{R}_{k-1}\ldots\mathbf{R}_1\end{cases} \tag{7.152}$$

We note that the product of lower triangular matrices is also a lower triangular matrix, so is the case with upper triangular matrices. Now, we can rewrite Equation 7.151 as follows:

$$\mathbf{T}_k\mathbf{A}_{k+1} = \mathbf{A}\mathbf{T}_k \tag{7.153}$$

In the following operations, we are using the triangular decomposition relation $\mathbf{A}_k = \mathbf{L}_k \mathbf{R}_k$.

$$
\begin{aligned}
\mathbf{T}_k\mathbf{U}_k &= (\mathbf{L}_1 \ldots \mathbf{L}_{k-1}\mathbf{L}_k)(\mathbf{R}_k\mathbf{R}_{k-1} \ldots \mathbf{R}_1) \\
&= (\mathbf{L}_1 \ldots \mathbf{L}_{k-1})\mathbf{A}_k(\mathbf{R}_{k-1} \ldots \mathbf{R}_1) \\
&= \mathbf{A}(\mathbf{L}_1 \ldots \mathbf{L}_{k-1})(\mathbf{R}_{k-1} \ldots \mathbf{R}_1) \\
&= \mathbf{A}(\mathbf{L}_1 \ldots \mathbf{L}_{k-2})\mathbf{A}_{k-1}(\mathbf{R}_{k-2} \ldots \mathbf{R}_1) \\
&= \mathbf{A}^2(\mathbf{L}_1 \ldots \mathbf{L}_{k-2})(\mathbf{R}_{k-2} \ldots \mathbf{R}_1) \\
&\;\;\vdots \\
&= \mathbf{A}^k
\end{aligned}
\tag{7.154}
$$

Wilkinson (1965) has shown that as $\lim_{k\to\infty} \mathbf{L}_k \to \mathbf{I}$, the lower triangular matrix $\mathbf{T}_k$ approaches an invariant matrix $\mathbf{T}$.

$$
\lim_{k\to\infty} \mathbf{T}_k \to \mathbf{T} \tag{7.155}
$$

Next, we define a matrix $\mathbf{U}$ such that $\mathbf{X} = \mathbf{T}\mathbf{U}$, where $\mathbf{X}$ is the matrix of the eigenvectors of A, $\mathbf{A}\mathbf{X} = \mathbf{X}\,\mathbf{\Lambda}$.

$$
\begin{aligned}
\mathbf{U} &= \mathbf{T}^{-1}\,\mathbf{X} \\
\mathbf{T} &= \mathbf{X}\mathbf{U}^{-1} \\
\mathbf{T}^{-1} &= \mathbf{U}\mathbf{X}^{-1}
\end{aligned}
\tag{7.156}
$$

$$
\begin{aligned}
\lim_{k\to\infty} \mathbf{A}_{k+1} &= \lim_{k\to\infty}\left(\mathbf{T}_k^{-1}\mathbf{A}\mathbf{T}_k\right) = \mathbf{T}^{-1}\mathbf{A}\mathbf{T} \\
&= \mathbf{U}\mathbf{X}^{-1}\mathbf{A}\mathbf{X}\mathbf{U}^{-1} \\
&= \mathbf{U}\mathbf{\Lambda}\mathbf{U}^{-1}
\end{aligned}
\tag{7.157}
$$

In the right-hand side of the above equation, $\mathbf{U}$ is a unit upper triangular matrix, so is $\mathbf{U}^{-1}$; and $\mathbf{\Lambda}$ is a diagonal matrix. Therefore, the result of this multiplication is also an upper triangular matrix of the following form:

$$
\mathbf{U}\mathbf{\Lambda}\mathbf{U}^{-1} = \begin{bmatrix}
\lambda_n & x & x & \cdots & x \\
 & \lambda_{n-1} & x & \cdots & \vdots \\
 & & \ddots & \ddots & x \\
 & & & \lambda_2 & x \\
 & & & & \lambda_1
\end{bmatrix}
\tag{7.158}
$$

This in principle illustrates the convergence property of the LR method. However, in the current form, this method is not of much practical value in computational mechanics because it destroys symmetry and the banded property of the stiffness matrices in finite element systems. Nevertheless, the basic idea behind this method serves as the prelude to some of the more useful methods such as the QR and QL methods, which will be discussed in Section 7.5.

7.5 QR METHOD

7.5.1 QR algorithm

We have observed in performing orthogonal transformation that, for any real matrix **A**, we can always find an orthogonal matrix $\mathbf{Q}^T$ such that the matrix product $\mathbf{Q}^T\mathbf{A}$ is an upper triangular matrix **R**.

$$\mathbf{Q}^T\mathbf{A} = \mathbf{R} \tag{7.159}$$

$$\mathbf{A} = \mathbf{QR} \tag{7.160}$$

This is usually called orthogonal decomposition of **A**. Since **Q** is an orthogonal matrix, then $\mathbf{Q}^{-1} = \mathbf{Q}^T$. We can see that this orthogonal decomposition can be realized by using Givens method or Householder method. In Givens method, we use a series of plane rotation matrices to obtain the matrix **Q**; and in Householder method, we use a series of reflection elementary matrices.

Similar to the LR method, for an $n \times n$ real matrix **A**, the basic idea behind the QR method for computing the eigenvalues of **A** can be described in the following iterative procedure:

$$\begin{aligned}
&\begin{cases} \mathbf{A}_1 = \mathbf{A} \\ \mathbf{A}_1 = \mathbf{Q}_1\mathbf{R}_1 \end{cases} \\
&\begin{cases} \mathbf{A}_2 = \mathbf{R}_1\mathbf{Q}_1 \\ \mathbf{A}_2 = \mathbf{Q}_2\mathbf{R}_2 \end{cases} \\
&\quad\vdots \\
&\begin{cases} \mathbf{A}_k = \mathbf{R}_{k-1}\mathbf{Q}_{k-1} \\ \mathbf{A}_k = \mathbf{Q}_k\mathbf{R}_k \end{cases} \\
&\quad\vdots
\end{aligned} \tag{7.161}$$

From the computations carried out in the above iterative process, the following recursive relation can be verified:

$$\begin{aligned}
\mathbf{A}_{k+1} &= \mathbf{R}_k\mathbf{Q}_k \\
&= (\mathbf{Q}_k^T\mathbf{Q}_k)\mathbf{R}_k\mathbf{Q}_k \\
&= \mathbf{Q}_k^T\mathbf{A}_k\mathbf{Q}_k
\end{aligned} \tag{7.162}$$

This implies that matrices $\mathbf{A}_{k+1}$ and $\mathbf{A}_k$ are similar and therefore, their eigenvalues are the same. There are also other interesting properties associated with the QR decomposition. We can observe that if **A** is a symmetric banded matrix, then the subsequently generated matrices, $\mathbf{A}_2, \ldots, \mathbf{A}_k$ are also symmetric banded matrices, and the bandwidth is not changed during the QR process. Specifically, if **A** is a symmetric tridiagonal matrix, then the subsequently generated matrices are also symmetric tridiagonal matrices. This is a very important property with practical implications.

Similar to the convergence property of LR method, it can be proven that the convergence property of the QR method is as follows, where x denotes nonzero terms in the upper triangular matrix:

$$\lim_{k\to\infty}\begin{cases} Q_k \to I \\ R_k \to \begin{bmatrix} \lambda_n & x & \cdots & x \\ & \ddots & \ddots & \vdots \\ & & \lambda_2 & x \\ & & & \lambda_1 \end{bmatrix} \end{cases} \tag{7.163}$$

Of course, if **A** is symmetric, then $\mathbf{R}_k$ is also symmetric, implying that it must be a diagonal matrix.

$$\lim_{k\to\infty} \mathbf{R}_k \to \mathrm{diag}[\lambda_n,\ldots,\lambda_1] \tag{7.164}$$

Now we briefly discuss two ways in proving the convergence property, which would provide some insight into the workings of this method. First, we describe Wilkinson's proof.

From the recursive relation in QR process, we can obtain the expanded relation as

$$\begin{aligned} \mathbf{A}_{k+1} &= \mathbf{Q}_k^T\mathbf{A}_k\mathbf{Q}_k \\ &= \mathbf{Q}_k^T\mathbf{Q}_{k-1}^T\mathbf{A}_{k-1}\mathbf{Q}_{k-1}\mathbf{Q}_k \\ &\ \ \vdots \\ &= \mathbf{Q}_k^T\cdots\mathbf{Q}_1^T\mathbf{A}\mathbf{Q}_1\cdots\mathbf{Q}_k \\ &= (\mathbf{Q}_1\cdots\mathbf{Q}_k)^T\mathbf{A}(\mathbf{Q}_1\cdots\mathbf{Q}_k) \end{aligned} \tag{7.165}$$

We define the orthogonal matrix $\mathbf{P}_k$ and upper triangular matrix $\mathbf{U}_k$.

$$\begin{cases} \mathbf{P}_k = \mathbf{Q}_1\cdots\mathbf{Q}_k \\ \mathbf{U}_k = \mathbf{R}_k\cdots\mathbf{R}_1 \end{cases} \tag{7.166}$$

Therefore, Equation 7.165 can be written as follows:

$$\mathbf{A}_{k+1} = \mathbf{P}_k^T\mathbf{A}\mathbf{P}_k \tag{7.167}$$

$$\mathbf{P}_k\mathbf{A}_{k+1} = \mathbf{A}\mathbf{P}_k \tag{7.168}$$

It follows that through successive substitution, we can arrive at the following relation:

$$\begin{aligned} \mathbf{P}_k\mathbf{U}_k &= (\mathbf{Q}_1\ldots\mathbf{Q}_k)(\mathbf{R}_k\ldots\mathbf{R}_1) \\ &= (\mathbf{Q}_1\ldots\mathbf{Q}_{k-1})\mathbf{A}_k(\mathbf{R}_{k-1}\ldots\mathbf{R}_1) \\ &= \mathbf{A}(\mathbf{Q}_1\ldots\mathbf{Q}_{k-1})(\mathbf{R}_{k-1}\ldots\mathbf{R}_1) \\ &\ \ \vdots \\ &= \mathbf{A}^k \end{aligned} \tag{7.169}$$

This relation means that $\mathbf{P}_k\ \mathbf{U}_k$ is the QR decomposition of matrix $\mathbf{A}^k$.

Now we perform spectral decomposition on $\mathbf{A}^k$.

$$\mathbf{A}^k = \mathbf{X}\Lambda^k\mathbf{X}^{-1} = \mathbf{X}\Lambda^k\mathbf{Y} \tag{7.170}$$

$\mathbf{X}$ = eigenvector matrix of $\mathbf{A}$, ($\mathbf{AX} = \mathbf{X}\Lambda$) and $\mathbf{Y}$ is the inverse of the eigenvector matrix. We also assume that matrix $\mathbf{A}$ has n distinct eigenvalues, that is, $\lambda_1 < \ldots < \lambda_n$. Next, we perform orthogonal decomposition on $\mathbf{X}$ and triangular decomposition on $\mathbf{Y}$.

$$\begin{cases} \mathbf{X} = \mathbf{QR} \\ \mathbf{Y} = \mathbf{X}^{-1} = \mathbf{LU} \end{cases} \tag{7.171}$$

$\mathbf{L}$ is a unit lower triangular matrix. We then substitute these relations in Equation 7.170.

$$\begin{aligned} \mathbf{A}^k &= \mathbf{X}\Lambda^k\mathbf{Y} = \mathbf{QR}\Lambda^k\mathbf{LU} \\ &= \mathbf{QR}(\Lambda^k\mathbf{L}\Lambda^{-k})\Lambda^k\mathbf{U} \end{aligned} \tag{7.172}$$

$$\Lambda^k\mathbf{L}\Lambda^{-k} = \begin{bmatrix} \lambda_1^k & & \\ & \ddots & \\ & & \lambda_n^k \end{bmatrix} \begin{bmatrix} 1 & & \\ \vdots & \ddots & \\ \cdots & \cdots & 1 \end{bmatrix} \begin{bmatrix} \lambda_1^{-k} & & \\ & \ddots & \\ & & \lambda_n^{-k} \end{bmatrix} \tag{7.173}$$

Obviously, $\Lambda^k\mathbf{L}\Lambda^{-k}$ is a unit lower triangular matrix with unit diagonal terms. The above matrix multiplication can be expressed as follows:

$$\Lambda^k\mathbf{L}\Lambda^{-k} = \mathbf{I} + \mathbf{E}_k \tag{7.174}$$

In this equation $\mathbf{E}$ is a strictly lower triangular matrix. We can verify the following relation:

$$\lim_{k\to\infty} \mathbf{E}_k \to 0 \tag{7.175}$$

Substituting Equation 7.174 into Equation 7.172, we arrive at the following relation:

$$\begin{aligned} \mathbf{A}^k &= \mathbf{QR}(\mathbf{I} + \mathbf{E}_k)\Lambda^k\mathbf{U} \\ &= \mathbf{Q}(\mathbf{I} + \mathbf{RE}_k\mathbf{R}^{-1})\mathbf{R}\Lambda^k\mathbf{U} \\ &= \mathbf{Q}(\mathbf{I} + \bar{\mathbf{E}}_k)\mathbf{R}\Lambda^k\mathbf{U} \end{aligned} \tag{7.176}$$

$$\bar{\mathbf{E}}_k = \mathbf{RE}_k\mathbf{R}^{-1} \tag{7.177}$$

We can verify the following expression based on Equation 7.175:

$$\lim_{k\to\infty} \bar{\mathbf{E}}_k \to 0 \tag{7.178}$$

Now the matrix $\mathbf{I} + \bar{\mathbf{E}}_k$ can be factorized by orthogonal decomposition.

$$\mathbf{I} + \bar{\mathbf{E}}_k = \bar{\mathbf{Q}}_k\bar{\mathbf{R}}_k \tag{7.179}$$

Since $\lim_{k\to\infty}(\mathbf{I}+\bar{\mathbf{E}}_k) \to \mathbf{I}$, then we have the following:

$$\lim_{k\to\infty}\begin{cases}\bar{\mathbf{Q}}_k \to \mathbf{I}\\ \bar{\mathbf{R}}_k \to \mathbf{I}\end{cases} \quad (7.180)$$

Finally, we arrive at the following expression:

$$\mathbf{A}^k = (\mathbf{Q}\bar{\mathbf{Q}}_k)(\bar{\mathbf{R}}_k\mathbf{R}\Lambda^k\mathbf{U}) = \mathbf{P}_k\mathbf{U}_k \quad (7.181)$$

We observe that the product of orthogonal matrices of $(\mathbf{Q}\bar{\mathbf{Q}}_k)$ is an orthogonal matrix, and the matrix $(\bar{\mathbf{R}}_k\mathbf{R}\Lambda^k\mathbf{U})$ is upper triangular. Since the orthogonal decomposition is unique, we have the following:

$$\mathbf{P}_k = \mathbf{Q}\bar{\mathbf{Q}}_k \quad (7.182)$$

$$\begin{aligned}\lim_{k\to\infty}\mathbf{P}_k &= \lim_{k\to\infty}\mathbf{Q}\bar{\mathbf{Q}}_k\\ &= \mathbf{Q}\lim_{k\to\infty}\bar{\mathbf{Q}}_k \to \mathbf{Q}\end{aligned} \quad (7.183)$$

Since $\mathbf{P}_k = \mathbf{Q}_1 \ldots \mathbf{Q}_k$, the above relation means that each additional orthogonal matrix calculated from QR algorithm would approach the identity matrix.

$$\lim_{k\to\infty}\mathbf{Q}_k \to \mathbf{I} \quad (7.184)$$

Recalling that $\mathbf{X} = \mathbf{Q}\,\mathbf{R}$, we have $\mathbf{Q} = \mathbf{X}\,\mathbf{R}^{-1}$.

$$\mathbf{Q}^T = \mathbf{Q}^{-1} = \mathbf{R}^{-T}\mathbf{X}^T = \mathbf{R}\mathbf{X}^{-1} \quad (7.185)$$

We can use this equation in the limit of Equation 7.167

$$\begin{aligned}\lim_{k\to\infty}\mathbf{A}_{k+1} &= \lim_{k\to\infty}\mathbf{P}_k^T\mathbf{A}\mathbf{P}_k \to \mathbf{Q}^T\mathbf{A}\mathbf{Q}\\ &= \mathbf{R}\mathbf{X}^{-1}\mathbf{A}\mathbf{X}\mathbf{R}^{-1}\\ &= \mathbf{R}\Lambda\mathbf{R}^{-1}\end{aligned} \quad (7.186)$$

Obviously, $\mathbf{R}\Lambda\mathbf{R}^{-1}$ is an upper triangular matrix with eigenvalues as diagonal elements and x denotes nonzero off-diagonal terms.

$$\mathbf{R}\Lambda\mathbf{R}^{-1} = \begin{bmatrix}\lambda_n & x & \cdots & x\\ & \ddots & \ddots & \vdots\\ & & \lambda_2 & x\\ & & & \lambda_1\end{bmatrix} \quad (7.187)$$

If matrix $\mathbf{A}$ is symmetric, then $\mathbf{R}\Lambda\mathbf{R}^{-1}$ is a diagonal matrix.

$$\mathbf{R}\Lambda\mathbf{R}^{-1} = \Lambda = \mathrm{diag}[\lambda_n, \ldots, \lambda_1] \quad (7.188)$$

$$\lim_{k\to\infty} \mathbf{A}_k = \Lambda \tag{7.189}$$

The next proof of the QR method is due to Parlett (1980), which shows the relationship between the QR method and the Power method. We start with the recursive relationship from QR algorithm.

$$\mathbf{P}_{k-1}\mathbf{A}_k = \mathbf{A}\mathbf{P}_{k-1} \tag{7.190}$$

The orthogonal decomposition of $\mathbf{A}_k$ is substituted into the above equation.

$$\mathbf{P}_{k-1}\mathbf{Q}_k\mathbf{R}_k = \mathbf{A}\mathbf{P}_{k-1} \tag{7.191}$$

We note that $\mathbf{P}_{k-1} = \mathbf{Q}_1 \ldots \mathbf{Q}_{k-1}$ which leads to $\mathbf{P}_{k-1}\,\mathbf{Q}_k = \mathbf{P}_k$ that is substituted in the above equation.

$$\mathbf{P}_k\mathbf{R}_k = \mathbf{A}\mathbf{P}_{k-1} \tag{7.192}$$

In this equation $\mathbf{R}_k$ is an upper triangular matrix with diagonal terms $r_{11}, \ldots, r_{nn}$. We next expand the equation above and look at its first column, noting that $\mathbf{e}_1 = [1, 0, \ldots, 0]$ is the first unit directional vector

$$(r_{11})_k\,\mathbf{P}_k\mathbf{e}_1 = \mathbf{A}\mathbf{P}_{k-1}\mathbf{e}_1 \tag{7.193}$$

The above relation can be compared with the basic equations in the power method, given below, to arrive at equivalent relations for the QR method.

$$\begin{cases} \overline{\mathbf{y}}_k = \mathbf{A}\mathbf{y}_{k-1} \\ \mathbf{y}_k = \dfrac{1}{\left\|\overline{\mathbf{y}}_k\right\|}\,\overline{\mathbf{y}}_k \end{cases} \tag{7.194}$$

$$\begin{cases} \mathbf{y}_{k-1} = \mathbf{P}_{k-1}\mathbf{e}_1 \\ \overline{\mathbf{y}}_k = (r_{11})_k\,\mathbf{P}_k\mathbf{e}_1 \end{cases} \tag{7.195}$$

We recall that columns of orthogonal matrices are unit vectors.

$$\left\|\mathbf{P}_k\mathbf{e}_1\right\|_2 = \left\|\mathbf{P}_{k-1}\mathbf{e}_1\right\|_2 = \left[\left(\mathbf{P}_{k-1}\mathbf{e}_1\right)^T\left(\mathbf{P}_{k-1}\mathbf{e}_1\right)\right]^{1/2} = 1 \tag{7.196}$$

From the convergence property of the power method, we can obtain the following convergence to the highest eigenvector of $\mathbf{A}$:

$$\lim_{k\to\infty} \mathbf{y}_k = \lim_{k\to\infty} \mathbf{P}_k\mathbf{e}_1 \to \mathbf{x}_n \tag{7.197}$$

$$\lim_{k\to\infty} \mathbf{P}_k^T\mathbf{x}_n \to \mathbf{e}_1 + O\left(\frac{\lambda_{n-1}}{\lambda_n}\right)^k \tag{7.198}$$

We can also show the convergence of the first diagonal term of $\mathbf{A}_k$ to the highest eigenvalue.

$$\begin{aligned}\lim_{k\to\infty} \mathbf{A}_{k+1}\mathbf{e}_1 &= \lim_{k\to\infty} \mathbf{P}_k^T \mathbf{A} \mathbf{P}_k \mathbf{e}_1 \\ &= \lim_{k\to\infty} \mathbf{P}_k^T \mathbf{A} \mathbf{x}_n \\ &= \lim_{k\to\infty} \lambda_n \mathbf{P}_k^T \mathbf{x}_n \to \lambda_n \mathbf{e}_1\end{aligned} \tag{7.199}$$

This shows that the fundamental process behind the QR method is the power method.

We examined the first column in the iterative process. We can see that the first column of matrix $\mathbf{P}_k$ converges to the highest eigenvector and the first column of $\mathbf{A}_k$ converges to all zeroes below diagonal with the highest eigenvalue on the diagonal.

Next, we will show the convergence to the lowest eigenvalue and eigenvector by examining the last column in the iterative process. But first we need to manipulate Equation 7.192 to put in a suitable form; we take its transpose and pre and postmultiply it with $\mathbf{P}_{k-1}$ and $\mathbf{P}_k$.

$$\mathbf{R}_k^T \mathbf{P}_k^T = \mathbf{P}_{k-1}^T \mathbf{A} \tag{7.200}$$

$$\mathbf{P}_{k-1}(\mathbf{R}_k^T \mathbf{P}_k^T)\mathbf{P}_k = \mathbf{P}_{k-1}(\mathbf{P}_{k-1}^T \mathbf{A})\mathbf{P}_k \tag{7.201}$$

$$\mathbf{P}_{k-1}\mathbf{R}_k^T = \mathbf{A}\mathbf{P}_k \tag{7.202}$$

We note that $\mathbf{R}_k^T$ is a lower triangular matrix. Now, we examine the last column of this equation, and note that $\mathbf{e}_n = [0, 0, \ldots, 1]$.

$$(r_{nn})_k \mathbf{P}_{k-1}\mathbf{e}_n = \mathbf{A}\mathbf{P}_k\mathbf{e}_n \tag{7.203}$$

Comparing the above equation with the basic equation in the inverse power method, given below, leads to the equivalent relations for the QR method.

$$\begin{cases} \mathbf{y}_{k-1} = \mathbf{A}\overline{\mathbf{y}}_k \\ \mathbf{y}_k = \dfrac{1}{\|\overline{\mathbf{y}}_k\|}\overline{\mathbf{y}}_k \end{cases} \tag{7.204}$$

$$\begin{cases} \mathbf{y}_{k-1} = (r_{nn})_k \mathbf{P}_{k-1}\mathbf{e}_n \\ \overline{\mathbf{y}}_k = \mathbf{P}_k\mathbf{e}_n \end{cases} \tag{7.205}$$

Using the convergence property of the inverse power method, we arrive at the following expression for the convergence of the QR to the first eigenpair:

$$\lim_{k\to\infty} \mathbf{y}_k = \lim_{k\to\infty} \mathbf{P}_k\mathbf{e}_n \to \mathbf{x}_1 \tag{7.206}$$

$$\lim_{k\to\infty} \mathbf{P}_k^T \mathbf{x}_1 \to \mathbf{e}_n + O\left(\frac{\lambda_1}{\lambda_2}\right)^k \tag{7.207}$$

$$\lim_{k\to\infty} \mathbf{A}_{k+1}\mathbf{e}_n = \lim_{k\to\infty} \mathbf{P}_k^T \mathbf{A} \mathbf{P}_k \mathbf{e}_n$$
$$= \lim_{k\to\infty} \mathbf{P}_k^T \mathbf{A} \mathbf{x}_1 \tag{7.208}$$
$$= \lim_{k\to\infty} \lambda_1 \mathbf{P}_k^T \mathbf{x}_1 \to \lambda_1 \mathbf{e}_n$$

In summary, the convergence property of QR method for symmetric matrix can be stated as follows:

$$\begin{cases} \lim_{k\to\infty} \mathbf{A}_k \to \Lambda = \text{diag}\left[\lambda_n, \ldots, \lambda_1\right] \\ \lim_{k\to\infty} \mathbf{P}_k \to \mathbf{X} = \left[\mathbf{x}_n | \ldots | \mathbf{x}_1\right] \end{cases} \tag{7.209}$$

At this juncture, we have illustrated the convergence property of QR method and the inner working of the power method behind the algorithm. Because of the use of orthogonal decomposition, the QR method has many numerical features that are important in computational mechanics, such as the conservation of symmetry and bandwidth of the original matrix. However, if we use the QR algorithm entirely in its original form, the convergence rate is not likely to be very fast. In practice, we usually use shift coupled with QR method to speed up the convergence process.

7.5.2 QR with shift

Shifting can be used in different ways in the QR method. If we have a reasonably good value of shift upfront that will speed up the convergence, then we shift the matrix and apply the QR iterations to the shifted matrix. In this case the value of the shift remains constant. In QR method we have the option of varying the value of the shift during the iterations and we also have the option of introducing it at any point during the iterations. Here, we will describe the general form of variable shift.

$$\begin{gathered} \begin{cases} \mathbf{A}_1 = \mathbf{A} \\ \mathbf{A}_1 - \mu_1 \mathbf{I} = \mathbf{Q}_1 \mathbf{R}_1 \end{cases} \\ \begin{cases} \mathbf{A}_2 = \mathbf{R}_1 \mathbf{Q}_1 + \mu_1 \mathbf{I} \\ \mathbf{A}_2 - \mu_2 \mathbf{I} = \mathbf{Q}_2 \mathbf{R}_2 \end{cases} \\ \vdots \\ \begin{cases} \mathbf{A}_k = \mathbf{R}_{k-1} \mathbf{Q}_{k-1} + \mu_{k-1} \mathbf{I} \\ \mathbf{A}_k - \mu_k \mathbf{I} = \mathbf{Q}_k \mathbf{R}_k \end{cases} \\ \vdots \end{gathered} \tag{7.210}$$

Here, we have started by shifting the initial matrix $\mathbf{A}_1$. Then we perform the orthogonal decomposition on the shifted matrix. In forming the next matrix $\mathbf{A}_2$ we add the shift to the right-hand side of the equation. This is for the purpose of maintaining the similarity transformations in the QR algorithm. We will demonstrate this for the kth step in iteration.

$$\begin{aligned} \mathbf{A}_k - \mu_k \mathbf{I} &= \mathbf{Q}_k \mathbf{R}_k \\ \mathbf{A}_{k+1} &= \mathbf{R}_k \mathbf{Q}_k + \mu_k \mathbf{I} \end{aligned} \quad (7.211)$$

From the first equation, we obtain the following relation and substitute it in the second equation:

$$\mathbf{R}_k = \mathbf{Q}_k^T(\mathbf{A}_k - \mu_k \mathbf{I}) \quad (7.212)$$

$$\begin{aligned} \mathbf{A}_{k+1} &= \mathbf{Q}_k^T(\mathbf{A}_k - \mu_k \mathbf{I})\mathbf{Q}_k + \mu_k \mathbf{I} \\ &= \mathbf{Q}_k^T \mathbf{A}_k \mathbf{Q}_k - \mu_k \mathbf{I} + \mu_k \mathbf{I} \\ &= \mathbf{Q}_k^T \mathbf{A}_k \mathbf{Q}_k \end{aligned} \quad (7.213)$$

This shows that $\mathbf{A}_{k+1}$ is similar to $\mathbf{A}_k$ and their eigenvalues are the same. By maintaining the similarity, we have the option of changing the shift or introducing it at any point or terminating it.

In practical implementation, the shift operation is not performed at the beginning of QR algorithm. Usually, we would carry out QR algorithm to certain extent such that the diagonal terms are getting close to certain eigenvalues. Then we employ the shift operation to speed up the convergence of the QR algorithm to those eigenvalues.

7.5.3 QL method

The idea behind the QL method is similar to the QR method. Let **A** be a real symmetric matrix, and we can perform the following orthogonal decomposition:

$$\mathbf{A} = \mathbf{QL} \quad (7.214)$$

where:

Q is an orthogonal matrix
L is a lower triangular matrix

We note that the **Q** matrix here is different from the **Q** matrix determined in the QR algorithm. Similar to the QR algorithm, the QL algorithm can be described in the following procedure:

$$\begin{aligned} &\begin{cases} \mathbf{A}_1 = \mathbf{A} \\ \mathbf{A}_1 = \mathbf{Q}_1 \mathbf{L}_1 \end{cases} \\ &\begin{cases} \mathbf{A}_2 = \mathbf{L}_1 \mathbf{Q}_1 \\ \mathbf{A}_2 = \mathbf{Q}_2 \mathbf{L}_2 \end{cases} \\ &\quad \vdots \\ &\begin{cases} \mathbf{A}_k = \mathbf{L}_{k-1} \mathbf{Q}_{k-1} \\ \mathbf{A}_k = \mathbf{Q}_k \mathbf{L}_k \end{cases} \\ &\quad \vdots \end{aligned} \quad (7.215)$$

In general, we have the following expressions similar to the QR method described in the previous section:

$$\begin{aligned} \mathbf{A}_{k+1} &= \mathbf{L}_k\mathbf{Q}_k = \mathbf{Q}_k^T\mathbf{A}_k\mathbf{Q}_k \\ &\vdots \\ &= \mathbf{Q}_k^T\ldots\mathbf{Q}_1^T\mathbf{A}\mathbf{Q}_1\ldots\mathbf{Q}_k \\ &= (\mathbf{Q}_1\ldots\mathbf{Q}_k)^T\mathbf{A}(\mathbf{Q}_1\ldots\mathbf{Q}_k) \end{aligned} \tag{7.216}$$

$$\mathbf{P}_k = \mathbf{Q}_1\ldots\mathbf{Q}_k \tag{7.217}$$

$$\mathbf{A}_{k+1} = \mathbf{P}_k^T\mathbf{A}\mathbf{P}_k \tag{7.218}$$

This indicates that similarity is maintained and $\mathbf{A}_{k+1}$ is similar to $\mathbf{A}$. Thus the symmetry of the original matrix as well as the eigenvalues of the matrix $\mathbf{A}$ is not altered during the QL process.

It can be proven, similar to the proof given for QR method in the previous section, that for symmetric matrix $\mathbf{A}$, the convergence property of the QL method is as follows:

$$\begin{cases} \lim\limits_{k\to\infty} \mathbf{A}_k \to \Lambda = \text{diag}[\lambda_1,\ldots,\lambda_n] \\ \lim\limits_{k\to\infty} \mathbf{P}_k \to \mathbf{X} = [\mathbf{x}_1|\ldots|\mathbf{x}_n] \end{cases} \tag{7.219}$$

This convergence characteristics can be demonstrated in a similar manner as with QR method. QL method, similar to QR method converges from both ends of the spectrum toward the middle.

7.5.4 Orthogonal decomposition techniques

In both QR and QL methods we need to decompose $\mathbf{A}_k$ by orthogonal decompositions.

$$\begin{aligned} \mathbf{A}_k &= \mathbf{Q}_k^{(R)}\mathbf{R}_k \\ \mathbf{A}_k &= \mathbf{Q}_k^{(L)}\mathbf{L}_k \end{aligned} \tag{7.220}$$

We have included a superscript in matrix $\mathbf{Q}$ to distinguish between the two methods. In both methods, the orthogonal decomposition is performed at each iteration. In fact, we can say that computing orthogonal decomposition of a matrix forms the core tasks of the QR and QL algorithms.

There are many ways to compute the orthogonal matrices. We will start with the QR method.

$$\mathbf{A} = \mathbf{Q}\mathbf{R} \tag{7.221}$$

$$\mathbf{A} = \left[\mathbf{a}_1 | \ldots | \mathbf{a}_n\right]$$
$$\mathbf{Q} = \left[\mathbf{q}_1 | \ldots | \mathbf{q}_n\right]$$
$$\mathbf{R} = \begin{bmatrix} r_{11} \cdots r_{1n} \\ \ddots \quad \vdots \\ r_{nn} \end{bmatrix} \tag{7.222}$$

$$\mathbf{a}_k = \sum_{j=1}^{k} r_{jk}\mathbf{q}_j \tag{7.223}$$

On the other hand, as matrix Q is orthogonal, we have the following relations:

$$\mathbf{R} = \mathbf{Q}^T\mathbf{A} \tag{7.224}$$

$$r_{jk} = \mathbf{q}_j^T\mathbf{a}_k \tag{7.225}$$

$$\begin{aligned} \mathbf{a}_k &- \sum_{j=1}^{k-1} r_{jk}\mathbf{q}_j + r_{kk}\mathbf{q}_k \\ &= \sum_{j=1}^{k-1} (\mathbf{q}_j^T\mathbf{a}_k)\mathbf{q}_j + r_{kk}\mathbf{q}_k \end{aligned} \tag{7.226}$$

Hence, the orthogonal vector can be determined as follows:

$$\mathbf{q}_k = \frac{1}{r_{kk}}\left(\mathbf{a}_k - \sum_{j=1}^{k-1}(\mathbf{q}_j^T\mathbf{a}_k)\mathbf{q}_j\right) = \frac{\bar{\mathbf{q}}_k}{r_{kk}} \tag{7.227}$$

Obviously, $\bar{\mathbf{q}}_k$ can be considered as a vector in the direction of $\mathbf{q}_k$. We obtain a unit vector in that direction through normalization.

$$\mathbf{q}_k = \frac{\bar{\mathbf{q}}_k}{\|\bar{\mathbf{q}}_k\|} \tag{7.228}$$

Therefore, we can determine the orthogonal matrix in the following procedure:

$$\begin{aligned} &\begin{cases} \bar{\mathbf{q}}_1 = \mathbf{a}_1 \\ \mathbf{q}_1 = \dfrac{\bar{\mathbf{q}}_1}{\|\bar{\mathbf{q}}_1\|} = \dfrac{\mathbf{a}_1}{\|\mathbf{a}_1\|} \end{cases} \\ &\begin{cases} \bar{\mathbf{q}}_2 = \mathbf{a}_2 - (\mathbf{q}_1^T\mathbf{a}_2)\mathbf{q}_1 = \mathbf{a}_2 - \dfrac{(\mathbf{a}_1^T\mathbf{a}_2)\mathbf{a}_1}{\|\mathbf{a}_1\|^2} \\ \mathbf{q}_2 = \dfrac{\bar{\mathbf{q}}_2}{\|\bar{\mathbf{q}}_2\|} \end{cases} \\ &\quad\vdots \end{aligned} \tag{7.229}$$

Obviously, this is the Gram–Schmidt orthogonalization procedure. To ensure numerical stability, we should use the modified Gram–Schmidt orthogonalization procedure discussed earlier in Chapter 1.

7.5.5 Computing R matrix in QR decomposition

From the equations just obtained above, we have the following relations:

$$\bar{\mathbf{q}}_k = \mathbf{a}_k - \sum_{j=1}^{k-1} r_{jk}\mathbf{q}_j \tag{7.230}$$

$$r_{jk} = \mathbf{q}_j^T \mathbf{a}_k \tag{7.231}$$

From Equations 7.227 and 7.228 we have the following:

$$r_{kk} = \|\bar{\mathbf{q}}_k\| \tag{7.232}$$

This shows that we can determine the elements in **R** matrix in the process of computing the Q matrix. On the other hand, since **A** = **QR**, then

$$\begin{aligned} \mathbf{A}^T\mathbf{A} &= \mathbf{R}^T\mathbf{Q}^T\mathbf{Q}\mathbf{R} \\ &= \mathbf{R}^T\mathbf{R} = \bar{\mathbf{L}}\bar{\mathbf{L}}^T \end{aligned} \tag{7.233}$$

in which $\bar{\mathbf{L}}\bar{\mathbf{L}}^T$ is the Cholesky decomposition of $\mathbf{A}^T\mathbf{A}$. This shows that when $\mathbf{A}^T\mathbf{A}$ is positive definite, **R** is equal to the Cholesky decomposition matrix $\bar{\mathbf{L}}^T$.

7.5.6 Stable orthogonalization methods

As has been discussed earlier, the Gram–Schmidt orthogonalization procedure is not stable, especially when the initial vectors are not strictly independent. That is the reason that it is not used in practical computations. The stable procedures for orthogonalization make use of Givens plane rotation matrices and Householder reflection matrices.

7.5.6.1 Householder method

In the Householder method, we use a series of reflection matrices to transform matrix **A** to an upper triangular matrix.

$$\mathbf{H}_{n-1}\ldots\mathbf{H}_2\mathbf{H}_1\mathbf{A} = \mathbf{U}_H \tag{7.234}$$

where $\mathbf{U}_H$ is the upper triangular matrix and $\mathbf{H}_j$ is the reflection matrix to reduce to zero all the terms below the diagonal in the jth column of $\mathbf{A}^{(j-1)}$. The determination of each reflection matrix has been discussed earlier in Chapter 3, which is not repeated here. After computing all the reflection matrices, we have

$$\begin{aligned} \mathbf{A} &= \mathbf{H}_1^T\mathbf{H}_2^T\ldots\mathbf{H}_{n-1}^T\mathbf{U}_H \\ &= \mathbf{Q}_H\mathbf{U}_H \end{aligned} \tag{7.235}$$

where $\mathbf{Q}_H$ is an orthogonal matrix.

7.5.6.2 Givens method

In Givens method, we use a series of plane rotation matrices to transform matrix **A** to an upper triangular matrix. But we can zero only one off-diagonal element at a time.

$$(\mathbf{G}_{n,n-1}) \dots (\mathbf{G}_{n,2} \dots \mathbf{G}_{3,2})(\mathbf{G}_{n,1} \dots \mathbf{G}_{2,1})\ \mathbf{A} = \mathbf{U}_G \tag{7.236}$$

$$(\mathbf{G}_{n-1} \dots \mathbf{G}_2\ \mathbf{G}_1)\ \mathbf{A} = \mathbf{U}_G \tag{7.237}$$

where $\mathbf{U}_G$ is the upper triangular matrix and $\mathbf{G}_j$ is the an orthogonal matrix that reduces to zero all the terms below the diagonal in the jth column of $\mathbf{A}^{(j-1)}$.

$$\begin{aligned}\mathbf{A} &= (\mathbf{G}_1^T\ \mathbf{G}_2^T\ \dots\ \mathbf{G}_{n-1}^T)\ \mathbf{U}_G \\ &= \mathbf{G}\ \mathbf{U}_G\end{aligned} \tag{7.238}$$

Givens method also has been discussed in detail earlier in Chapter 3 and will not be repeated here.

In general, the Householder orthogonalization is stable and computationally more efficient than Givens method. In practice, therefore, we usually use Householder method. It can be observed that if we use QR or QL method directly for solving eigenvalues and eigenvectors of a full symmetric matrix, the computational efforts involved in performing orthogonal decomposition are quite time consuming and of enormous scale especially when the dimensionality of the problem is very large. Therefore, QR or QL methods tend to be used for obtaining all eigenvalues and eigenvectors of small matrices. For real symmetric matrices, to reduce the computational efforts, normally orthogonal similarity transformation is used to reduce the original matrix **A** to its tridiagonal form, and then perform QR or QL process to obtain the eigenvalues and eigenvectors. In so doing, the effectiveness of the QR method is maximized and the computational expenses involved in the orthogonal decomposition process in the QR iteration process are reduced.

7.5.7 Tridiagonalization of real symmetric matrices

By tridiagonalization, we mean to perform orthogonal similarity transformations on a real symmetric matrix **A** such that it is reduced to the following tridiagonal form:

$$\mathbf{Q}^T\mathbf{A}\mathbf{Q} = \begin{bmatrix} \alpha_1 & \beta_1 & & \\ \beta_1 & \alpha_2 & \ddots & \\ & \ddots & \ddots & \beta_{n-1} \\ & & \beta_{n-1} & \alpha_n \end{bmatrix} \tag{7.239}$$

We can see that the tridiagonalization process can be accomplished through the use of either Givens plane rotation matrix or Householder reflection matrix. First, we illustrate the use of reflection matrices. We start with a real symmetric matrix **A** and perform an orthogonal similarity transformation to reduce to zero the last n – 2 elements in the first row and first column. The resulting matrix after this transformation is as follows:

$$\mathbf{A}_1 = \mathbf{Q}_1^T\mathbf{A}\mathbf{Q}_1 \tag{7.240}$$

$$\mathbf{A}_1 = \begin{bmatrix} 1 & 0 \\ 0 & \mathbf{R}_1 \end{bmatrix} \begin{bmatrix} a_{11} & \bar{\mathbf{b}}_1^T \\ \bar{\mathbf{b}}_1 & \bar{\mathbf{A}}_1 \end{bmatrix} \begin{bmatrix} 1 & 0 \\ 0 & \mathbf{R}_1^T \end{bmatrix} = \begin{bmatrix} a_{11} & \bar{\mathbf{b}}_1^T \mathbf{R}_1^T \\ \mathbf{R}_1 \bar{\mathbf{b}}_1 & \mathbf{R}_1 \bar{\mathbf{A}}_1 \mathbf{R}_1^T \end{bmatrix} \tag{7.241}$$

$\mathbf{R}_1$ = an $(n - 1) \times (n - 1)$ reflection matrix, $\bar{\mathbf{b}}_1$ = column vector consisting of the last $n - 1$ elements of the first column of $\mathbf{A}$, and $\bar{\mathbf{A}}_1$ = the remaining matrix after removing the first row and column. The following relation satisfies the condition that the last $n - 2$ elements in the first row and column be zero:

$$\mathbf{R}_1 \bar{\mathbf{b}}_1 = \beta_1 \bar{\mathbf{e}}_1^{(1)} = \begin{Bmatrix} \beta_1 \\ 0 \\ \vdots \\ 0 \end{Bmatrix}_{n-1} \tag{7.242}$$

$\bar{\mathbf{e}}_1^{(1)}$ is a unit vector of size $(n - 1)$ with 1 as the first term and all other terms zero. We recall that the reflection transformation preserves the length of a vector, leading to the following relation:

$$\beta_1 = \left\| \bar{\mathbf{b}}_1 \right\| \tag{7.243}$$

The reflection matrix can be determined as follows:

$$\mathbf{R}_1 = \mathbf{I}_{n-1} - \frac{2}{\mathbf{u}_1^T \mathbf{u}_1} \mathbf{u}_1 \mathbf{u}_1^T \tag{7.244}$$

$$\mathbf{u}_1 = \bar{\mathbf{b}}_1 + \text{sign}(a_{11}) \left\| \bar{\mathbf{b}}_1 \right\| \bar{\mathbf{e}}_1^{(1)} \tag{7.245}$$

After completing the reduction of the first row and column to tridiagonal form we arrive at the following:

$$\mathbf{A}_1 = \begin{bmatrix} a_{11} & \beta_1 & \\ \beta_1 & \bar{a}_{22} & \bar{\mathbf{b}}_2^T \\ & \bar{\mathbf{b}}_2 & \bar{\mathbf{A}}_2 \end{bmatrix} \tag{7.246}$$

Now, we can reduce to zero the last $n - 3$ terms in the second row and column of $\mathbf{A}_1$ through similar operations to obtain $\mathbf{A}_2$. We can continue to complete the process of tridiagonalization. We see that with the use of Householder reflection matrices, we can transform a real symmetric matrix into its tridiagonal form in $n - 2$ steps.

We can also reduce a real symmetric matrix into a tridiagonal matrix using orthogonal similarity transformation with Givens plane rotation matrices. At each stage, plane rotation matrices are used to reduce to zero the last $n - (j + 1)$ terms in the jth row and column of the matrix $\mathbf{A}_j$. Givens method has been described in detail in Chapter 3 and will not be repeated here.

Because of the symmetry of matrix $\mathbf{A}$, the tridiagonalization process with Givens method requires about $4n^3/3$ multiplications and $n^2/2$ square root operations. Whereas, with

Householder tridiagonalization process, it requires $2n^3/3$ multiplications and $(n-2)$ square root operations. This shows that the Householder tridiagonalization method is more efficient than the Givens method. That is one of the reasons Householder tridiagonalization method is often used in practice.

7.5.8 Wilkinson shift

In the practical use of QR or QL method for eigenvalue solution of real symmetric matrices, we first perform tridiagonalization process on matrix **A** and then compute all eigenvalue and eigenvectors of the tridiagonal matrix. Usually we use shift with QR or QL. In our early discussion on shift, we observed that Rayleigh quotient would provide an optimal value for shift. However, with tridiagonal matrix, there are exceptions. It is possible that with Rayleigh quotient shift, in some rare cases, the off-diagonal terms in the tridiagonal matrix may not be reduced to zero during a QR or QL process. This means that Rayleigh quotient shift coupled with QR or QL may result in numerical instability in some rare cases.

A stable shift, called Wilkinson shift (Wilkinson, 1965), guarantees the convergence of the QR or QL process applied to a tridiagonal matrix **T** of the following form:

$$\mathbf{T} = \begin{bmatrix} \alpha_1 & \beta_1 & & \\ \beta_1 & \alpha_2 & \ddots & \\ & \ddots & \ddots & \beta_{n-1} \\ & & \beta_{n-1} & \alpha_n \end{bmatrix} \tag{7.247}$$

Wilkinson shift is taken as the eigenvalue of the second order principal submatrix closer to α_1. The two eigenvalues of the second-order principal submatrix can be computed from the following equation:

$$\det\begin{bmatrix} \alpha_1-\lambda & \beta_1 \\ \beta_1 & \alpha_2-\lambda \end{bmatrix} = 0 \tag{7.248}$$

The eigenvalue closer to α_1 that is taken as the Wilkinson shift is given in the following equation:

$$\mu = \alpha_1 - \mathrm{sign}(\delta)\left(\frac{\beta_1^2}{|\delta| + \sqrt{\delta^2+\beta_1^2}}\right)$$
$$\delta = \frac{1}{2}(\alpha_2-\alpha_1) \tag{7.249}$$

It can be proven that QR or QL method with Wilkinson shift always converges for any symmetric tridiagonal matrices, and the rate of convergence is at least of second order.

7.6 DIRECT SOLUTION OF GENERALIZED EIGENVALUE PROBLEM

We have discussed numerical schemes that transform the generalized eigenvalue problem to standard form. Then we use standard eigenvalue solution methods to compute the eigenvalues and eigenvectors for the transformed matrix. In this section we will discuss that many solution methods can be directly applied to the generalized eigenvalue problem in the following form:

$$\mathbf{Ax} = \lambda\mathbf{Bx} \tag{7.250}$$

The matrices **A** and **B** are assumed to be symmetric and positive definite. Let **R** = matrix of eigenvectors of **B**, ($\mathbf{R}^T\mathbf{R} = \mathbf{I}$), and $\mathbf{D}^2 = \text{diag}\left(d_{ii}^2\right)$ the eigenvalues of **B**, then we can perform spectral decomposition of **B**.

$$\begin{aligned} \mathbf{BR} &= \mathbf{RD}^2 \\ \mathbf{B} &= \mathbf{RD}^2\mathbf{R}^T \end{aligned} \tag{7.251}$$

$$\begin{aligned} \mathbf{A} - \lambda\mathbf{B} &= \mathbf{A} - \lambda\mathbf{RDDR}^T \\ &= \mathbf{RD}(\mathbf{D}^{-1}\mathbf{R}^T\mathbf{ARD}^{-1} - \lambda\mathbf{I})\mathbf{DR}^T \\ &= \mathbf{RD}(\bar{\mathbf{A}} - \lambda\mathbf{I})\mathbf{DR}^T \end{aligned} \tag{7.252}$$

$$\bar{\mathbf{A}} = \mathbf{D}^{-1}\mathbf{R}^T\mathbf{ARD}^{-1} \tag{7.253}$$

By definition, the eigenvalues of the generalized problem are the roots of the characteristic polynomial.

$$\begin{aligned} \det(\mathbf{A} - \lambda\mathbf{B}) &= \det\left[\mathbf{RD}(\bar{\mathbf{A}} - \lambda\mathbf{I})\mathbf{DR}^T\right] \\ &= \det(\mathbf{D}^2)\det(\bar{\mathbf{A}} - \lambda\mathbf{I}) = 0 \end{aligned} \tag{7.254}$$

Because $\det(\mathbf{D}^2)$ is a scalar multiple, then the above relation means that the roots of $\det(\mathbf{A} - \lambda\mathbf{B})$ are the same as those of $\det(\bar{\mathbf{A}} - \lambda\mathbf{I})$ but the later has been transformed to a standard form. The eigenvalues of the two problems are the same, but their eigenvectors are different. This indeed is a way to transform a generalized eigenvalue problem to a standard form.

To determine the relationship between eigenvectors of the transformed and original systems, we can use all the standard methods to compute all eigenvalues and eigenvectors from transformed problem.

$$\bar{\mathbf{A}}\bar{\mathbf{x}}_k = \lambda_k\bar{\mathbf{x}}_k \tag{7.255}$$

Since the matrix in this equation is symmetric and positive definite, then its eigenvectors are an orthogonal set of vectors. Now, we can substitute for the transformed matrix.

$$(\mathbf{D}^{-1}\mathbf{R}^T\mathbf{ARD}^{-1})\bar{\mathbf{x}}_k = \lambda_k\bar{\mathbf{x}}_k \tag{7.256}$$

$$\begin{aligned} \mathbf{A}(\mathbf{RD}^{-1}\bar{\mathbf{x}}_k) &= \lambda_k\mathbf{RD}\bar{\mathbf{x}}_k \\ &= \lambda_k\mathbf{RD}(\mathbf{DR}^T\mathbf{RD}^{-1})\bar{\mathbf{x}}_k \\ &= \lambda_k\mathbf{B}(\mathbf{RD}^{-1}\bar{\mathbf{x}}_k) \end{aligned} \tag{7.257}$$

$$\mathbf{Ax}_k = \lambda_k\mathbf{Bx}_k \tag{7.258}$$

$$\mathbf{x}_k = \mathbf{R}\mathbf{D}^{-1}\bar{\mathbf{x}}_k \tag{7.259}$$

This gives the relationship between the eigenvectors of the transformed system and the original generalized system. Since the eigenvectors for the transformed system constitute an orthogonal set, then we have the following relation:

$$\begin{aligned}\bar{\mathbf{x}}_j^T\bar{\mathbf{x}}_k &= (\mathbf{x}_j^T\mathbf{R}\mathbf{D})(\mathbf{D}\mathbf{R}^T\mathbf{x}_k)\\ &= \mathbf{x}_j^T\mathbf{B}\mathbf{x}_k = \delta_{ij}\end{aligned} \tag{7.260}$$

This shows that the eigenvectors of the original generalized system are orthogonal with respect to space spanned by vectors in matrix **B** or we say that eigenvectors are **B** orthogonal.

7.6.1 Rayleigh quotient

With the transformed standard form obtained, we can use Rayleigh quotient to compute some eigenvalues. By definition, we have the following:

$$\begin{aligned}\rho &= \frac{\bar{\mathbf{x}}^T\bar{\mathbf{A}}\bar{\mathbf{x}}}{\bar{\mathbf{x}}^T\bar{\mathbf{x}}}\\ &= \frac{(\mathbf{x}^T\mathbf{R}\mathbf{D})(\mathbf{D}^{-1}\mathbf{R}^T\mathbf{A}\mathbf{R}\mathbf{D}^{-1})(\mathbf{D}\mathbf{R}^T\mathbf{x})}{(\mathbf{x}^T\mathbf{R}\mathbf{D})(\mathbf{D}\mathbf{R}^T\mathbf{x})}\\ &= \frac{\mathbf{x}^T\mathbf{A}\mathbf{x}}{\mathbf{x}^T\mathbf{B}\mathbf{x}}\end{aligned} \tag{7.261}$$

This is the Rayleigh quotient for generalized eigenvalue problems. In structural dynamics, **A** is the structural stiffness matrix and **B** the mass matrix. In modeling structural systems with finite element method, it is most likely that we may introduce extra constraints into the system such that the resulting structural model would be stiffer than the real structure. The mass matrix is always symmetric and at least positive semi-definite. The mass matrix can be determined consistent with the finite element approximation or through a simple lumping mass scheme. Lumping mass to few degrees of freedom usually does not affect the lower modes, unless some uncharacteristic mass lumping scheme is used. We can thus intuitively infer that the eigenvalue predicted from the Rayleigh quotient would give an upper bound on the real eigenvalue of the original system.

If we introduce a shift μ, we saw from previous discussion that the eigenvalues are preserved between the transformed system and original generalized system, Equation 7.254. This mean that if matrix $\bar{\mathbf{A}}$ is shifted by $\mu\mathbf{I}$, then matrix **A** must be shifted by $\mu\mathbf{B}$ such that the relation in Equation 7.254 can be maintained.

$$\det[\mathbf{A} - (\lambda + \mu)\mathbf{B}] = \det(\mathbf{D}^2)\det\left[\bar{\mathbf{A}} - (\lambda + \mu)\mathbf{I}\right] \tag{7.262}$$

This relationship describing the effect of shift can also be directly derived as follows.

$$\begin{aligned}\bar{\mathbf{A}} - \mu\mathbf{I} &= \mathbf{D}^{-1}\mathbf{R}^T\mathbf{A}\mathbf{R}\mathbf{D}^{-1} - \mu\mathbf{I}\\ &= \mathbf{D}^{-1}\mathbf{R}^T(\mathbf{A} - \mu\mathbf{R}\mathbf{D}\mathbf{D}\mathbf{R}^T)\mathbf{R}\mathbf{D}^{-1}\\ &= \mathbf{D}^{-1}\mathbf{R}^T(\mathbf{A} - \mu\mathbf{B})\mathbf{R}\mathbf{D}^{-1}\end{aligned} \tag{7.263}$$

$$(\bar{\mathbf{A}} - \mu\mathbf{I})\bar{\mathbf{x}} = \lambda\bar{\mathbf{x}}$$

$$\mathbf{D}^{-1}\mathbf{R}^{T}(\mathbf{A} - \mu\mathbf{B})\mathbf{R}\mathbf{D}^{-1}\bar{\mathbf{x}} = \lambda\bar{\mathbf{x}} \tag{7.264}$$

$$(\mathbf{A} - \mu\mathbf{B})\mathbf{R}\mathbf{D}^{-1}\bar{\mathbf{x}} = \lambda\mathbf{R}\mathbf{D}\bar{\mathbf{x}}$$

$$(\mathbf{A} - \mu\mathbf{B})\mathbf{R}\mathbf{D}^{-1}\bar{\mathbf{x}} = \lambda\mathbf{R}\mathbf{D}(\mathbf{D}\mathbf{R}^{T}\mathbf{R}\mathbf{D}^{-1})\bar{\mathbf{x}} \tag{7.265}$$

$$(\mathbf{A} - \mu\mathbf{B})\mathbf{x} = \lambda\mathbf{B}\mathbf{x} \tag{7.266}$$

This is the shifted eigenvalue problem in a generalized form.

7.6.2 Power method and the inverse power method

To use some of the iterative methods, we can mathematically transform the generalized eigenvalue problem into the following form:

$$\mathbf{B}^{-1}\mathbf{A}\mathbf{x} = \lambda\mathbf{x} \tag{7.267}$$

Therefore, if we use the power method to compute the largest eigen-pairs, we are iteratively carrying out the following computation:

$$\bar{\mathbf{y}}_k = \mathbf{B}^{-1}\mathbf{A}\mathbf{y}_{k-1} \tag{7.268}$$

In practice, instead of computing the inverse of **B**, unless it is a diagonal matrix from lumping mass, we usually solve the following system of equations:

$$\mathbf{B}\,\bar{\mathbf{y}}_k = \mathbf{A}\mathbf{y}_{k-1} \tag{7.269}$$

Hence the power method for generalized eigenvalue problems can be implemented in the following procedure:

$$\begin{aligned}
&\mathbf{y}_0, \text{ initial vector}\\
&\mathbf{B}\,\bar{\mathbf{y}}_1 = \mathbf{A}\mathbf{y}_0, \text{ solve for } \bar{\mathbf{y}}_1\\
&\mathbf{y}_1 = \frac{1}{\|\bar{\mathbf{y}}_1\|}\bar{\mathbf{y}}_1\\
&\quad\vdots\\
&\mathbf{B}\,\bar{\mathbf{y}}_k = \mathbf{A}\mathbf{y}_{k-1}, \text{ solve for } \bar{\mathbf{y}}_k\\
&\mathbf{y}_k = \frac{1}{\|\bar{\mathbf{y}}_k\|}\bar{\mathbf{y}}_k\\
&\quad\vdots
\end{aligned} \tag{7.270}$$

The convergence is the same as in the standard problem discussed earlier. If $\mathbf{y}_0^T\mathbf{x}_n \neq 0$, then the method converges as follows.

$$\begin{cases} \lim_{k\to\infty} \mathbf{y}_k \to \mathbf{x}_n \\ \lim_{k\to\infty} \|\bar{\mathbf{y}}_k\|_2 \to \lambda_n \end{cases} \tag{7.271}$$

The inverse power method can also be directly used to compute the lowest eigen-pairs. We solve the following system of equations:

$$\mathbf{A}\ \bar{\mathbf{y}}_k = \mathbf{B}\mathbf{y}_{k-1} \tag{7.272}$$

With this calculations, the basic steps are as follows.

$$\begin{aligned} &\mathbf{A}\ \bar{\mathbf{y}}_k = \mathbf{B}\mathbf{y}_{k-1}, \text{ solve for } \bar{\mathbf{y}}_k \\ &\mathbf{y}_k = \frac{1}{\|\bar{\mathbf{y}}_k\|}\bar{\mathbf{y}}_k \end{aligned} \tag{7.273}$$

If $\mathbf{y}_0^T\mathbf{x}_1 \neq 0$, the method converges as follows.

$$\begin{cases} \lim_{k\to\infty} \mathbf{y}_k \to \mathbf{x}_1 \\ \lim_{k\to\infty} \|\bar{\mathbf{y}}_k\|_2 \to \lambda_1 \end{cases} \tag{7.274}$$

Similarly, we can use the deflation process coupled with the power method and inverse power method to compute other eigenvalues and eigenvectors. However, it can be seen that both methods involve the solution of system of equations at each iteration. This alone would make the solution process computationally costly. A more practical method for solving generalized eigenvalue problems is called the *subspace iteration* method, which will be discussed in Section 7.7.

7.7 SUBSPACE ITERATION METHOD: THE RAYLEIGH–RITZ ANALYSIS

7.7.1 Basic concept

Assuming that A is an n × n symmetric and positive definite matrix with distinct eigenvalues $\lambda_1 < \lambda_2 < \ldots < \lambda_n$, we have the standard eigenvalue problem.

$$\mathbf{A}\ \mathbf{x}_j = \lambda_j\mathbf{x}_j, \quad j = 1,\ldots, n \tag{7.275}$$

The Rayleigh quotient is defined as

$$\rho(\mathbf{y}) = \frac{\mathbf{y}^T\mathbf{A}\mathbf{y}}{\mathbf{y}^T\mathbf{y}} \tag{7.276}$$

and has the property that $\lambda_1 < \rho(\mathbf{y}) < \lambda_n$. From our earlier discussion, it is observed that the smallest eigenvalue can be determined by minimizing $\rho(\mathbf{y})$ in the n-dimensional vector space.

$$\lambda_1 = \min\left[\rho(\mathbf{y})\right] \quad \text{for } \mathbf{y} \in R^n \tag{7.277}$$

The corresponding eigenvector $\mathbf{x}_1$ can be computed using any of the eigenvalue solution methods described earlier. Now if we restrict that vector $\mathbf{y}$ lies in the subspace orthogonal to $\mathbf{x}_1$, that is, $\mathbf{y}^T\mathbf{x}_1 = 0$, then we have the next lowest eigenvalue as the minimum Rayleigh quotient.

$$\lambda_2 = \min[\rho(\mathbf{y})] \quad \text{for } \mathbf{y} \in S^{n-1} = \text{span } [\mathbf{x}_1]^{\perp} \tag{7.278}$$

The corresponding eigenvector is $\mathbf{x}_2$ which can be readily obtained. Similarly, if we set the constraints that $\mathbf{y}^T\mathbf{x}_1 = 0$ and $\mathbf{y}^T\mathbf{x}_2 = 0$, then

$$\lambda_3 = \min[\rho(\mathbf{y})] \quad \text{for } \mathbf{y} \in S^{n-2} = \text{span } [\mathbf{x}_1, \mathbf{x}_2]^{\perp} \tag{7.279}$$

This process defines the Rayleigh–Ritz minimization problem and shows that we can look at the solution of eigenvalue problems as a minimization problem.

Through natural extrapolation, the same type of minimization in an m-dimensional subspace within an n-dimensional vector space, where $m < n$, forms the basis of the Rayleigh–Ritz analysis. In the Rayleigh–Ritz method, we start with m initial vectors as our initial guess of the first m eigenvectors,

$$\mathbf{Y} = [\mathbf{y}_1 \,|\, \ldots \,|\, \mathbf{y}_m] \in R^{n\times m} \tag{7.280}$$

and assume that these m vectors are mutually orthogonal, $\mathbf{Y}^T\mathbf{Y} = \mathbf{I}$. The initial subspace spanned by the m vectors is

$$S^m = \text{span}[\mathbf{y}_1 | \ldots | \mathbf{y}_m] \tag{7.281}$$

We define a vector in this subspace that is a linear combination of vectors $\mathbf{y}_j$.

$$\bar{\mathbf{x}} = \sum_{j=1}^{m} \xi_j \mathbf{y}_j = \mathbf{Y}\mathbf{z} \tag{7.282}$$

$$\mathbf{z} = \begin{Bmatrix} \xi_1 \\ \vdots \\ \xi_m \end{Bmatrix} \in R^m \tag{7.283}$$

The Rayleigh quotient for this vector is defined as follows.

$$\begin{aligned} \rho(\bar{\mathbf{x}}) &= \frac{\bar{\mathbf{x}}^T\mathbf{A}\bar{\mathbf{x}}}{\bar{\mathbf{x}}^T\bar{\mathbf{x}}} = \frac{\mathbf{z}^T\mathbf{Y}^T\mathbf{A}\mathbf{Y}\mathbf{z}}{\mathbf{z}^T\mathbf{Y}^T\mathbf{Y}\mathbf{z}} \\ &= \frac{\mathbf{z}^T\bar{\mathbf{A}}\mathbf{z}}{\mathbf{z}^T\mathbf{z}} \end{aligned} \tag{7.284}$$

$$\bar{\mathbf{A}} = \mathbf{Y}^T\mathbf{A}\mathbf{Y} \in R^{m\times m} \tag{7.285}$$

We note that $\bar{\mathbf{A}}$ is a $m \times m$ matrix. Because $m < n$, now we are dealing with a smaller matrix in computing the Rayleigh quotient.

Next, we minimize the Raleigh quotient with respect to $\mathbf{z}$. This means that the minimization takes place in the subspace S^m.

$$\frac{\partial[\rho(\bar{\mathbf{x}})]}{\partial \mathbf{z}} = \frac{(2\bar{\mathbf{A}}\mathbf{z})(\mathbf{z}^T\mathbf{z}) - (2\mathbf{z})(\mathbf{z}^T\bar{\mathbf{A}}\mathbf{z})}{(\mathbf{z}^T\mathbf{z})^2} = 0 \tag{7.286}$$

$$(2\bar{\mathbf{A}}\mathbf{z})(\mathbf{z}^T\mathbf{z}) - (2\mathbf{z})(\mathbf{z}^T\bar{\mathbf{A}}\mathbf{z}) = 0 \tag{7.287}$$

$$\bar{\mathbf{A}}\mathbf{z} - \left(\frac{\mathbf{z}^T\bar{\mathbf{A}}\mathbf{z}}{\mathbf{z}^T\mathbf{z}}\right)\mathbf{z} = 0 \tag{7.288}$$

$$\bar{\mathbf{A}}\mathbf{z} = \rho\mathbf{z} \tag{7.289}$$

This relation shows that minimization of $\rho(\mathbf{x})$ with respect to $\mathbf{z}$ is equivalent to solving a reduced order eigenvalue problem. We arrive at the following expression by substituting for $\bar{\mathbf{A}}$ from Equation 7.285 into Equation 7.289 and taking advantage of the orthogonality of $\mathbf{Y}$.

$$\mathbf{A}(\mathbf{Y}\mathbf{z}) = \rho(\mathbf{Y}\mathbf{z}) \tag{7.290}$$

Comparing with the relation that $\mathbf{A}\,\mathbf{x} = \lambda\,\mathbf{x}$, we can conclude that if $\rho = \lambda$, then $\mathbf{Y}\,\mathbf{z} = \mathbf{x}$. The vectors $\mathbf{Y}\,\mathbf{z}_j$ are usually referred to as *Ritz vectors* and $(\rho_j, \mathbf{Y}\,\mathbf{z}_j)$ are called *Ritz-pairs*.

Now we perform spectral decomposition of $\bar{\mathbf{A}}$.

$$\bar{\mathbf{A}}\mathbf{Z} = \mathbf{Z}\bar{\Lambda} \tag{7.291}$$

$$\begin{cases} \bar{\Lambda} = \mathrm{diag}[\rho_1, \ldots, \rho_m] \\ \mathbf{Z} = [\mathbf{z}_1 \mid \ldots \mid \mathbf{z}_m] \end{cases} \tag{7.292}$$

Then we can expect that $\{\rho_1, \ldots, \rho_m\}$ are approximations to the eigenvalues $\{\lambda_1, \ldots, \lambda_m\}$; and the Ritz vectors $\{\mathbf{Y}\,\mathbf{z}_1, \ldots, \mathbf{Y}\,\mathbf{z}_m\}$, are approximations to the eigenvectors $\{\mathbf{x}_1, \ldots, \mathbf{x}_m\}$. This process of the Rayleigh–Ritz analysis can also be viewed as a method of reduction of dimensionality.

7.7.2 Subspace iteration method

With the idea introduced in the Rayleigh–Ritz analysis, we can use it to obtain the approximate solutions to eigenvalues and eigenvectors in a reduced dimensionality. This process can be implemented by starting with a subspace of initial vectors that are assumed to be mutually orthogonal.

$$\mathbf{Y}_0 = \left[\mathbf{y}_1^{(0)} | \ldots | \mathbf{y}_m^{(0)}\right] \in \mathbf{R}^{n \times m} \tag{7.293}$$

Then, we perform similarity transformation to obtain a matrix of reduced dimensionality and solve the reduced dimension eigenvalue problem.

$$\bar{\mathbf{A}} = \mathbf{Y}^T\mathbf{A}\mathbf{Y} \tag{7.294}$$

$$\bar{\mathbf{A}}\mathbf{z} = \rho\mathbf{z} \tag{7.295}$$

We obtain approximations to eigenvalues and eigenvectors in Ritz pairs $(\rho_j, \mathbf{Y}\,\mathbf{z}_j)$. If the currently obtained Ritz vectors are not accurate enough, we use them to repeat the iteration and continue until the desired accuracy level is reached.

The standard subspace iteration method uses the idea behind the Rayleigh–Ritz analysis and is in fact the blocked power method in an m-dimensional subspace. In this case, instead of using a single vector as trial vector, we start with a subspace of vectors. The algorithm can be described in the following procedure:

$$
\begin{aligned}
&\mathbf{Y}_0 \in \mathrm{R}^{n\times m}, \quad m < n \\
&\text{for } j = 1, 2, \dots \\
&\quad \bar{\mathbf{Y}}_j = \mathbf{A}\mathbf{Y}_{j-1} \\
&\quad \bar{\mathbf{Y}}_j = \mathbf{Q}_j\mathbf{R}_j, \quad \text{orthogonal decomposition} \\
&\quad \mathbf{Y}_j = \mathbf{Q}_j
\end{aligned}
\tag{7.296}
$$

The orthogonal decomposition can be accomplished by using Householder decomposition method. The convergence property of this method is that $\mathbf{Y}_j$ converges to the m highest eigenvectors of **A**, and the rate of convergence of outer eigenvectors is faster than the inner eigenvectors. In other words, the method converges to eigenvector n faster than to eigenvector n–1, and so on.

Similarly, we can also extend the basic subspace iteration method with $\mathbf{A}^{-1}$ to obtain some lowest eigenvectors of **A**. The algorithm can be described as

$$
\begin{aligned}
&\mathbf{Y}_0 \in \mathrm{R}^{n\times m}, \quad m < n \\
&\text{for } j = 1, 2, \dots \\
&\quad \mathbf{Y}_{j-1} = \mathbf{A}\bar{\mathbf{Y}}_j, \qquad \text{solve for } \bar{\mathbf{Y}}_j \\
&\quad \bar{\mathbf{Y}}_j = \mathbf{Q}_j\mathbf{R}_j, \quad \text{orthogonal decomposition} \\
&\quad \mathbf{Y}_j = \mathbf{Q}_j
\end{aligned}
\tag{7.297}
$$

This algorithm is equivalent to the inverse power method, but in an m-dimensional subspace. As expected, the convergence property is that it converges to the m lowest eigenvectors of **A**. Again, similar to direct subspace iteration, convergence to the first eigenvector is faster than to the second eigenvector and convergence to the second eigenvector is faster than to the third eigenvector, and so on. At certain stage of computation, we need to conduct Sturm sequence check to ensure that the method converges to the first m lowest eigenvectors. In structural dynamics, we are most interested in the first few (lowest) eigenvectors of the structural systems. Therefore, this is the algorithm used in structural analysis.

7.7.3 Subspace iteration with the Rayleigh–Ritz analysis

With the use of the Rayleigh–Ritz analysis, the subspace iteration method can be described in the following procedure:

$$\begin{aligned}
&\mathbf{Y}_0 \in R^{n\times m}, \quad m < n \\
&\text{for } j = 1, 2, \ldots \\
&\quad \bar{\mathbf{Y}}_j = \mathbf{A}\mathbf{Y}_{j-1} \\
&\quad \bar{\mathbf{Y}}_j = \mathbf{Q}_j\mathbf{R}_j, \quad \text{orthogonal decomposition} \\
&\quad \bar{\mathbf{A}}_j = \mathbf{Q}_j^T\mathbf{A}\mathbf{Q}_j \in R^{m\times m} \\
&\bar{\mathbf{A}}_j\mathbf{Z}_j = \mathbf{Z}_j\bar{\mathbf{\Lambda}}_j \\
&\quad \mathbf{Y}_j = \mathbf{Q}_j\mathbf{Z}_j
\end{aligned} \tag{7.298}$$

We note that the last three steps are the Rayleigh–Ritz procedure. Because $\bar{\mathbf{A}}_j$ is of reduced dimensionality, its eigenvalues and eigenvectors can be computed using QR or Jacobi method. Because the block version of the power method is used in this algorithm, the convergence of this method has the pattern that the Ritz pairs $(\bar{\lambda}_j, \mathbf{y}_j)$ converge to the highest eigen-pairs $(\lambda_j, \mathbf{x}_j)$.

If we are interested in the lowest eigen-pairs, then we can replace the first step in the iteration of the above algorithm by the following:

$$\mathbf{A}\bar{\mathbf{Y}}_j = \mathbf{Y}_{j-1}, \text{ solve for } \bar{\mathbf{Y}}_j \tag{7.299}$$

Now the Ritz-pairs $(\bar{\lambda}_j, \mathbf{y}_j)$ converge to the lowest eigen-pairs $(\lambda_1, \mathbf{x}_1)$. The resulting algorithm couples the block inverse power method with the Rayleigh–Ritz analysis. This is the practical subspace iteration algorithm most frequently used in computational mechanics, especially in structural dynamics. For structural systems, the distribution of eigenvalues is not random, rather, it follows an orderly pattern. In general, the lowest few eigenvalues are reasonably well separated; whereas the highest few eigenvalues are close to each other. It can be expected that convergence to the lowest few eigenvalues should be faster than the convergence to the highest eigenvalues.

In order to implement subspace iteration methods, there are issues that need to be addressed, such as, how to determine the value of m or the number of vectors in the subspace and the initial subspace of trial vectors. Of course, the value of m should be larger than the number of eigenvalues that we intend to compute. For computing k eigenvalues, often the following rule of thumb is used:

$$m = \max[k + 8, 2k] \tag{7.300}$$

This empirical value usually gives a reasonable estimation on the value of m. However, there is no fixed rule for selecting the trial vectors in $\mathbf{Y}_0$. In general, it should contain most of the information that pertains to the first few lower eigenvectors. And in structural analysis, practical experience shows that the following procedure works fine in selecting the $\mathbf{Y}_0$ matrix:

$$\mathbf{Y}_0 = \left[\mathbf{y}_1^{(0)} | \ldots | \mathbf{y}_m^{(0)}\right]$$

$$\mathbf{y}_1^{(0)} = \begin{Bmatrix} 1 \\ 1 \\ \vdots \\ \vdots \\ \vdots \\ 1 \\ 1 \end{Bmatrix}, \quad \mathbf{y}_2^{(0)} = \begin{Bmatrix} 0 \\ \vdots \\ 0 \\ 1 \\ 0 \\ \vdots \\ 0 \end{Bmatrix}, \ldots, \quad \mathbf{y}_m^{(0)} = \begin{Bmatrix} 0 \\ 0 \\ 1 \\ 0 \\ \vdots \\ \vdots \\ 0 \end{Bmatrix} \tag{7.301}$$

The vector $\mathbf{y}_1^{(0)}$ consists of n ones which are to excite all the eigenvectors of the structural system; and vectors $\mathbf{y}_2^{(0)}$ to $\mathbf{y}_m^{(0)}$ are all unit vectors and the unit term in these vectors resides at a unique position where we have max (B_{jj}/A_{jj}) within the kth column. In this case, **B** is the mass matrix **M** and **A** is the stiffness matrix **K** of the structural system.

7.7.4 Subspace iteration for solution of generalized eigenvalue problem

The subspace iteration method can be directly used for solution of a generalized eigenvalue problem in structural dynamic problems. First, we derive the Rayleigh–Ritz procedure for the problem.

$$\mathbf{AX} = \mathbf{BX\Lambda} \tag{7.302}$$

In this equation $\mathbf{\Lambda} = \mathrm{diag}[\lambda_1, \ldots, \lambda_n]$ and $\mathbf{X} = \left[\mathbf{x}_1 | \ldots | \mathbf{x}_n\right]$ are the eigenvalue and eigenvector matrices, respectively. Objective is to reduce this to an m × m, m < n, generalized eigenvalue problem. The procedure is similar to the standard eigenvalue problem discussed earlier in Equations 7.280 through 7.291. We start with selecting m initial trial vectors as our first guess of the first eigenvectors and define a vector as a linear combination of the trial vectors.

$$\mathbf{Y} = \left[\mathbf{y}_1 | \ldots | \mathbf{y}_m\right] \in R^{n\times m} \tag{7.303}$$

$$\overline{\mathbf{x}} = \sum_{j=1}^{m} \xi_j \mathbf{y}_j = \mathbf{Yz}$$

$$\mathbf{z} = \begin{Bmatrix} \xi_1 \\ \vdots \\ \xi_m \end{Bmatrix} \in R^m \tag{7.304}$$

We have derived earlier that the Rayleigh quotient for generalized eigenvalue problem is as follows.

$$\rho(\bar{\mathbf{x}}) = \frac{\bar{\mathbf{x}}^T \mathbf{A}\bar{\mathbf{x}}}{\bar{\mathbf{x}}^T \mathbf{B}\bar{\mathbf{x}}} = \frac{\mathbf{z}^T\mathbf{Y}^T\mathbf{A}\mathbf{Y}\mathbf{z}}{\mathbf{z}^T\mathbf{Y}^T\mathbf{B}\mathbf{Y}\mathbf{z}} = \frac{\mathbf{z}^T\bar{\mathbf{A}}\mathbf{z}}{\mathbf{z}^T\bar{\mathbf{B}}\mathbf{z}} \tag{7.305}$$

$$\left\{\begin{aligned} \bar{\mathbf{A}} &= \mathbf{Y}^T\mathbf{A}\mathbf{Y} \\ \bar{\mathbf{B}} &= \mathbf{Y}^T\mathbf{B}\mathbf{Y} \end{aligned}\right. \in \mathrm{R}^{m\times m} \tag{7.306}$$

We note that the selected initial trial vectors need not necessarily be **B**-orthogonal.

Since the eigenvalue solution process is a minimization process on the Rayleigh quotient, we minimize the Rayleigh quotient above with respect to the coefficient vector **z**.

$$\frac{\partial\|\rho(\bar{\mathbf{x}})\|}{\partial \mathbf{z}} = \frac{(2\bar{\mathbf{A}}\mathbf{z})(\mathbf{z}^T\bar{\mathbf{B}}\mathbf{z}) - (2\bar{\mathbf{B}}\mathbf{z})(\mathbf{z}^T\bar{\mathbf{A}}\mathbf{z})}{(\mathbf{z}^T\bar{\mathbf{B}}\mathbf{z})^2} = 0 \tag{7.307}$$

This leads to the following reduced dimension generalized eigenvalue problem:

$$\bar{\mathbf{A}}\mathbf{z} - \rho\bar{\mathbf{B}}\mathbf{z} = 0 \tag{7.308}$$

This relation shows that the minimization of Rayleigh quotient with respect to **z** is equivalent to solving a reduced dimension generalized eigenvalue problem that is expressed below in matrix form.

$$\bar{\mathbf{A}}\mathbf{Z} = \bar{\mathbf{B}}\mathbf{Z}\bar{\Lambda} \tag{7.309}$$

$$\left\{\begin{aligned} \bar{\Lambda} &= \mathrm{diag}\left[\rho_1,\ldots,\rho_m\right] \\ \mathbf{Z} &= \left[\mathbf{z}_1|\ldots|\mathbf{z}_m\right] \end{aligned}\right. \tag{7.310}$$

We can expect that the Ritz-pairs (ρ_j, $\mathbf{Y}\,\mathbf{z}_j$) are approximations to the eigen-pairs (λ_j, $\mathbf{x}_j$). This process can be iteratively implemented so that the desired accuracy on the Ritz-pairs can be satisfied.

Similar to standard eigenvalue problem, the subspace iteration method with the Rayleigh–Ritz process for computing the first few eigenvectors can be described in the following procedure:

$$Y_0 = \left[\mathbf{y}_1^{(0)} | \ldots | \mathbf{y}_m^{(0)} \right] \in R^{n\times m}, \ m < n$$

for j = 1, 2,...

$$\begin{aligned} &\mathbf{R}_j = \mathbf{B}\mathbf{Y}_{j-1} \\ &\mathbf{A}\overline{\mathbf{Y}}_j = \mathbf{R}_j, \text{ solve for } \overline{\mathbf{Y}}_j \\ &\overline{\mathbf{A}}_j = \overline{\mathbf{Y}}_j^T\mathbf{R}_j \\ &\overline{\mathbf{B}}_j = \overline{\mathbf{Y}}_j^T\mathbf{B}\overline{\mathbf{Y}}_j \\ &\overline{\mathbf{A}}_j\mathbf{Z}_j = \overline{\mathbf{B}}_j\mathbf{Z}_j\overline{\Lambda}_j \\ &\mathbf{Y}_j = \overline{\mathbf{Y}}_j\mathbf{Z}_j \end{aligned} \tag{7.311}$$

It can be easily seen that the Ritz-pairs obtained from this method, $(\rho_j, \mathbf{y}_j)$, converge to the lowest eigen-pairs $(\lambda_j, \mathbf{x}_j)$. In this algorithm, due to the use of the Rayleigh–Ritz analysis, the solution of the generalized eigenvalue problem in reduced dimensionality can be obtained by using QR with shift or Jacobi method.

7.8 LANCZOS METHOD

7.8.1 Simple Lanczos algorithm

The Lanczos algorithm for solution of eigenvalue problem was proposed in 1950 (Lanczos, 1950). It was based on the observation that any symmetric and positive definite matrix can be reduced to tridiagonal form through finite numbers of orthogonal similarity transformations. Lanczos algorithm is a *direct* method of reduction of a symmetric and positive definite matrix to tridiagonal form. It basically constructs a set of orthogonal vectors $\mathbf{Q} = [\mathbf{q}_1 | \ldots | \mathbf{q}_n]$, such that through orthogonal similarity transformations, this matrix of orthogonal vectors would tridiagonalize a real symmetric matrix. Since this tridiagonal matrix maintains the same eigenvalues as the original matrix, this would result in substantial saving in computational efforts. Early on, we have discussed that the QR algorithm would be very efficient in computing the eigenvalues and eigenvectors of a tridiagonal symmetric matrix. Now the question is how we can determine the Q matrix that directly accomplishes the tridiagonalization process.

Let **A** be a symmetric and positive definite $n \times n$ matrix and **Q** an orthogonal matrix, then we have the following:

$$\mathbf{T} = \mathbf{Q}^T\mathbf{A}\mathbf{Q} \tag{7.312}$$

$$\mathbf{Q}^T\mathbf{Q} = \mathbf{I} \tag{7.313}$$

This means that the column vectors of matrix Q are unit vectors and are mutually orthogonal. We also assume that matrix T is a tridiagonal matrix of the following general form:

$$\mathbf{T} = \begin{bmatrix} \alpha_1 & \beta_1 & & \\ \beta_1 & \alpha_2 & \ddots & \\ & \ddots & \ddots & \beta_{n-1} \\ & & \beta_{n-1} & \alpha_n \end{bmatrix} \tag{7.314}$$

Lanczos algorithm allows direct computation of $\mathbf{q}_j$, α_j and β_j. From the original matrix equation, we have the following and we can equate the jth column from both sides of the equation:

$$\mathbf{AQ} = \mathbf{QT} \tag{7.315}$$

$$[\mathbf{A}][\ldots|\mathbf{q}_j|\ldots] = [\ldots|\mathbf{q}_{j-1}|\mathbf{q}_j|\mathbf{q}_{j+1}|\ldots]\begin{bmatrix} \ddots & \ddots & & & \\ \ddots & \ddots & \beta_{j-1} & & \\ & \ddots & \alpha_j & \ddots & \\ & & \beta_j & \ddots & \ddots \\ & & & \ddots & \ddots \end{bmatrix} \tag{7.316}$$

After performing matrix multiplications, we obtain the following relation:

$$\mathbf{Aq}_j = \beta_{j-1}\mathbf{q}_{j-1} + \alpha_j\mathbf{q}_j + \beta_j\mathbf{q}_{j+1} \tag{7.317}$$

This is the basic equation of Lanczos algorithm. Pre-multiplying the above equation by $\mathbf{q}_j^T$ yields the following:

$$\mathbf{q}_j^T\mathbf{Aq}_j = \beta_{j-1}\mathbf{q}_j^T\mathbf{q}_{j-1} + \alpha_j\mathbf{q}_j^T\mathbf{q}_j + \beta_j\mathbf{q}_j^T\mathbf{q}_{j+1} \tag{7.318}$$

Because $\mathbf{q}_j^T\mathbf{q}_k = \delta_{jk}$, then

$$\alpha_j = \mathbf{q}_j^T\mathbf{Aq}_j \tag{7.319}$$

Once α_j is determined, and since $\mathbf{q}_j$, $\mathbf{q}_{j-1}$ and β_{j-1} are known quantities, we can determine β_j and $\mathbf{q}_{j+1}$.

$$\mathbf{r}_j = \beta_j\mathbf{q}_{j+1} = \mathbf{Aq}_j - \alpha_j\mathbf{q}_j - \beta_{j-1}\mathbf{q}_{j-1} \tag{7.320}$$

$$\|\mathbf{r}_j\| = \beta_j\left(\mathbf{q}_{j+1}^T\mathbf{q}_{j+1}\right) = \beta_j \tag{7.321}$$

$$\mathbf{q}_{j+1} = \frac{\mathbf{r}_j}{\|\mathbf{r}_j\|} = \frac{\mathbf{r}_j}{\beta_j} \tag{7.322}$$

With these equations, we arrive at the *simple Lanczos algorithm* for direct tridiagonalization of a symmetric and positive definite matrix, which is described in the following procedure:

$$
\begin{aligned}
&\mathbf{r}_0 = \text{initial vector} \\
&\beta_0 = \|\mathbf{r}_0\| \\
&\text{for } j = 1, 2, \ldots \\
&\qquad \mathbf{q}_j = \frac{1}{\beta_{j-1}} \mathbf{r}_{j-1} \\
&\qquad \mathbf{a} = \mathbf{A}\mathbf{q}_j \\
&\qquad \alpha_j = \mathbf{q}_j^T \mathbf{a} \; \left(= \mathbf{q}_j^T \mathbf{A} \mathbf{q}_j\right) \\
&\qquad \mathbf{r}_j = \mathbf{a} - \alpha_j \mathbf{q}_j - \beta_{j-1} \mathbf{q}_{j-1} \\
&\qquad \left(\text{for } j = 1, \mathbf{r}_j = \mathbf{a} - \alpha_j \mathbf{q}_j\right) \\
&\qquad \beta_j = \|\mathbf{r}_j\|
\end{aligned}
\tag{7.323}
$$

It should be noted that because the Q matrix is composed of unit orthogonal basis vectors in the n-dimensional space, there are infinite number of vectors that satisfy the condition. Therefore, the Q matrix is not unique. It seems that the simple Lanczos algorithm will keep computing values of α_j, β_j, $\mathbf{q}_j$, even when j > n. However, in order to complete the tridiagonalization process we are only required to compute up to α_n, β_{n-1}, $\mathbf{q}_n$. Now the question is when should the Lanczos process be terminated?

We can observe from above procedure that in exact mathematics, when $\beta_n = 0$, then the reduction to tridiagonal form is complete. In finite precision mathematics, because of the presence of round-off error, however, complete reduction to tridiagonal form is not possible. Hence, $\beta_n \neq 0$, and the method can continue indefinitely, computing more and more Lanczos vectors. We are actually looking at the scenario that there could only be n orthogonal vectors in the n-dimensional space, but the Lanczos algorithm in finite precision mathematics generates more than n vectors. Therefore, the Lanczos vectors computed in finite precision mathematics cannot be precisely orthogonal. As a matter of fact, in finite precision mathematics, simple Lanczos algorithm cannot maintain the orthogonality of Lanczos vectors at all. The main question is related to the impact of the loss of orthogonality on the practical use of the algorithm. To shed light on this problem of loss of orthogonality, first we need to discuss some of the fundamental properties of Lanczos algorithm.

7.8.2 Lanczos algorithm for eigenvalue solution

Suppose that in carrying out the simple Lanczos algorithm for tridiagonalization, we have performed k (< n) Lanczos steps to obtain k Lanczos vectors as $\mathbf{Q}_k = [\mathbf{q}_1 | \ldots | \mathbf{q}_k]$, and we have obtained the k × k tridiagonal matrix.

$$\mathbf{T}_k = \mathbf{Q}_k^T \mathbf{A} \mathbf{Q}_k \tag{7.324}$$

$$\mathbf{T}_k = \begin{bmatrix} \alpha_1 & \beta_1 & & & \\ \beta_1 & \ddots & \ddots & & \\ & \ddots & \ddots & \beta_{k-1} \\ & & \beta_{k-1} & \alpha_k \end{bmatrix}_{k\times k} \tag{7.325}$$

Obviously, because of the orthogonal similarity transformation performed in the tridiagonalization process, we can compute the eigenvalues of $\mathbf{T}_k$ as approximations to eigenvalues of $\mathbf{A}$, and the Ritz vectors $\mathbf{y}_j = \mathbf{Q}_k\, \mathbf{z}_j$ are approximations to eigenvectors of $\mathbf{A}$. This is in fact the basic idea behind the Lanczos algorithm for eigenvalue solutions.

Now we perform spectral decomposition on $\mathbf{T}_k$ and use Equation 7.324 to arrive at the Ritz vectors $\mathbf{Y}_k$.

$$\mathbf{T}_k\mathbf{Z}_k = \mathbf{Z}_k\bar{\mathbf{\Lambda}}_k \tag{7.326}$$

$$(\mathbf{Q}_k^T\mathbf{A}\mathbf{Q}_k)\mathbf{Z}_k = \mathbf{Z}_k\bar{\mathbf{\Lambda}}_k \Rightarrow \mathbf{A}(\mathbf{Q}_k\mathbf{Z}_k) = (\mathbf{Q}_k\mathbf{Z}_k)\bar{\mathbf{\Lambda}}_k \tag{7.327}$$

$$\mathbf{Y}_k = \mathbf{Q}_k\mathbf{Z}_k \tag{7.328}$$

$$[\mathbf{Y}_k]_{n\times k} = [\mathbf{Q}_k]_{n\times k}[\mathbf{Z}_k]_{k\times k} \tag{7.329}$$

It can be shown that the extreme Ritz pairs converge to the extreme eigen-pairs of $\mathbf{A}$.

$$\begin{aligned} &\bar{\lambda}_1 \to \lambda_1,\ \mathbf{y}_1 \to \mathbf{x}_1 \\ &\bar{\lambda}_2 \to \lambda_2,\ \mathbf{y}_2 \to \mathbf{x}_2 \\ &\quad\vdots \\ &\bar{\lambda}_{k-1} \to \lambda_{n-1},\ \mathbf{y}_{k-1} \to \mathbf{x}_{n-1} \\ &\bar{\lambda}_k \to \lambda_n,\ \mathbf{y}_k \to \mathbf{x}_n \end{aligned} \tag{7.330}$$

It is very important to note that the convergence of Lanczos algorithm Ritz pairs is dependent on maintaining a *reasonable* orthogonality of Lanczos vectors. In exact mathematics, by orthogonality, we mean the following:

$$\mathbf{Q}_k^T\mathbf{Q}_k - \mathbf{I}_k = 0 \tag{7.331}$$

However, in finite precision mathematics, because of the presence of round-off error, we cannot expect precise orthogonality.

$$\mathbf{Q}_k^T\mathbf{Q}_k - \mathbf{I}_k \neq 0 \tag{7.332}$$

$$\kappa_k = \left\| \mathbf{Q}_k^T\mathbf{Q}_k - \mathbf{I}_k \right\| \tag{7.333}$$

This shows that the magnitude of κ_k should be a *reasonable* measure of the loss of orthogonality. In other words, if there is a severe loss of orthogonality, then the size of off-diagonal terms in the matrix $Q_k^TQ_k$ keeps on increasing as the iteration progresses. It should be emphasized that loss of orthogonality in Lanczos algorithm is due to round-off error in finite precision operations within a computer. It is not a problem of algorithm itself.

Because of the damaging effect of round-off error in computing the Lanczos vectors, in practice it was discovered that simple Lanczos algorithm does not converge at all because matrix loses its orthogonality very fast, which renders the algorithm unstable for eigenvalue solutions. Therefore, in the 1950s and 1960s, Lanczos algorithm was seldom used in practice. Until 1971, the publication of Paige's theorem (Paige, 1971) revived the algorithm for eigenvalue solutions.

7.8.3 Paige's theorem and its consequences

After k Lanczos steps, we have computed k Lanczos vectors $\mathbf{Q}_k = [\mathbf{q}_1 | \ldots | \mathbf{q}_k]$, and the terms in the tridiagonal matrix.

$$\mathbf{T}_k = \begin{bmatrix} \alpha_1 & \beta_1 & & \\ \beta_1 & \ddots & \ddots & \\ & \ddots & \ddots & \beta_{k-1} \\ & & \beta_{k-1} & \alpha_k \end{bmatrix}_{k\times k} \tag{7.334}$$

Now we consider the original orthogonal similarity transformation in the following form:

$$\mathbf{AQ} = \mathbf{QT} \tag{7.335}$$

$$[\mathbf{A}][\ldots | \mathbf{q}_k | \ldots] = [\ldots | \mathbf{q}_{k-1} | \mathbf{q}_k | \mathbf{q}_{k+1} | \ldots]\begin{bmatrix} \ddots & \ddots & & & \\ \ddots & \ddots & \beta_{k-1} & & \\ & \beta_{k-1} & \alpha_k & \beta_k & \\ & & \beta_k & \ddots & \ddots \\ & & & \ddots & \ddots \end{bmatrix} \tag{7.336}$$

Through direct matrix operations, the kth column in the above equation can be computed as follows which is the basic Lanczos equation.

$$\mathbf{A}\mathbf{q}_k = \beta_{k-1}\mathbf{q}_{k-1} + \alpha_k\mathbf{q}_k + \beta_k\mathbf{q}_{k+1} \tag{7.337}$$

We can also write this equation in the following matrix form:

$$[\mathbf{A}]_{n\times n}[\mathbf{Q}_k]_{n\times k} = [\mathbf{Q}_k]_{n\times k}[\mathbf{T}_k]_{k\times k} + [0 | \beta_k\mathbf{q}_{k+1}]_{n\times k} \tag{7.338}$$

We examine the last term in the equation above, which is an n × k matrix.

$$\begin{aligned}[0 | \beta_k\mathbf{q}_{k+1}]_{n\times k} &= \beta_k\{\mathbf{q}_{k+1}\}_{n\times 1}[0,\ldots,0,1]_{1\times k} \\ &= \beta_k\mathbf{q}_{k+1}\mathbf{e}_k^T\end{aligned} \tag{7.339}$$

$$\mathbf{AQ}_k = \mathbf{Q}_k\mathbf{T}_k + \beta_k\mathbf{q}_{k+1}\mathbf{e}_k^T \tag{7.340}$$

We arrive at the following expressions by substituting in the above equation from Equation 7.320:

$$\mathbf{A}\mathbf{Q}_k = \mathbf{Q}_k\mathbf{T}_k + \mathbf{r}_k\mathbf{e}_k^T \tag{7.341}$$

$$\mathbf{r}_k = \left[\mathbf{A}\mathbf{Q}_k - \mathbf{Q}_k\mathbf{T}_k\right]\mathbf{e}_k \tag{7.342}$$

Now we conduct spectral decomposition on $\mathbf{T}_k$ and determine the Ritz pairs of matrix **A**.

$$\mathbf{T}_k\mathbf{z}_j = \bar{\lambda}_j\mathbf{z}_j \tag{7.343}$$

$$\mathbf{y}_j = \mathbf{Q}_k\mathbf{z}_j \tag{7.344}$$

As the Ritz pairs $(\bar{\lambda}_j, \mathbf{y}_j)$ converge to eigen-pairs $(\lambda_j, \mathbf{x}_j)$, a good measure of error can be expressed as the following norm:

$$\left\|\mathbf{A}\mathbf{y}_j - \bar{\lambda}_j\mathbf{y}_j\right\| \tag{7.345}$$

This norm would approach zero as the Ritz pairs converge to eigen pair

$$\lim_{(\bar{\lambda}_j, \mathbf{y}_j)\to(\lambda_j, \mathbf{x}_j)} \left\|\mathbf{A}\mathbf{y}_j - \bar{\lambda}_j\mathbf{y}_j\right\| \to 0 \tag{7.346}$$

Now we evaluate this norm using the relations obtained earlier.

$$\begin{aligned}
\left\|\mathbf{A}\mathbf{y}_j - \bar{\lambda}_j\mathbf{y}_j\right\| &= \left\|\mathbf{A}\mathbf{Q}_k\mathbf{z}_j - \bar{\lambda}_j\mathbf{Q}_k\mathbf{z}_j\right\| \\
&= \left\|\mathbf{A}\mathbf{Q}_k\mathbf{z}_j - \mathbf{Q}_k\mathbf{T}_k\mathbf{z}_j\right\| \\
&= \left\|(\mathbf{A}\mathbf{Q}_k - \mathbf{Q}_k\mathbf{T}_k)\mathbf{z}_j\right\| = \left\|\left(\mathbf{r}_k\mathbf{e}_k^T\right)\mathbf{z}_j\right\| \\
&= \left\|\beta_k\mathbf{q}_{k+1}\mathbf{e}_k^T\mathbf{z}_j\right\| \\
&= \left\|\beta_k\mathbf{e}_k^T\mathbf{z}_j\right\|
\end{aligned} \tag{7.347}$$

Hence, the convergence condition of the Lanczos algorithm becomes

$$\lim_{(\bar{\lambda}_j, \mathbf{y}_j)\to(\lambda_j, \mathbf{x}_j)} \left\|\mathbf{A}\mathbf{y}_j - \bar{\lambda}_j\mathbf{y}_j\right\| = \left\|\beta_k\mathbf{e}_k^T\mathbf{z}_j\right\| \to 0 \tag{7.348}$$

The convergence criteria can be based on the following values:

$$\beta_{jk} = \left|\beta_k\mathbf{e}_k^T\mathbf{z}_j\right| = \left|\beta_k\left[0,\ldots,0,1\right]\begin{Bmatrix} z_{j1} \\ \vdots \\ z_{jk} \end{Bmatrix}\right| = \beta_k z_{jk} \tag{7.349}$$

Therefore, to achieve convergence, either we should have $\beta_j \to 0$ or the last term of the vector $\mathbf{z}_j$ should be zero, $z_{jk} \to 0$. This shows that significant departure from orthogonality between Lanczos vectors, $\mathbf{q}_j$, can occur even when β_j is of small value, which indicates that the criterion of $\beta_j \to 0$ is not essential in measuring convergence. Hence the convergence criterion for the Ritz pairs to converge is that the last term in the vector $\mathbf{z}_j$ should be approaching zero. This is the basic result of Paige's theorem.

This theorem also yields insight into the reasons behind loss of orthogonality between Lanczos vectors. It shows that loss of orthogonality of Lanczos vectors does not occur in a random way. Instead, orthogonality of Lanczos vectors start deteriorating as soon as some of the Ritz vectors start converging; and new Lanczos vectors have significant components along the converged Ritz vectors. In addition, simple Lanczos algorithm starts finding duplicate eigenvalues. Due to Paige's theorem, it is obvious that Lanczos vectors computed from the algorithm should be checked frequently and orthogonalized if necessary. In practice, there are Lanczos algorithms that use either full orthogonalization of Lanczos vectors or selective orthogonalization.

7.8.4 Lanczos algorithm with full orthogonalization

With the use of orthogonalization or re-orthogonalization, the computed Lanczos vectors are orthogonalized once there is a loss of orthogonality. In full orthogonalization, during each Lanczos step, we orthogonalize the current computed Lanczos vector with respect to all Lanczos vectors computed in previous steps. We can use modified Gram–Schmidt procedure or Householder method. Simple orthogonalization can be carried out as follows in the jth Lanczos step.

$$\mathbf{r}_j = \mathbf{r}_j - \sum_{k=1}^{j-1} \left(\mathbf{q}_k^T \mathbf{r}_j\right)\mathbf{q}_k \tag{7.350}$$

However, as can be observed from numerical experimentation, the use of full orthogonalization schemes does not guarantee orthogonality of Lanczos vectors to working precision. This is because the source of loss of orthogonality lies with the round-off error. On the other hand, because new Lanczos vector has components in the converged Ritz vectors, it seems that carrying out re-orthogonalization of Lanczos vectors with each other is not a good idea.

7.8.5 Lanczos algorithm with selective orthogonalization

The main idea behind selective orthogonalization is the realization that the most recently computed Lanczos vector, $\mathbf{q}_{k+1}$ or $\mathbf{r}_k$ tends to have a component in the direction of any converged Ritz vectors. Therefore, orthogonalizing the newly computed Lanczos vector $\mathbf{q}_{k+1}$ against the converged Ritz vectors $\mathbf{y}_k$ will accomplish the same results obtained from full orthogonalization. This is an interesting idea because not only will it get rid of the converged Ritz vector components in the newly computed Lanczos vector, but the number of converged Ritz vectors is also fewer than the number of computed Lanczos vectors at that iteration. As a consequence, by maintaining reasonable linear independence of Lanczos vectors, it results in an efficient algorithm that works reasonably well.

One way to carry out the selective orthogonalization is to orthogonalize $\mathbf{q}_{j+1}$ or $\mathbf{r}_j$ with respect to $\mathbf{y}_k^{(j)}$ which have converged, when the following conditions are satisfied:

$$\begin{cases} \beta_{jk} \le \dfrac{\varepsilon \|\mathbf{A}\|}{\left(\kappa/\sqrt{j}\right)} = \sqrt{\varepsilon j}\,\|\mathbf{A}\| \\ \mathbf{y}_k^{(j)T}\mathbf{r}_j \ge \kappa\sqrt{j}\,\left\|\mathbf{y}_k^{(j)}\right\|\|\mathbf{r}_j\| \end{cases} \tag{7.351}$$

ε = relative precision of arithmetic; $\kappa = \sqrt{\varepsilon}$; j = iteration number; and $\kappa_j = \left\|\mathbf{Q}_j^T\mathbf{Q}_j - \mathbf{I}_j\right\|$, with $\kappa_j < \kappa$. This scheme was proposed by Parlett and Scott (1979). It basically monitors the loss of orthogonality by considering the measurement $\kappa_j = \left\|\mathbf{Q}_j^T\mathbf{Q}_j - \mathbf{I}_j\right\|$. When a preset bound signals loss of orthogonality, it then expands the set of good Ritz vectors by solving the eigenvalue problem $\mathbf{T}_k\mathbf{z}_j = \bar{\lambda}_j\mathbf{z}_j$. Obviously, this scheme does not require the eigenvalue solution with $\mathbf{T}_k$ at each iteration; it calls for eigenvalue solution only when it experiences loss of orthogonality among Lanczos vectors.

Another way to implement the selective orthogonalization scheme is to compute the eigenvalue solution of $\mathbf{T}_k\mathbf{z}_j = \bar{\lambda}_j\mathbf{z}_j$ at each iteration. And monitor the converged Ritz vectors by checking the last term in the vector $\mathbf{z}_j$ which is the eigenvector of $\mathbf{T}_k$. This procedure is computationally more intensive.

7.8.6 Lanczos step for lowest few eigenvalues and eigenvectors

Like the power method, the standard Lanczos step converges to the highest few eigen-pairs of the original matrix **A**. To compute the lowest few eigen-pairs, we need to make some modifications on the algorithm. From the Lanczos decomposition in Equation 7.315, we have the following:

$$\mathbf{A}^{-1}\mathbf{Q} = \mathbf{Q}\mathbf{T}^{-1} \tag{7.352}$$

It can be observed that if **A** and **T** are symmetric, then $\mathbf{A}^{-1}$ and $\mathbf{T}^{-1}$ would remain symmetric and the tridiagonal form of **T** should be preserved. Let $\mathbf{T}^{-1} = \bar{\mathbf{T}}$ then

$$\mathbf{A}^{-1}\mathbf{Q} = \mathbf{Q}\bar{\mathbf{T}} \tag{7.353}$$

$$\bar{\mathbf{T}} = \begin{bmatrix} \ddots & \ddots & & & \\ \ddots & \ddots & \bar{\beta}_{k-1} & & \\ & \bar{\beta}_{k-1} & \bar{\alpha}_k & \bar{\beta}_k & \\ & & \bar{\beta}_k & \ddots & \ddots \\ & & & \ddots & \ddots \end{bmatrix} \tag{7.354}$$

Similarly, from the matrix Equation 7.352, we can obtain the following relation:

$$\mathbf{A}^{-1}\mathbf{q}_k = \bar{\beta}_{k-1}\mathbf{q}_{k-1} + \bar{\alpha}_k\mathbf{q}_k + \bar{\beta}_k\mathbf{q}_{k+1} \tag{7.355}$$

Using orthogonality of Lanczos vectors, we can obtain the following relations:

$$\begin{cases} \bar{\alpha}_j = \mathbf{q}_j^T \mathbf{A}^{-1} \mathbf{q}_j = (\mathbf{A}^{-1} \mathbf{q}_j)^T \mathbf{q}_j \\ \mathbf{r}_j = \mathbf{A}^{-1} \mathbf{q}_j - \bar{\alpha}_j \mathbf{q}_j - \bar{\beta}_{j-1} \mathbf{q}_{j-1} \\ \bar{\beta}_j = (\mathbf{r}_j^T \mathbf{r}_j)^{1/2} \\ \mathbf{q}_{j+1} = \dfrac{1}{\bar{\beta}_j} \mathbf{r}_j \end{cases} \tag{7.356}$$

With this Lanczos step, we can expect that the computed Ritz pairs would approximate the lowest eigen-pairs of matrix **A**.

7.8.7 Practical Lanczos algorithms

From our discussion in this section, we observe that a practical Lanczos algorithm should have selective orthogonalization by orthogonalizing its Lanczos vectors with respect to computed Ritz vectors. In the following, we will describe two practical algorithms. The first algorithm converges to higher eigen-pairs; and the second algorithm converges to the lower eigen-pairs.

Algorithm #1

Initialization

$\mathbf{r} \leftarrow \mathbf{r}_0 \neq 0$

$\beta_0 \leftarrow (\mathbf{r}^T\mathbf{r})^{1/2}$

for j = 1, 2, ...

$\mathbf{q}_j \leftarrow \dfrac{1}{\beta_{j-1}} \mathbf{r}$

$\mathbf{r} \leftarrow \mathbf{A}\mathbf{q}_j$

$\alpha_j \leftarrow \mathbf{r}^T\mathbf{q}_j$

$\mathbf{r} \leftarrow \mathbf{r} - \alpha_j \mathbf{q}_j - \beta_{j-1} \mathbf{q}_{j-1}$

$\left(\text{for } j = 1, \mathbf{r} \leftarrow \mathbf{r} - \alpha_j \mathbf{q}_j\right)$

Perform selective orthogonalization, if necessary

$\beta_j \leftarrow (\mathbf{r}^T\mathbf{r})^{1/2}$

Compute Ritz pairs

In this algorithm, the "compute Ritz pairs" procedure involves the following eigenvalue solution process:

$$\begin{aligned} \mathbf{T}_j\mathbf{z}_i^{(j)} &= \bar{\lambda}_i\mathbf{z}_i^{(j)}, \quad i = 1,\ldots, j \\ \mathbf{y}_i^{(j)} &= \mathbf{Q}_j\mathbf{z}_i^{(j)}, \quad i = 1,\ldots, j \end{aligned} \tag{7.357}$$

Because $\mathbf{T}_j$ is a small matrix, its eigenvalues and eigenvectors can be readily computed using QR or Jacobi method. This algorithm computes the highest few eigen-pairs.

Algorithm #2

Initialization

$\mathbf{r} \leftarrow \mathbf{r}_0 \neq 0$

$\beta_0 \leftarrow (\mathbf{r}^T\mathbf{r})^{1/2}$

for $j = 1, 2, \ldots$

$\mathbf{q}_j \leftarrow \dfrac{1}{\beta_{j-1}}\mathbf{r}$

$\mathbf{r} \leftarrow$ solution of $\mathbf{Br} = \mathbf{q}_j$

$\alpha_j \leftarrow \mathbf{r}^T\mathbf{q}_j$

$\mathbf{r} \leftarrow \mathbf{r} - \alpha_j\mathbf{q}_j - \beta_{j-1}\mathbf{q}_{j-1}$

(for $j = 1$, $\mathbf{r} \leftarrow \mathbf{r} - \alpha_j\mathbf{q}_j$)

Perform selective orthogonalization, if necessary

$\beta_j \leftarrow (\mathbf{r}^T\mathbf{r})^{1/2}$

Compute Ritz pairs

In this algorithm, the "compute Ritz pairs" procedure is the same as in Algorithm 1. This algorithm converges to the lowest few eigen-pairs. It can be observed that the only difference between the two algorithms is in the computation of vector **r**. These two algorithms have close resemblance to power and inverse power methods discussed earlier.

In Algorithm 2, the solution of $\mathbf{Ar} = \mathbf{q}_j$ can be obtained using pre-conditioned conjugate gradient method or other direct solution methods such as Gauss-elimination method. However, if preconditioned conjugate gradient method is used in the solution process, with finite element analysis, there is no need to store and formulate the structural stiffness matrix. This feature would make the Lanczos algorithm well suited for computing eigen-pairs of very large structural problems. There are other interesting properties associated with the use of conjugate gradient method in the solution of $\mathbf{Ar} = \mathbf{q}_j$. After performing selective orthogonalization, the vector **r** is orthogonal to the converged Ritz vectors, which means **r** is almost orthogonal to $\mathbf{q}_j$. As more Ritz vectors reach convergence, the dimensionality of search space for vector **r** which is complement to the subspace spanned by $\mathbf{q}_j$ decreases. Therefore, we can expect faster and faster convergence rate as the Lanczos iteration progresses.

The Lanczos algorithm is well suited for solving very large eigenvalue problems. In algorithm 1, each iteration involves matrix vector multiplication using matrix **A**. Even with algorithm 2, the solution of system of equations can be solved with preconditioned conjugate gradient method that does not require the explicit formulation of the structural stiffness matrix. All other computations require only vector manipulations. The Ritz vectors can be computed readily from standard eigenvalue solution routines.

7.8.8 Lanczos algorithm for generalized eigenvalue problems

The Lanczos algorithm for generalized eigenvalue problems, $\mathbf{Ax} = \lambda\ \mathbf{Bx}$, can be readily derived. Similar to standard eigenvalue problem, we can have two algorithms: the first converges to the largest few eigen-pairs and the second algorithm converges to the lowest few eigen-pairs. We can describe the first algorithm in the following procedure:

Intialization

$\mathbf{r} \leftarrow \mathbf{r}_0 \neq 0$

$\beta_0 \leftarrow \left(\mathbf{r}^T\mathbf{Br}\right)^{1/2}$

for $j=1,2,\ldots$

$$\mathbf{q}_j \leftarrow \frac{1}{\beta_{j-1}}\mathbf{r}$$

$\mathbf{r} \leftarrow$ solution of $\mathbf{Br} = \mathbf{Aq}_j$

$\alpha_j \leftarrow \mathbf{r}^T\mathbf{Bq}_j$

$\mathbf{r} \leftarrow \mathbf{r} - \alpha_j\mathbf{q}_j - \beta_{j-1}\mathbf{q}_{j-1}$

$\left(\text{for } j=1,\ \mathbf{r} \leftarrow \mathbf{r} - \alpha_j\mathbf{q}_j\right)$

Perform selective orthogonalization, if necessary

$\beta_j \leftarrow \left(\mathbf{r}^T\mathbf{Br}\right)^{1/2}$

Compute Ritz pairs

In the second algorithm, the vector **r** is computed from solving the following system of equations:

$$\mathbf{Ar} = \mathbf{Bq}_j \tag{7.358}$$

The rest of the algorithm is the same as the first algorithm

Chapter 8

Direct integration of dynamic equations of motion

8.1 INTRODUCTION

In dynamic analysis of linear structural systems, after the spatial discretization with finite elements, finite difference, or any other discretization method, we arrive at the dynamic equation of motion in the form of a second-order ordinary differential equation.

$$\mathbf{M}\ddot{\mathbf{x}}(t) + \mathbf{C}\dot{\mathbf{x}}(t) + \mathbf{K}\mathbf{x}(t) = \mathbf{p}(t) \tag{8.1}$$

where:

- $\mathbf{M}$ is the structural mass matrix
- $\mathbf{C}$ is the damping matrix
- $\mathbf{K}$ is the linear structural stiffness matrix
- $\ddot{\mathbf{x}}(t)$ is the vector of accelerations at the degrees of freedom
- $\dot{\mathbf{x}}(t)$ is the vector of velocities at the degrees of freedom
- $\mathbf{x}(t)$ is the vector of displacements
- $\mathbf{p}(t)$ is the external force vector

The dynamic equilibrium equation of a linear dynamic system can also be written in terms of the inertia force vector, the damping force vector, and the elastic force or internal resisting force vector.

$$\mathbf{f}_I(t) + \mathbf{f}_D(t) + \mathbf{f}_E(t) = \mathbf{p}(t) \tag{8.2}$$

For a geometrically nonlinear structural system under incremental Lagrangian formulation, the dynamic equation of motion can be derived as follows:

$$\mathbf{M}\ddot{\mathbf{x}}(t) + \mathbf{C}\dot{\mathbf{x}}(t) + \mathbf{K}_t\Delta\mathbf{x}(t) = \mathbf{p}(t) - \mathbf{I}(t - \Delta t) \tag{8.3}$$

$\mathbf{I}(t)$ is the internal resisting force vector that is normally computed by assembling the element contributions, and $\mathbf{K}_t$ is the tangent stiffness matrix. Even for nonlinear systems, usually, the mass matrix $\mathbf{M}$ and damping matrix $\mathbf{C}$ are assumed to remain constant. This is because there are many factors contributing to the damping property of structural systems and our understanding of the characteristics of damping is fairly limited. On the other hand, experience shows that even though this is a simplified model for the mass and damping matrices, it is a reasonable assumption.

Another type of time-dependent equation of motion we encounter in engineering modeling is called diffusion equation, which arises from the modeling of problems such as viscous flow, thermal flow, and flow through porous media. We also encounter these problems in viscoelastic or nonlinear time-dependent material behavior. The diffusion equation describes the *transient* behavior of a system as a first-order ordinary differential equation. They are also referred to as quasi-static problems.

$$\mathbf{C}\dot{\mathbf{x}}(t) + \mathbf{K}\mathbf{x}(t) = \mathbf{p}(t) \tag{8.4}$$

Obviously, problems described by these dynamic system of equations are initial value problems. In practice, unless the system is extremely simple, solutions of initial value problems are obtained using numerical integration methods. In computational mechanics, specifically structural dynamics and earthquake engineering, these initial value problems are solved using basically two approaches: the *mode superposition method* and *direct integration methods*. Within the realm of direct integration methods, there are *explicit* integration methods and *implicit* integration methods. Direct integration methods can be used for solving dynamic equation of motion of linear and nonlinear systems. However, the mode superposition method applies to linear dynamic systems only. Our discussion in this chapter focuses on the solution methods for dynamic responses of linear systems. Specifically, we are dealing with structural dynamic problems. Issues associated with obtaining dynamic responses of geometrically nonlinear systems will be also discussed.

8.2 MODE SUPERPOSITION METHOD

Considering the dynamic equation of motion of a linear structural system, Equation 8.1, in a general case, matrices **M**, **C**, and **K** are not all diagonal. Although they are all symmetric matrices, their bandwidth is not small in most cases, which means that the system of equations is highly coupled. It is obvious that if the system of equations of motion can be decoupled, then the solution process can be reduced to obtaining the solution of individual ordinary differential equations. The system response can be obtained by summing up the responses of the individual ordinary differential equations. This is in fact the basic idea behind the mode superposition method. We can show that this decoupling process can be accomplished through coordinate transformations.

Suppose we transform the displacement vector to generalized displacements by following linear transformation:

$$\mathbf{x}(t) = \mathbf{Q}\ \mathbf{y}(t) \tag{8.5}$$

where:

$\mathbf{x}(t)$ refers to spatial coordinates
$\mathbf{y}(t)$ is the generalized coordinates
$\mathbf{Q}$ is an $n \times n$ nonsingular square transformation matrix of order n

At the present, we do not specify the property of matrix **Q**, which will become evident later. Now, we substitute the above coordinate transformation equation into the original dynamic equation of motion, Equation 8.1, and premultiply it by $\mathbf{Q}^T$.

$$\left(\mathbf{Q}^T\mathbf{M}\mathbf{Q}\right)\ddot{\mathbf{y}}(t) + \left(\mathbf{Q}^T\mathbf{C}\mathbf{Q}\right)\dot{\mathbf{y}}(t) + \left(\mathbf{Q}^T\mathbf{K}\mathbf{Q}\right)\mathbf{y}(t) = \mathbf{Q}^T\mathbf{p}(t) \tag{8.6}$$

$$\bar{\mathbf{M}}\ddot{\mathbf{y}}(t) + \bar{\mathbf{C}}\ \dot{\mathbf{y}}(t) + \bar{\mathbf{K}}\ \mathbf{y}(t) = \bar{\mathbf{p}}(t) \tag{8.7}$$

$$\begin{cases} \bar{M} = Q^T M Q \\ \bar{C} = Q^T C Q \\ \bar{K} = Q^T K Q \end{cases} \tag{8.8}$$

After transformation $\bar{M}$, $\bar{C}$, and $\bar{K}$ are the mass, damping, and stiffness matrices in the new coordinate system. Since our objective is to decouple the original system of equation, then we should expect that after the transformation, $\bar{M}$, $\bar{C}$, and $\bar{K}$ become diagonal matrices. This requires that the transformation matrix Q should be orthogonal because only through orthogonal similarity transformation can we transform a full symmetric matrix to a diagonal matrix. Now the question is what kind of Q matrix would accomplish the above tasks. To determine the Q matrix, first, we conduct some analysis on the free vibration response of an idealized undamped system.

8.2.1 Undamped free vibration analysis

In theory, there can be many different transformation matrices Q, which through similarity transformation would reduce a symmetric matrix to its diagonal form. However, in practice, an effective transformation matrix can be obtained from the displacement solution of equilibrium equation for undamped free vibration. In this special case, there is no physical damping present in the system and no external dynamic loading acting on it. We do need to have initial conditions (either displacement or acceleration) to set the structure in motion. We set $\mathbf{C} = 0$ and $\mathbf{p}(t) = 0$ in the dynamic equation of motion for the undamped free vibration of the system.

$$\mathbf{M}\ddot{\mathbf{x}}(t) + \mathbf{K}\mathbf{x}(t) = 0 \tag{8.9}$$

We can represent the dynamic response of a structure as a linear combination of its mode shapes.

$$\mathbf{x}(t) = \sum_{j=1}^{n} \boldsymbol{\phi}_j \, y_j(t) = \boldsymbol{\Phi}\mathbf{y}(t) \tag{8.10}$$

$$\begin{cases} \boldsymbol{\Phi} = [\boldsymbol{\phi}_1 | \cdots | \boldsymbol{\phi}_n] \\ \mathbf{y}(t) = [y_1(t), \ldots, y_n(t)]^T \end{cases} \tag{8.11}$$

$$\ddot{\mathbf{x}}(t) = \boldsymbol{\Phi}\ddot{\mathbf{y}}(t) \tag{8.12}$$

where:

$\boldsymbol{\phi}_j$ is the vibration mode shape
y_j is the mode amplitude function

Now we consider the case of single mode response of the system. Let ω_j, $\boldsymbol{\phi}_j$ be the jth natural frequency and corresponding vibration mode, respectively. The amplitude function can be assumed to be of the following general form:

$$y_j(t) = A_j \cos\omega_j t + B_j \sin\omega_j t \tag{8.13}$$

$$\ddot{y}_j(t) = -\omega_j^2 \, y_j(t) \tag{8.14}$$

Substituting the relations for the single-mode response into the equation of motion for undamped free vibration yields the following:

$$\left(-\omega_j^2 \mathbf{M}\phi_j + \mathbf{K}\phi_j\right) y_j(t) = 0 \tag{8.15}$$

For nontrivial solution we arrive at the following generalized eigenvalue problem:

$$\mathbf{K}\phi_j = \omega_j^2 \mathbf{M}\phi_j \tag{8.16}$$

$$\mathbf{K\Phi} = \mathbf{M\Phi\Omega} \tag{8.17}$$

$$\Omega = \begin{bmatrix} \omega_1^2 & & \\ & \ddots & \\ & & \omega_n^2 \end{bmatrix} \tag{8.18}$$

From properties of generalized eigenvalue problem, we know that eigenvectors are M-orthonormal.

$$\begin{cases} \mathbf{\Phi}^T \mathbf{M\Phi} = \mathbf{I} \\ \mathbf{\Phi}^T \mathbf{K\Phi} = \Omega \end{cases} \tag{8.19}$$

This shows that if we take $\mathbf{Q} = \mathbf{\Phi}$, which is the eigenvector matrix obtained from undamped free vibration analysis, then the equation of motion for undamped free vibration becomes decoupled. This means that the following matrix equation is a decoupled system of equations:

$$\left(\mathbf{\Phi}^T \mathbf{M\Phi}\right)\ddot{\mathbf{y}}(t) + \left(\mathbf{\Phi}^T \mathbf{K\Phi}\right)\mathbf{y}(t) = 0 \tag{8.20}$$

$$\begin{Bmatrix} \ddot{y}_1(t) \\ \vdots \\ \ddot{y}_n(t) \end{Bmatrix} + \begin{bmatrix} \omega_1^2 & & \\ & \ddots & \\ & & \omega_n^2 \end{bmatrix} \begin{Bmatrix} y_1(t) \\ \vdots \\ y_n(t) \end{Bmatrix} = 0 \tag{8.21}$$

With the system of equations of motion decoupled, we can then solve the equation of motion for each mode and add the individual responses together to obtain the system response. The solution for a second-order ordinary differential equation can be routinely obtained.

8.2.2 Decoupling of dynamic equation of motion

Using $\mathbf{\Phi}$, the eigenvector matrix obtained from undamped free vibration analysis, as the transformation matrix, we have the transformation from spatial coordinates to generalized coordinates.

$$\begin{cases} \mathbf{x}(t) = \mathbf{\Phi y}(t) \\ \dot{\mathbf{x}}(t) = \mathbf{\Phi}\dot{\mathbf{y}}(t) \\ \ddot{\mathbf{x}}(t) = \mathbf{\Phi}\ddot{\mathbf{y}}(t) \end{cases} \tag{8.22}$$

Next, we substitute these relations into the dynamic equation of motion,

$$\mathbf{M}\left(\Phi\ddot{\mathbf{y}}(t)\right)+\mathbf{C}\left(\Phi\dot{\mathbf{y}}(t)\right)+\mathbf{K}\left(\Phi\mathbf{y}(t)\right)=\mathbf{p}(t) \tag{8.23}$$

$$\left(\Phi^T\mathbf{M}\Phi\right)\ddot{\mathbf{y}}(t)+\left(\Phi^T\mathbf{C}\Phi\right)\dot{\mathbf{y}}(t)+\left(\Phi^T\mathbf{K}\Phi\right)\mathbf{y}(t)=\Phi^T\mathbf{p}(t) \tag{8.24}$$

$$\bar{\mathbf{M}}\ddot{\mathbf{y}}(t)+\bar{\mathbf{C}}\dot{\mathbf{y}}(t)+\bar{\mathbf{K}}\mathbf{y}(t)=\bar{\mathbf{p}}(t) \tag{8.25}$$

$$\begin{cases}\bar{\mathbf{M}}=\Phi^T\mathbf{M}\Phi=\mathbf{I}\\ \bar{\mathbf{C}}=\Phi^T\mathbf{C}\Phi\\ \bar{\mathbf{K}}=\Phi^T\mathbf{K}\Phi=\Omega\end{cases} \tag{8.26}$$

We observe that the transformed mass matrix and stiffness matrix are diagonalized or decoupled. However, the question is whether the transformed damping matrix would be diagonalized or decoupled as well. Obviously, if the matrix is decoupled, then we have uncoupled the system of dynamic equations of motion of structural system into individual equations of motion of each mode. This indicates the decoupling of the transformed damping matrix is the key in accomplishing this decoupling procedure.

To decouple the generalized damping matrix, we have to understand what contributes to damping in a structural system. Those factors include material property, joint friction, and so on, which are often complicated phenomena to model. It is in fact difficult to delineate the parameters and their contributions to damping in mathematical models. Therefore, in practice, the concept of *modal damping* is used, which means assigning percentage of damping ratio for each mode. This is a reasonable assumption which complies with observations on the behavior of structures. With modal damping employed, we decouple the generalized damping matrix. With damping ratio for the jth mode assumed to be ξ_j, we achieve decoupling of the generalized damping matrix.

$$\bar{\mathbf{C}}=\Phi^T\mathbf{C}\Phi=\begin{bmatrix}2\xi_1\omega_1 & & \\ & \ddots & \\ & & 2\xi_n\omega_n\end{bmatrix} \tag{8.27}$$

Usually, higher modes have higher damping ratio. In structural control, attempt is made to reduce the response of lower modes, and shift the energy to higher modes where it can be dissipated with higher damping.

Another way to decouple the generalized damping matrix is through the use of proportional damping which assumes that the damping matrix is a linear combination of the mass and stiffness matrices.

$$\mathbf{C}=\alpha\mathbf{M}+\beta\mathbf{K} \tag{8.28}$$

α and β are two constants. There are some intuitive reasons behind this assumption. In the above relation, the first part ($\alpha\mathbf{M}$) can be considered as the damping contribution from fluid–structure interaction, and the constant α can be considered as related to the viscosity of the fluid; and the second part ($\beta\mathbf{K}$) represents viscoelastic behavior. All these factors can contribute to the damping characteristics of a structural system. With proportional damping, we can achieve decoupling of the generalized damping matrix.

$$\begin{aligned}\bar{C} &= \Phi^T C \Phi \\ &= \alpha \Phi^T M \Phi + \beta \Phi^T K \Phi \\ &= \alpha I + \beta \Omega\end{aligned} \tag{8.29}$$

Therefore, with the generalized damping matrix diagonalized, we have uncoupled the system of equation of motion into n ordinary differential equations, one for each mode.

With modal damping, the uncoupled system of equation of motion and the uncoupled equation of motion for jth mode is as follows:

$$\begin{Bmatrix} \ddot{y}_1(t) \\ \vdots \\ \ddot{y}_n(t) \end{Bmatrix} + \begin{bmatrix} 2\xi_1\omega_1 & & \\ & \ddots & \\ & & 2\xi_n\omega_n \end{bmatrix} \begin{Bmatrix} \dot{y}_1(t) \\ \vdots \\ \dot{y}_n(t) \end{Bmatrix} + \begin{bmatrix} \omega_1^2 & & \\ & \ddots & \\ & & \omega_n^2 \end{bmatrix} \begin{Bmatrix} y_1(t) \\ \vdots \\ y_n(t) \end{Bmatrix} = \begin{Bmatrix} \phi_1^T p(t) \\ \vdots \\ \phi_n^T p(t) \end{Bmatrix} \tag{8.30}$$

$$\ddot{y}_j(t) + 2\xi_j\omega_j\dot{y}_j(t) + \omega_j^2 y_j(t) = \phi_j^T \mathbf{p}(t) \tag{8.31}$$

For proportional damping, we can obtain equation of motion for the jth mode as follows:

$$\ddot{y}_j(t) + \left(\alpha + \beta\omega_j^2\right)\dot{y}_j(t) + \omega_j^2 y_j(t) = \phi_j^T \mathbf{p}(t) \tag{8.32}$$

These are scalar ordinary differential equations and their closed form or numerical solutions can be readily obtained. With the solution of n uncoupled ordinary differential equations obtained, we then determine the response of structural system from superposition of modal responses.

$$\mathbf{x}(t) = \sum_{j=1}^{n} \phi_j \, y_j(t) \tag{8.33}$$

This procedure for obtaining dynamic response of structural system through decoupling of the dynamic equation of motion is thus called mode superposition method. It should be mentioned that this method can only be applied to linear structural systems.

8.2.3 Base motion

Dynamic excitation to structural systems can be of different forms from various sources. One unique type of dynamic excitation is from base motion. Two examples are shown in Figure 8.1, a structure subjected to earthquake ground motion and a truck driving over bumpy road. In this case, if we let $\mathbf{x}_r$ represent the displacement of the structural system relative to the ground and $\mathbf{x}_g$ as the ground motion, then the dynamic equation of motion of a structural system subjected to base motion can be derived as follows:

$$\mathbf{M}\ddot{\mathbf{x}}_r(t) + \mathbf{C}\dot{\mathbf{x}}_r(t) + \mathbf{K}\mathbf{x}_r(t) = \mathbf{M}\mathbf{L}\ddot{\mathbf{x}}_g(t) \tag{8.34}$$

The right-hand side of the equation represents the force exerted on the structure by the ground acceleration, such as the recorded earthquake acceleration time histories. In the case of the earthquake ground motion, the vector **L** consists of elements of value 1 at the horizontal degrees of freedom and zeroes at all the other degrees of freedom. In the case of the truck going over bumpy road, the values of 1 would be at the vertical degrees of freedom.

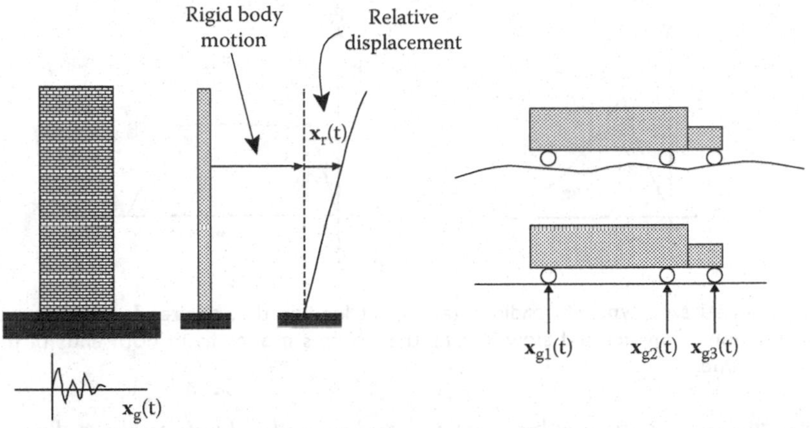

Figure 8.1 Examples of base motion: seismic excitation and vehicle over bumpy road.

We note that in arriving at formulating these dynamic problems, we have made the assumption that stress waves propagating through the structural system are ignored. With structures subjected to near source earthquakes, where bursts of motions occur due to fault movement, this formulation may not be appropriate to capture the whole response of structures.

After decoupling the system of equation of motion, we have the following generalized external load for the mode number j:

$$\phi_j^T p(t) = \left(\phi_j^T ML\right) \ddot{x}_g(t) \tag{8.35}$$

The scalar quantity $\left(\phi_j^T ML\right)$ is called *modal participation factor.* This is an important concept in structural dynamics. A typical relationship between modal participation factor and the natural frequency is as shown in Figure 8.2. We observe from this figure that in most cases in structural dynamics, keeping only the first (lowest) few modes would be sufficient to obtain reasonable structural response.

The main motivation behind modal analysis is the fact that in most cases only a few modes of a structural system are activated significantly and have dominant contribution to the response of the structural system. The rest of the modes, having insignificant contribution to

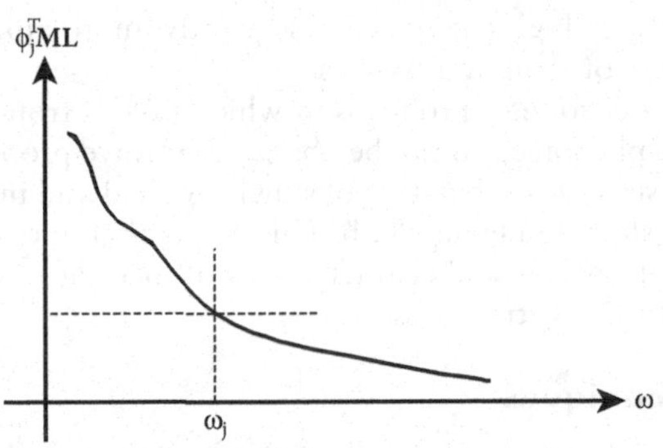

Figure 8.2 A typical relationship for modal participation factor.

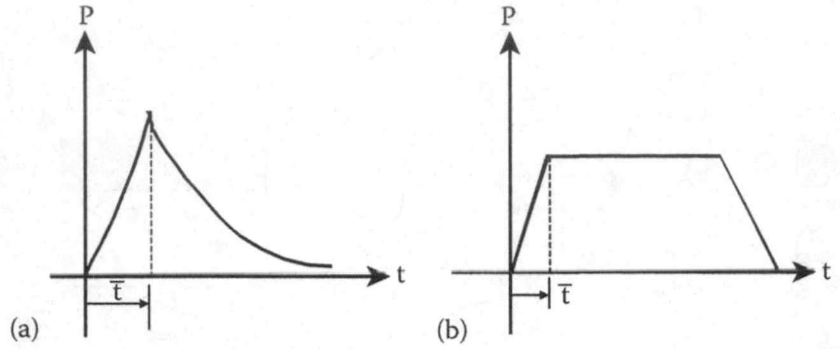

Figure 8.3 Characteristics of dynamic loadings: (a) impact loading that excites few higher modes and (b) combination of impact and slow loading that excites modes from both ends of the spectrum. $\bar{t}$ = rise time.

the structural response, can thus be neglected and excluded from analysis. By so doing, we can obtain reasonably accurate response of structures without the highly time-consuming task of computing all modes. However, the question as to which modes would be excited in a particular dynamics problem depends on many factors including the stiffness of the structure, damping property, and the nature of loading.

Of particular interest is the class of problems in which mode shapes from either extreme ends of the eigenvalue spectrum are activated. Often the modes from the middle part of the spectrum can be ignored, since they contribute little to the structural response. As has been discussed in Chapter 7, these extreme eigenpairs can be readily computed using efficient iterative methods for solving eigenvalue problems, such as the power method, inverse power method, or subspace iterations. In general, for most structural dynamic problems, only a few of the lower modes are activated. These include earthquake excitation and wind loading on tall buildings, and problems associated with slowly applied loading conditions. For all these cases, we only need to solve for the lowest few modal responses. For most structures (buildings), the natural frequencies fall into the range of $\omega = 0.1 - 10$ Hz, or the natural period $T = 10 - 0.1$ sec. Even in case the damping matrix cannot be diagonalized, then we only need to solve a $k \times k$ coupled equations (k is the number modes included in computing the structural response).

When wave propagation is involved, usually, higher modes of the structure are activated. This includes cases where structures are subject to impact, blast-and-shock loadings. A typical impact loading can be represented as a P–t curve in Figure 8.3a, in which the *rise time* is typically very short. The parameter rise time has a determining effect on the cluster of higher modes that could be activated by the impact loading. In general, the shorter the rise time, the higher the order of modes activated. This means that under impact loading, higher modes can carry very high energies. Subsequently, more modes need to be included in computing the response of structural system.

It is not uncommon to encounter problems in which modes from both ends of the spectrum are activated. This phenomenon can be observed in wave propagation problem and in cases where loading consists of combination of slowly applied and impact-type loading such as the loading condition shown in Figure 8.3b. This shows that understanding of fundamental behavior of structural systems and characteristics of loading is essential in carrying out reasonable dynamic analysis of structures.

8.2.4 Reduction techniques

It has been observed that lower natural frequencies and mode shapes are not sensitive to mass distributions. Accordingly, dynamic property of structural systems can be explored by appropriate mass lumping method such that the dimensionality of the problem can be drastically reduced. For example, in the dynamic analysis of frame structures, if we lump the mass at the structural joints, then we can dramatically reduce the dynamic degrees of freedom of the system. In structural dynamics, basically we neglect the rotational inertia of the lumped mass and ignore axial deformation modes, which essentially contribute to higher mode responses. In applying a *mass lumping* method by neglecting the mass at certain type of dynamic degrees of freedom which contribute to higher mode responses, we are in fact performing partitioning on the original system of dynamic equations of motion. With undamped free vibration, we partition the equation of motion in the following way:

$$\begin{bmatrix} \mathbf{M} & \\ & 0 \end{bmatrix} \begin{Bmatrix} \ddot{\mathbf{x}}_1 \\ \ddot{\mathbf{x}}_2 \end{Bmatrix} + \begin{bmatrix} \mathbf{K}_{11} & \mathbf{K}_{12} \\ \mathbf{K}_{21} & \mathbf{K}_{22} \end{bmatrix} \begin{Bmatrix} \mathbf{x}_1 \\ \mathbf{x}_2 \end{Bmatrix} = 0 \tag{8.36}$$

We can reduce the dimensionality of the system by only maintaining the lumped mass degrees of freedom.

$$\mathbf{M}\ddot{\mathbf{x}}_1 + \bar{\mathbf{K}}\mathbf{x}_1 = 0 \tag{8.37}$$

$$\bar{\mathbf{K}} - \mathbf{K}_{11} - \mathbf{K}_{12}\mathbf{K}_{22}^{-1}\mathbf{K}_{21} \tag{8.38}$$

The reduced stiffness matrix is obtained from static condensation; determining $\mathbf{x}_2$ from the second equation and substituting in the first equation. In practice, this reduced system of equations of motion can have as few as 10% of the total degrees of freedom.

Another method for carrying out dimensionality reduction is called *Guyan reduction* (Guyan, 1965). We partition the system of equation of motion in undamped free vibration.

$$\begin{bmatrix} \mathbf{M}_{11} & \mathbf{M}_{12} \\ \mathbf{M}_{21} & \mathbf{M}_{22} \end{bmatrix} \begin{Bmatrix} \ddot{\mathbf{x}}_1 \\ \ddot{\mathbf{x}}_2 \end{Bmatrix} + \begin{bmatrix} \mathbf{K}_{11} & \mathbf{K}_{12} \\ \mathbf{K}_{21} & \mathbf{K}_{22} \end{bmatrix} \begin{Bmatrix} \mathbf{x}_1 \\ \mathbf{x}_2 \end{Bmatrix} = 0 \tag{8.39}$$

In this method, we assume that we have a corresponding static problem, where we perform the reduction in the structural degrees of freedom. This leads to the following relations:

$$\mathbf{K}_{21}\mathbf{x}_1 + \mathbf{K}_{22}\mathbf{x}_2 = 0 \tag{8.40}$$

$$\mathbf{x}_2 = -\mathbf{K}_{22}^{-1}\mathbf{K}_{21}\mathbf{x}_1 \tag{8.41}$$

$$\begin{Bmatrix} \mathbf{x}_1 \\ \mathbf{x}_2 \end{Bmatrix} = \begin{bmatrix} \mathbf{I} \\ -\mathbf{K}_{22}^{-1}\mathbf{K}_{21} \end{bmatrix} \{\mathbf{x}_1\} \tag{8.42}$$

$$\begin{Bmatrix} \ddot{\mathbf{x}}_1 \\ \ddot{\mathbf{x}}_2 \end{Bmatrix} = \begin{bmatrix} \mathbf{I} \\ -\mathbf{K}_{22}^{-1}\mathbf{K}_{21} \end{bmatrix} \{\ddot{\mathbf{x}}_1\} \tag{8.43}$$

With the relations obtained so far, substituting Equations 8.42 and 8.43 into the original system of equations, Equation 8.39 and premultiplying with $[\mathbf{I}, -(\mathbf{K}_{22}^{-1}\mathbf{K}_{21})^T]$, we can obtain the reduced set of equation of motion.

$$\bar{\mathbf{M}}\ddot{\mathbf{x}}_1 + \bar{\mathbf{K}}\mathbf{x}_1 = 0 \tag{8.44}$$

$$\bar{\mathbf{M}} = \left[\mathbf{I}, -\left(\mathbf{K}_{22}^{-1}\mathbf{K}_{21}\right)^{T}\right]\begin{bmatrix}\mathbf{M}_{11} & \mathbf{M}_{12} \\ \mathbf{M}_{21} & \mathbf{M}_{22}\end{bmatrix}\begin{bmatrix}\mathbf{I} \\ -\mathbf{K}_{22}^{-1}\mathbf{K}_{21}\end{bmatrix} \tag{8.45}$$

$$= \mathbf{M}_{11} - \mathbf{M}_{12}\mathbf{K}_{22}^{-1}\mathbf{K}_{21} - \mathbf{K}_{12}\mathbf{K}_{22}^{-1}\mathbf{M}_{21} + \mathbf{K}_{12}\mathbf{K}_{22}^{-1}\mathbf{M}_{22}\mathbf{K}_{22}^{-1}\mathbf{K}_{21}$$

$$\bar{\mathbf{K}} = \left[\mathbf{I}, -\left(\mathbf{K}_{22}^{-1}\mathbf{K}_{21}\right)^{T}\right]\begin{bmatrix}\mathbf{K}_{11} & \mathbf{K}_{12} \\ \mathbf{K}_{21} & \mathbf{K}_{22}\end{bmatrix}\begin{bmatrix}\mathbf{I} \\ -\mathbf{K}_{22}^{-1}\mathbf{K}_{21}\end{bmatrix} \tag{8.46}$$

$$= \mathbf{K}_{11} - \mathbf{K}_{12}\mathbf{K}_{22}^{-1}\mathbf{K}_{21}$$

Depending on the partitioning method used, the resulting reduced system of equations of motion can be of a fraction of the dimensionality of the original system. This would lead to speed up in the solution process for determining the dynamic response of structures.

8.2.5 Solution of diffusion equation

The idea of mode superposition can be readily applied to obtain the solution of the diffusion equation.

$$\mathbf{C}\dot{\mathbf{x}}(t) + \mathbf{K}\mathbf{x}(t) = \mathbf{p}(t) \tag{8.47}$$

The basic procedure involved in the decoupling of the above system of equation of motion is essentially the same as in the dynamic problems. To carry out modal decomposition process, we consider responses of the structural system without external loading and represent the displacement response as a linear combination of its mode shapes. First, we consider the single-mode response.

$$\begin{cases} x(t) = \phi\, y(t) \\ \dot{x}(t) = \phi\, \dot{y}(t) \end{cases} \tag{8.48}$$

With a single-mode response, we can assume the general form of modal amplitude function as follows:

$$\begin{cases} y(t) = Ae^{-\lambda t} \\ \dot{y}(t) = -\lambda A e^{-\lambda t} = -\lambda y(t) \end{cases} \tag{8.49}$$

We ignore the external force in Equation 8.47 and substitute the above single-mode response.

$$(-\lambda \mathbf{C}\phi + \mathbf{K}\phi)\, y(t) = 0 \tag{8.50}$$

$$\mathbf{K}\phi = \lambda \mathbf{C}\phi \tag{8.51}$$

This is a generalized eigenvalue problem. This shows that the mode shapes are the eigenvectors of the above generalized eigenvalue problem.

$$\Omega = \begin{bmatrix} \lambda_1 & & \\ & \ddots & \\ & & \lambda_2 \end{bmatrix} \tag{8.52}$$

$$\Phi = \left[\phi_1 | \cdots | \phi_n\right]$$

Now, we carry out the coordinate transformation and substitute into the original diffusion equation and premultiply by $\mathbf{\Phi}^T$ to arrive at the uncoupled equations.

$$\begin{cases} \mathbf{x}(t) = \mathbf{\Phi}\ \mathbf{y}(t) \\ \dot{\mathbf{x}}(t) = \mathbf{\Phi}\ \dot{\mathbf{y}}(t) \end{cases} \tag{8.53}$$

$$(\mathbf{\Phi}^T\mathbf{C}\mathbf{\Phi})\dot{\mathbf{y}}(t) + (\mathbf{\Phi}^T\mathbf{K}\mathbf{\Phi})\mathbf{y}(t) = \mathbf{\Phi}^T\mathbf{p}(t) \tag{8.54}$$

$$\begin{cases} \mathbf{\Phi}^T\mathbf{C}\mathbf{\Phi} = \mathbf{I} \\ \mathbf{\Phi}^T\mathbf{K}\mathbf{\Phi} = \mathbf{\Omega} \end{cases} \tag{8.55}$$

$$\begin{Bmatrix} \dot{y}_1(t) \\ \vdots \\ \dot{y}_n(t) \end{Bmatrix} + \begin{bmatrix} \lambda_1 & & \\ & \ddots & \\ & & \lambda_n \end{bmatrix} \begin{Bmatrix} y_1(t) \\ \vdots \\ y_n(t) \end{Bmatrix} = \begin{Bmatrix} \phi_1^T \mathbf{p}(t) \\ \vdots \\ \phi_n^T \mathbf{p}(t) \end{Bmatrix} \tag{8.56}$$

This is the uncoupled diffusion equation. For this first-order ordinary differential equation, the solution can be readily obtained for each mode. The system response can be obtained by summing up all the modal responses.

8.3 DIRECT INTEGRATION METHODS

In the previous section, we have discussed the solution of dynamic equation of motions using mode superposition method, which involves the solution of a generalized eigenvalue problems in computing mode shapes of the structural system in undamped free vibration, and the transformation leading to the uncoupling of the original system of equations of motion. Because of the use of superposition principle, the method can only be used for linear structural systems. In most cases in practice the response of the structural systems are usually determined through direct integration using a step-by-step numerical procedure. In direct numerical integration, the original system of equations, such as the dynamic equation of motion or the diffusion equation, are not transformed into a different form. Instead of obtaining a closed form solution in the continuous time domain, the time domain is first discretized into the time steps Δt. Then the displacement, velocity, and acceleration vectors at each discrete time station are computed numerically. Suppose that we use a constant time step, then at any time $t_k = k\Delta t$, we obtain a set of discrete vectors for displacement, velocity, and acceleration.

$$\begin{cases} \mathbf{x}_k = \mathbf{x}(t_k) \\ \dot{\mathbf{x}}_k = \dot{\mathbf{x}}(t_k) \\ \ddot{\mathbf{x}}_k = \ddot{\mathbf{x}}(t_k) \end{cases} \tag{8.57}$$

These should be reasonable approximations to the exact solution at those discrete time stations. By continuing this process forward step by step, we can then obtain results at all the discrete time stations in the time duration of interest, which can be considered as reasonable approximations to the exact solution in continuous time. This means that to describe the step-by-step numerical solution process, we need to derive recursive equations relating known information on the previous k time steps to the current time step.

Depending on the specifics of each functions, or the way the time space is discretized, we can arrive at different integration methods. This shows the basic idea behind direct integration for the solution of initial value problems; that is, through certain time discretization methods, the original system of continuous differential equations is transformed into a system of difference (algebraic) equations at discrete time stations. The results obtained from solving the system of difference (algebraic) equations are approximations to the exact solutions at the discrete time stations.

In practice, we usually use a one-step method to carry out the direct numerical integration process. A one-step method uses only the information at the immediate previous time station to compute solutions at the current time station. In structural dynamics, especially in evaluating the response of very large and complex structural systems under dynamic short-term loading such as explosion and shocks, where a short response history will be computed, we tend to use one-step methods to achieve computational efficiency.

8.3.1 Convergence, consistency, and stability

In this chapter we will discuss a number of methods for direct integration of equation of motion. All direct integration methods have to be convergent. This means that in the limit, as the time step goes to zero, the solution from the direct integration method should converge to the exact solution of the differential equation of motion. Convergent methods should satisfy the following condition at time $t_k = k\Delta t$:

$$\lim_{\substack{k\to\infty \\ \Delta t\to 0}} \mathbf{x}_k \to \mathbf{x}(t_k) \tag{8.58}$$

The convergence of the direct integration methods is not usually studied directly. We normally study the *stability* and *consistency* characteristics of the numerical integration methods. This is due to the *Lax equivalence theorem* (Lax and Richtmyer, 1956), also referred to as *Lax–Richtmyer theorem*, which states that any numerical integration method which is consistent and stable is also convergent.

Basically, *consistency* means that the equation in the direct integration should be consistent with the differential equation of motion. In the limit, as the time step goes to zero, the equations used in the discrete direct integration method should converge to the original differential equation. In other words, the resulting difference (algebraic) equation should be consistent with the original physical representation. In structural dynamics, the first-order difference operator should be consistent with velocity and the second-order difference operator should be consistent with accelerations, and so forth.

Stability is an important and existential characteristic of any direct integration method; for a method to be viable, it has to be stable. *Stability* means that small perturbation at the initial condition results in small changes in the system response. Small perturbations are always present in numerical computation in the form of round-off error. If the round-off errors grow and dominate the response of the system, then the solution becomes useless and the method is unstable. In the stable methods the round-off errors do not grow. We can see that stability is a practical issue in computation where round-off errors are always present.

In terms of stability, the direct integration methods fall into two categories; they are either *conditionally stable* or *unconditionally stable*. Conditional stability means the method is stable as long as we use a time step below a certain limit. The limit on the time step depends on the properties of the system being analyzed. In structural dynamics the limit on the time step depends on the highest natural frequency of the structural system. When there are no limits on time step, then the method is unconditionally stable; the method remains stable as we use larger and larger time step. In unconditionally stable methods there are other considerations that limit the size of the time step. Normally, numerical errors increase when we use larger time steps. Time step has to be chosen carefully to control the numerical errors, even when the stability is not the main concern in unconditionally stable methods.

Direct integration methods in structural dynamics can also be classified as *explicit* methods and *implicit* methods. From computational point of view, the difference between these classes of methods involves the stiffness matrix, **K**. Implicit methods require the solution of a system of equations with the stiffness matrix every time step. In explicit methods the stiffness matrix is multiplied by a vector at every step. A primary difference is that with explicit methods, the dynamic equation of motion is formed at the beginning of the time step, t_k. However, with implicit method, the equation of motion is formed at the end of the time step, t_{k+1}. This argument will be evident in our derivation of some direct solution methods for solving structural dynamic problems.

It should be mentioned that the direct integration methods developed for the solution of the dynamic equation of motion in structural dynamics and for the diffusion equation in transient analysis are specialized methods and based primarily on ideas originated from physical reasoning and engineering principles coupled with basic mathematical concepts in temporal discretization; whereas direct integration methods for initial value ordinary differential equations are general methods based on mathematical concepts in temporal discretization. Specifically, in direct integration of dynamic equation of motion, the dynamic equilibrium condition is satisfied at discrete time stations; and we make assumptions on the variation of displacements, velocities, and accelerations within each time interval. Although the final algorithms may have different forms, there is always an underlying connection between these "engineering-inspired" methods and established mathematical methods. In the following discussions, for starter, we will first describe some of the basic mathematical concepts in deriving numerical integration methods.

8.3.2 Basic concepts in temporal discretization

To illustrate the basic ideas in temporal discretization, we consider an initial value problem represented in the following first-order differential equation:

$$\begin{cases} \dot{y} = f(y, t) \\ y(t_0) = y_0 \end{cases} \tag{8.59}$$

In this equation t is a time-like variable, and f(y, t) can be a nonlinear function. This is a very general representation of initial value problems. We know that if the function f(y, t) is continuous and satisfies the Lipschitz condition that

$$|f(y, t) - f(\bar{y}, t)| \leq L|y - \bar{y}| \tag{8.60}$$

where $\bar{y}$ is any point in the domain, then the initial value problem has a unique solution y = y(t). In most cases with ordinary differential equations, it is very difficult, if not impossible, to obtain analytical solutions. Therefore, in practice, we use numerical methods to obtain discrete solutions at discrete time stations. This requires the use of temporal discretization methods to transform the system of differential equations into a system of difference equations. In this case, we are dealing with only first-order derivative, as higher order ordinary differential equation can always be transformed into the general form expressed in Equation 8.59.

There are many ways to carry out the temporal discretization process. For instance, from the definition of first-order derivative, we have approximately,

$$\dot{y}(t_k) = \lim_{\Delta t \to 0} \frac{y_{k+1} - y_k}{\Delta t} \approx \frac{y_{k+1} - y_k}{\Delta t} \tag{8.61}$$

With this discretization method, we can transform the above differential equation to a discrete difference equation.

$$\begin{cases} \dfrac{y_{k+1} - y_k}{\Delta t} = f(y_k, t_k), \; k = 0, 1, \ldots \\ y_0 = y_0 \end{cases} \tag{8.62}$$

Discrete values for y_k at different time stations can be readily computed from this algebraic equation. It can be seen that this discretization method is not very accurate, but we also do not know how accurate it is. For more advanced methods, we can consider Taylor series and assume that the function f(y, t) is sufficiently differentiable. Using Taylor series, we can express the discrete solution from t_k to t_{k+1} as follows:

$$y_{k+1} = y_k + \Delta t \, \dot{y}_k + \frac{\Delta t^2}{2} \ddot{y}_k + \ldots \tag{8.63}$$

Using the original differential equation and truncating the series, approximately, we have

$$y_{k+1} = y_k + \Delta t \, \phi(y_k, t_k; \Delta t) \tag{8.64}$$

$$\phi(y_k, t_k; \Delta t) = f(y, t) + \frac{\Delta t}{2} f^{(1)}(y, t) + \cdots + \frac{\Delta t^{p-1}}{p!} f^{(p-1)}(y, t) \tag{8.65}$$

In which $f^{(j)}$ denotes jth order derivative of f(y, t). Obviously, by selecting the value of p, we can derive different numerical integration methods with varying degrees of accuracy. Taking p = 1, we arrive at the following, which is same as in Equation 8.62:

$$\begin{cases} y_{k+1} = y_k + \Delta t \, f(y_k, t_k) \\ y_0 = y_0 \end{cases} \tag{8.66}$$

This relation is usually called *Euler formula* whose geometric interpretation is shown in Figure 8.4. If the initial value is known, then we can use the above equation to compute discrete function values at each time station in a step-by-step manner as follows:

$$\begin{aligned} y_1 &= y_0 + \Delta t \, f(y_0, t_0) \\ y_2 &= y_1 + \Delta t \, f(y_1, t_1) \\ &\vdots \end{aligned} \tag{8.67}$$

As illustrated in Figure 8.4, this corresponds to using piece-wise linear function within each time step as approximation to the exact solution of the first-order differential equation. And the gradient is taken at the time station t_k.

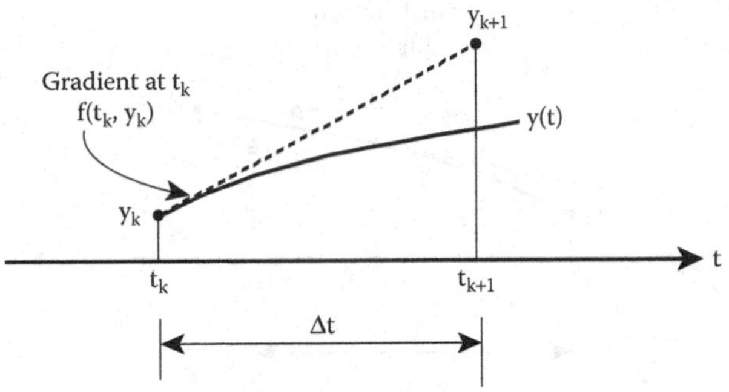

Figure 8.4 Illustration of Euler formula (forward difference).

To analyze the accuracy in approximation of Euler formula, we assume $y_k = y(t_k)$ and compare the Equation 8.66 with the following relation obtained from the Taylor series expansion:

$$y(t_{k+1}) = y(t_k) + \Delta t\ \dot{y}(t_k) + \frac{(\Delta t)^2}{2}\ \ddot{y}(\xi_k) \tag{8.68}$$

to arrive at the following:

$$y(t_{k+1}) - y_{k+1} = \frac{(\Delta t)^2}{2}\ \ddot{y}(\xi_k) \tag{8.69}$$

in which $t_k < \xi_k < t_{k+1}$. This shows that the truncation error with the use of Euler formula is $O[(\Delta t)^2]$. In this case, if the exact solution for y(t) is a first degree polynomial, then the result obtained from Euler formula is exact. Therefore, the Euler formula is a first-order method in numerical integration. Similarly, if the truncation error for a numerical integration method is $O[(\Delta t)^{p+1}]$, then the method would give accurate results for a solution of pth order polynomial. Of course, to obtain higher accuracy, higher-order derivatives need to be computed.

We note that the Euler formula is an explicit method because the differential equation is evaluated at the time station t_k, at the beginning of the time step.

Next, we write the differential equation at the end of the time step, at time station t_{k+1}.

$$\dot{y}(t_{k+1}) = f\left[y(t_{k+1}),\ t_{k+1}\right] \tag{8.70}$$

The backward difference is defined as follows:

$$\dot{y}(t_{k+1}) = \frac{y(t_{k+1}) - y(t_k)}{\Delta t} \tag{8.71}$$

With the use of backward difference for temporal discretization, we can obtain the *backward Euler formula*.

$$y_{k+1} = y_k + \Delta t\ f(y_{k+1},\ t_{k+1}) \tag{8.72}$$

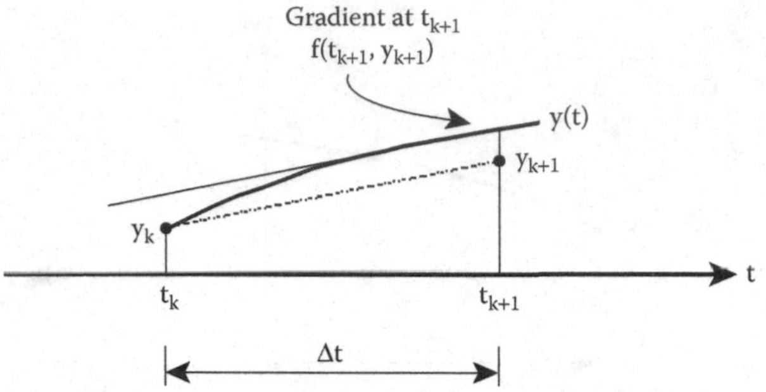

Figure 8.5 Illustration of backward Euler formula (backward difference).

This is an implicit method. The geometric interpretation of this formula is illustrated in Figure 8.5. This equation can only be solved with an iterative approach by first computing a starting $y_{k+1}^{(0)}$ value for using the Euler formula and iterating with the backward Euler formula. It can be shown that the truncation error with backward Euler formula is

$$y(t_{k+1}) - y_{k+1} = -\frac{(\Delta t)^2}{2}\,\ddot{y}(\eta_k) \tag{8.73}$$

in which $t_k < \eta_k < t_{k+1}$.

It can be observed that if we combine the Euler formula with backward Euler formula through averaging, we will obtain a method of potentially higher accuracy because a major part of the truncation errors would be canceled out. Such a method is called the *trapezoidal rule*, which can be expressed as follows:

$$y_{k+1} = y_k + \frac{\Delta t}{2}\,[f(y_k,\ t_k) + f(y_{k+1},\ t_{k+1})] \tag{8.74}$$

This method is implicit so that the solution process is iterative. We usually use Euler formula to compute a starting value for y_{k+1} and then iterate until we have convergence.

$$\begin{cases} y_{k+1}^{(0)} = y_k + \Delta t\ f(y_k,\ t_k) \\ y_{k+1}^{(j+1)} = y_k + \dfrac{\Delta t}{2}\left[f(y_k,\ t_k) + f(y_{k+1}^{(j)},\ t_{k+1})\right] \end{cases} \tag{8.75}$$

The superscript in the parenthesis is the iteration number. This method requires many iterations and the computation of functional values at each step. The geometric interpretation of trapezoidal rule as well as the forward and backward Euler rules is illustrated in Figure 8.6. Clearly, the trapezoidal rule is more likely to give higher level of accuracy.

To reduce the computational effort in using trapezoidal rule, instead of carrying out iteration to achieve convergence, we can use Euler formula to compute an initial approximation for y_{k+1} as a *predicator*, and then use the trapezoidal rule once to compute an improved value

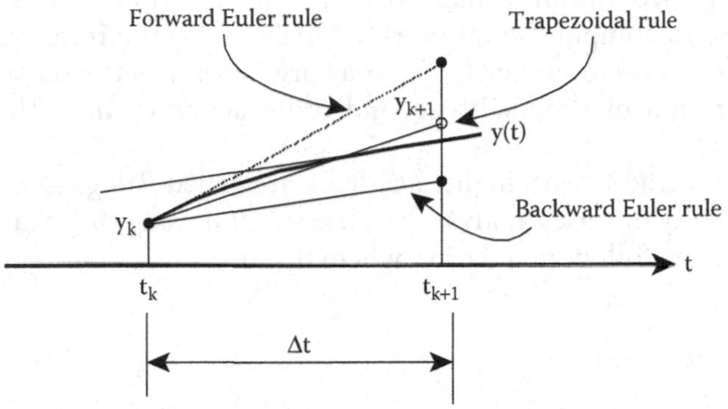

Figure 8.6 Illustration of the trapezoidal rule in temporal discretization.

for y_{k+1}, which is called a *corrector*. By so doing, we obtain a *predictor–corrector* method referred to as the *modified Euler method* as follows:

$$\begin{cases} y_{pre} = y_k + \Delta t\ f(y_k,\ t_k) \\ y_{cor} = y_k + \Delta t\ f(y_{pre},\ t_{k+1}) \\ y_{k+1} = \dfrac{1}{2}\left(y_{pre} + y_{cor}\right) \end{cases} \tag{8.76}$$

This method is expected to have a higher accuracy in approximation than the standard Euler formula.

At this juncture, we observe that all the methods described so far are single-step methods because we use only information at y_k to compute y_{k+1}. Next, we use the central difference formula to approximate the first derivative.

$$\dot{y}(t_k) = \frac{1}{2\Delta t}\left[y(t_{k+1}) - y(t_{k-1})\right] \tag{8.77}$$

Then, we obtain the *two-step Euler formula*.

$$y(t_{k+1}) = y(t_{k-1}) + 2\Delta t\ f(y_k,\ t_k) \tag{8.78}$$

This formula is of higher (second-order) accuracy and its truncation error can be shown to be as follows:

$$y(t_{k+1}) - y_{k+1} = -\frac{(\Delta t)^3}{3}\ y^{(3)}(\xi_k) \tag{8.79}$$

Obviously, with two-step method, we need to use information at y_{k-1} and y_k to compute y_{k+1}. This shows that in addition to the initial value y_0, we also need to use a single-step method to compute a starting value for y_1, in order to start using the two-step Euler formula.

That is, we need two initial values to start the iteration process. This additional information and extra computational work contributes to the improved accuracy of the algorithm. Using the two-step Euler formula as predictor and the trapezoidal rule as corrector, we will obtain a method with second-order accuracy in both the predictor and the corrector.

Another group of methods with higher accuracy are called *Runge–Kutta* methods, which use the basic idea in Taylor series analysis. Let $h = \Delta t$, and using the mean value theorem for derivatives, we have the following relation where $0 < \theta < 1$:

$$\frac{y(t_{k+1}) - y(t_k)}{h} = \dot{y}(t_k + \theta h) \tag{8.80}$$

$$\begin{aligned} y(t_{k+1}) &= y(t_k) + h\, f\left[t_k + \theta h,\ y(t_k + \theta h)\right] \\ &= y(t_k) + h\, \bar{\kappa} \end{aligned} \tag{8.81}$$

$\bar{\kappa}$ is called the average gradient in the interval $[t_k, t_{k+1}]$. This relation shows that any algorithm that computes the average gradient would result in a new temporal discretization method. We note that in Euler formula, the gradient at t_k is taken as the average gradient. Obviously, this is not an accurate approximation. A modified Euler formula can be expressed as follows.

$$\begin{aligned} & y_{k+1} = y_k + \frac{h}{2}(\kappa_1 + \kappa_2) \\ & \begin{cases} \kappa_1 = f(t_k,\ y_k) \\ \kappa_2 = f(t_{k+1},\ y_k + h\,\kappa_1) \end{cases} \end{aligned} \tag{8.82}$$

This shows that $\bar{\kappa}$ is taken as the average gradients at t_k and t_{k+1}, and k_2 is predicted from information at t_k. This observation indicates that a more accurate formulation may be obtained by computing more gradient values within the interval and using their weighted average. This is the basic idea behind Runge–Kutta group of methods.

For example, to derive the second-order Runge–Kutta method, we consider a point $t_{k+\xi}$ within the interval $[t_k, t_{k+1}]$.

$$t_{k+\xi} = t_k + \xi h,\ 0 < \xi \leq 1 \tag{8.83}$$

We use the weighted average of gradients at t_k and $t_{k+\xi}$.

$$\begin{aligned} & y_{k+1} = y_k + h(\lambda_1 \kappa_1 + \lambda_2 \kappa_2) \\ & \begin{cases} \kappa_1 = f(t_k,\ y_k) \\ \kappa_2 = f(t_{k+\xi},\ y_k + \xi h\,\kappa_1) \end{cases} \end{aligned} \tag{8.84}$$

It can be shown that if $\lambda_1 + \lambda_2 = 1$ and $\xi = 1/2$, then we obtain a family of methods which are of second-order accuracy. This family of methods are called second-order Runge–Kutta methods. A special version of the second-order Runge–Kutta method can be expressed as follows:

$$y_{k+1} = y_k + \Delta t\, \kappa_2$$

$$\begin{cases} \kappa_1 = f\left(t_k,\, y_k\right) \\ \kappa_2 = f\left(t_k + \dfrac{\Delta t}{2},\, y_k + \kappa_1 \dfrac{\Delta t}{2}\right) \end{cases} \tag{8.85}$$

The same idea can be extended to generate other higher order Runge–Kutta methods, which will not be discussed here.

So far, we have discussed some of the basic temporal discretization methods for the direct solution of initial value ordinary differential equations. They are general methods and may not be efficient and useful for the solution of most computational mechanics problems, especially in structural dynamics. For problems in computational mechanics, we need some specialized methods that will be discussed later in this chapter.

8.3.3 Direct solution of diffusion equation

Earlier we discussed the modal analysis of the diffusion equation by uncoupling it. Here, we will discuss the direct solution of the following diffusion equation without uncoupling it:

$$\mathbf{C}\dot{\mathbf{x}}(t) + \mathbf{K}\mathbf{x}(t) = \mathbf{p}(t) \tag{8.86}$$

The temporal discretization method is developed by assuming linear variation of $\mathbf{x}$ within a time step from t_j to t_{j+1} as shown in Figure 8.7. Let $\alpha = \tau/\Delta t$, where $\alpha \in [0, 1]$, which is the normalized time. Then, we have the following variation within the time step:

$$\mathbf{x}_{j+\alpha} = \left(1-\alpha\right)\mathbf{x}_j + \alpha\mathbf{x}_{j+1} \tag{8.87}$$

The first derivative within the time step is a constant.

$$\dot{\mathbf{x}}_{j+\alpha} = \frac{1}{\Delta t}\left(\mathbf{x}_{j+1} - \mathbf{x}_j\right) \tag{8.88}$$

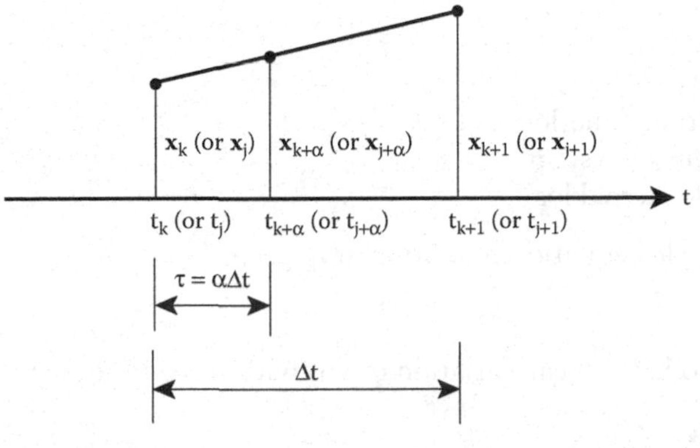

Figure 8.7 Assumption of linear variation of displacement within a time step for direct integration of diffusion equations.

The discretized system of equations is obtained after substituting the above equations in the equation of motion.

$$\frac{1}{\Delta t}\mathbf{C}(\mathbf{x}_{j+1}-\mathbf{x}_j)+(1-\alpha)\mathbf{K}\mathbf{x}_j+\alpha\mathbf{K}\mathbf{x}_{j+1}=\mathbf{p}(t_{j+\alpha}) \tag{8.89}$$

$$\left[\frac{1}{\Delta t}\mathbf{C}+\alpha\mathbf{K}\right]\mathbf{x}_{j+1}=\mathbf{p}(t_{j+\alpha})+\left[\frac{1}{\Delta t}\mathbf{C}-(1-\alpha)\mathbf{K}\right]\mathbf{x}_j \tag{8.90}$$

After choosing a value for α, we can use this equation in a time stepping manner to determine the discrete solution at each time station step by step.

If we choose $\alpha = 0$, then we obtain an explicit method called the forward difference method, where the equation of motion is satisfied at the beginning of the time step. We note that $\mathbf{C}$ is assumed to be a diagonal matrix.

$$\frac{1}{\Delta t}\mathbf{C}\mathbf{x}_{j+1}=\mathbf{p}(t_j)+\left[\frac{1}{\Delta t}\mathbf{C}-\mathbf{K}\right]\mathbf{x}_j \tag{8.91}$$

If we choose $\alpha = 1$, then the equation of motion is satisfied at the end of the time step and we have an implicit method called backward difference method.

$$\left[\frac{1}{\Delta t}\mathbf{C}+\mathbf{K}\right]\mathbf{x}_{j+1}=\mathbf{p}(t_{j+1})+\frac{1}{\Delta t}\mathbf{C}\mathbf{x}_j \tag{8.92}$$

Similarly, when $\alpha = 1/2$, the equation of motion is satisfied at the middle of the time step, and we have the mid-difference method or the Crank–Nicolson method, (Crank and Nicolson, 1947), which is an implicit method.

$$\left[\frac{1}{\Delta t}\mathbf{C}+\frac{1}{2}\mathbf{K}\right]\mathbf{x}_{j+1}=\mathbf{p}(t_{j+\frac{1}{2}})+\left[\frac{1}{\Delta t}\mathbf{C}-\frac{1}{2}\mathbf{K}\right]\mathbf{x}_j \tag{8.93}$$

This shows that by assuming linear variation of $\mathbf{x}$ in the time step, and consequently constant velocity within the interval, we have obtained a family of direct integration methods for the diffusion equations. Obviously, this temporal discretization method automatically satisfies the *consistency* condition. Now we discuss the stability characteristics of this family of methods.

We can study the stability characteristics of a direct integration method by considering *single-mode* response of the system. First, we need to uncouple the original system of equations of motion using mode superposition method described earlier. The uncoupled equation of motion is as follows:

$$\dot{y}(t)+\lambda y(t)=p(t) \tag{8.94}$$

where:

$y(t)$ is the amplitude function
λ is the eigenvalue of system
$p(t)$ is the modal external load

We write this uncoupled equation of motion at $t_{k+\alpha}$.

$$\dot{y}_{k+\alpha}+\lambda y_{k+\alpha}=p_{k+\alpha} \tag{8.95}$$

We note that we also have linear variation of $y(t)$ within the time step.

$$\begin{cases} y_{k+\alpha}=(1-\alpha)y_k+\alpha y_{k+1} \\ \dot{y}_{k+\alpha}=\dfrac{1}{\Delta t}(y_{k+1}-y_k) \end{cases} \tag{8.96}$$

Substituting these relations into the uncoupled equation of motion yields the following:

$$y_{k+1} = \frac{1-(1-\alpha)\lambda\Delta t}{1+\alpha\lambda\Delta t} y_k + \frac{\Delta t}{1+\alpha\lambda\Delta t} p_{k+\alpha} \tag{8.97}$$

$$y_{k+1} = R^{\alpha} y_k + F^{\alpha} p_{k+\alpha} \tag{8.98}$$

$$\begin{cases} R^{\alpha} = \dfrac{1-(1-\alpha)\lambda\Delta t}{1+\alpha\lambda\Delta t} \\ F^{\alpha} = \dfrac{\Delta t}{1+\alpha\lambda\Delta t} \end{cases} \tag{8.99}$$

We can ignore the external load, since it does not affect the stability of integration method.

$$\begin{aligned} y_{k+1} &= R^{\alpha} y_k \\ &= \left(R^{\alpha}\right)^2 y_{k-1} \\ &\ \vdots \\ &= \left(R^{\alpha}\right)^k y_0 \end{aligned} \tag{8.100}$$

This indicates that the stability of the method requires satisfying the following condition:

$$\left|R^{\alpha}\right| = \left|\frac{1-(1-\alpha)\lambda\Delta t}{1+\alpha\lambda\Delta t}\right| \le 1 \tag{8.101}$$

$$-1 \le \frac{1-(1-\alpha)\lambda\Delta t}{1+\alpha\lambda\Delta t} \le 1 \tag{8.102}$$

$$\begin{cases} \lambda\Delta t \ge 0 \\ (1-2\alpha)\lambda\Delta t \le 2 \end{cases} \tag{8.103}$$

Since the stiffness matrix is assumed to be symmetric and positive definite, its eigenvalues are positive. Then, the first condition is always satisfied. Therefore, the stability criterion requires that the second condition above be satisfied.

$$(1-2\alpha)\lambda\Delta t \le 2 \tag{8.104}$$

We observe that if $\alpha \ge 1/2$, then the above relation is always satisfied. This indicates that all the direct integration methods for diffusion equation with $\alpha \ge 1/2$ are unconditionally stable. However, if $\alpha < 1/2$, then the family of direct integration methods are conditionally stable. With conditional stability, the time step length is constrained by the following relation:

$$\Delta t \le \frac{2}{(1-2\alpha)\lambda} \tag{8.105}$$

With this relation, we can determine the stability condition of different methods. For example, When $\alpha = 0$, $R = 1 - \lambda\Delta t$, which corresponds to the forward difference method that is conditionally stable, with the following condition on the time step:

$$\lambda\Delta t \leq 2 \;\Rightarrow\; \Delta t \leq \frac{2}{\lambda} \tag{8.106}$$

As a conditionally stable explicit method, the forward difference method ($\alpha = 0$) has its step length constrained in order to maintain stability. In multidegree-of-freedom systems, such as finite element models, we should use a step length that satisfies the condition that $\Delta t \leq 2/\lambda_{max}$, where λ_{max} is the largest eigenvalue of the system. For nonlinear problems, with the progression of deformations, the eigenvalues of the system keep changing. Therefore, it is preferable that we use unconditionally stable numerical integration methods whose stability is independent of the size of step length.

If we take $\alpha = 1/4$, then the corresponding method is also conditionally stable and its step length is limited by $\Delta t \leq 4/\lambda_{max}$. This indicates that as the value of α increases toward the threshold for unconditional stability, the constraint on the step length becomes less stringent.

Now, we consider the backward difference method, $\alpha = 1$ and the mid-difference method, $\alpha = 1/2$. In both cases we have $|R| < 1$. This means that both these methods are unconditionally stable. This result can also be directly obtained from the stability condition obtained earlier. Since in both cases $\alpha \geq 1/2$, the methods are unconditionally stable.

For this simple case described in Equation 8.94, the exact (homogeneous) solution can be obtained as follows:

$$y(t) = Ae^{-\lambda t} \tag{8.107}$$

$$\begin{aligned} y_{k+1} &= Ae^{-\lambda t_{k+1}} \\ &= Ae^{-\lambda(t_k+\Delta t)} \\ &= e^{-\lambda\Delta t}Ae^{-\lambda t_k} \\ &= e^{-\lambda\Delta t}y_k \end{aligned} \tag{8.108}$$

$$R^{\alpha} = e^{-\lambda\Delta t} \tag{8.109}$$

The stability conditions of the family of methods discussed is illustrated in Figure 8.8. We can see that the mid-difference method ($\alpha = 1/2$) is closer to the exact solution than the method corresponding to $\alpha = 1$. It can be expected that the mid-difference method can result in a higher level of accuracy in the solution obtained.

8.3.4 Explicit second central difference method

In its simplest form, the second central difference method is applied to the undamped dynamic equations of motion.

$$\mathbf{M}\ddot{\mathbf{x}}(t) + \mathbf{K}\mathbf{x}(t) = \mathbf{p}(t) \tag{8.110}$$

This method is based on the use of the second central difference operator. Consider the time axis divided into equal time steps Δt, separated by discrete time stations t_k. The objective of direct integration is to compute the system variables, displacements, velocities, and

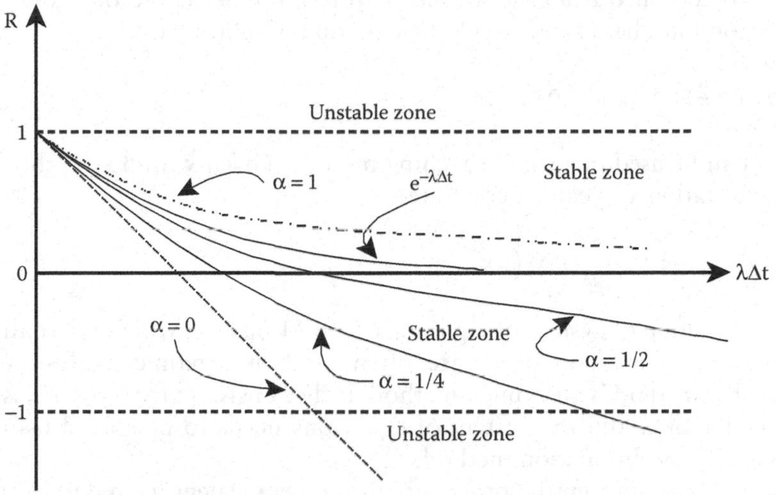

Figure 8.8 Stability conditions of direct integration methods for diffusion equation.

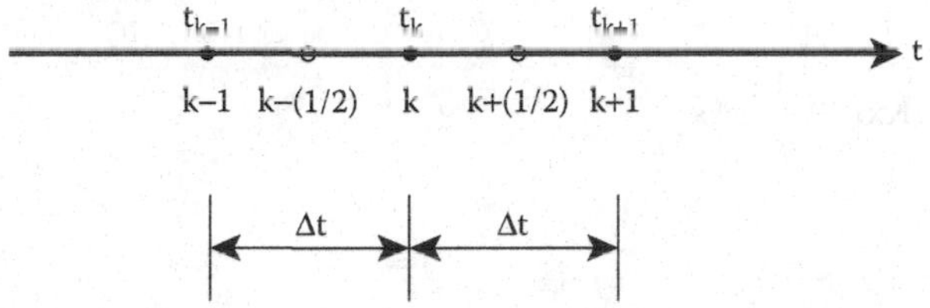

Figure 8.9 Temporal discretization for explicit second central difference method.

accelerations, at these discrete time stations. We take a typical time station t_k and develop the second central difference operator around this time station. This operator relates the acceleration at t_k to the displacements at t_k as well as the time stations before and after that, as illustrated in Figure 8.9. First, the equation of motion is satisfied at time station t_k.

$$\mathbf{M}\ddot{\mathbf{x}}_k + \mathbf{K}\mathbf{x}_k = \mathbf{p}_k \tag{8.111}$$

The velocity within each time interval and acceleration at t_k can be determined.

$$\begin{cases} \dot{\mathbf{x}}_{k-\frac{1}{2}} = \dfrac{1}{\Delta t}\left(\mathbf{x}_k - \mathbf{x}_{k-1}\right) \\ \dot{\mathbf{x}}_{k+\frac{1}{2}} = \dfrac{1}{\Delta t}\left(\mathbf{x}_{k+1} - \mathbf{x}_k\right) \end{cases} \tag{8.112}$$

$$\begin{aligned} \ddot{\mathbf{x}}_k &= \frac{1}{\Delta t}\left(\dot{\mathbf{x}}_{k+\frac{1}{2}} - \dot{\mathbf{x}}_{k-\frac{1}{2}}\right) \\ &= \frac{1}{\Delta t^2}\left(\mathbf{x}_{k+1} - 2\mathbf{x}_k + \mathbf{x}_{k-1}\right) \end{aligned} \tag{8.113}$$

Acceleration is determined from the second central difference operator. Now we substitute the above relation into the discrete dynamic equation of motion at t_k.

$$\frac{1}{\Delta t^2}\mathbf{M}\left(\mathbf{x}_{k+1} - 2\mathbf{x}_k + \mathbf{x}_{k-1}\right) + \mathbf{K}\mathbf{x}_k = \mathbf{p}_k \tag{8.114}$$

This equation can be used in a time marching method. Given $\mathbf{x}_k$ and $\mathbf{x}_{k-1}$, the displacement at the next time station $\mathbf{x}_{k+1}$ can be computed.

$$\frac{1}{\Delta t^2}\mathbf{M}\mathbf{x}_{k+1} = \mathbf{p}_k - \mathbf{K}\mathbf{x}_k + \frac{1}{\Delta t^2}\mathbf{M}\left(2\mathbf{x}_k - \mathbf{x}_{k-1}\right) \tag{8.115}$$

This requires a solution of system of equations with **M** matrix, which is usually a diagonal matrix because lump mass matrices are often used in dynamic analysis of structural systems. Thus, the method is an explicit method. If the consistent mass matrix is formulated, then **M** is not diagonal and the system of equations needs to be solved using a solution method such as Gauss-elimination method.

In practical implementation, the procedure uses an equivalent form derived from the following relations:

$$\frac{1}{\Delta t}\mathbf{M}\left(\dot{\mathbf{x}}_{k+\frac{1}{2}} - \dot{\mathbf{x}}_{k-\frac{1}{2}}\right) + \mathbf{K}\mathbf{x}_k = \mathbf{p}_k \tag{8.116}$$

$$\frac{1}{\Delta t}\mathbf{M}\dot{\mathbf{x}}_{k+\frac{1}{2}} = \mathbf{p}_k - \mathbf{K}\mathbf{x}_k + \frac{1}{\Delta t}\mathbf{M}\dot{\mathbf{x}}_{k-\frac{1}{2}} \tag{8.117}$$

$$\mathbf{x}_{k+1} = \mathbf{x}_k + \Delta t\, \dot{\mathbf{x}}_{k+\frac{1}{2}} \tag{8.118}$$

It should be noted that the solution of $\mathbf{x}_{k+1}$ is computed based on satisfying the dynamic equilibrium equation at time t_k. Because of this operation, the second central difference method is called an explicit integration method. It is also interesting to note that this method does not require the explicit formulation of the linear stiffness matrix because the matrix–vector multiplication, $\mathbf{K}\mathbf{x}_k$, is equal to the internal resisting force vector, which can be computed from direct assembly of element contributions. The procedure for computing the internal resisting force vector has been discussed earlier in Chapter 2.

When damping is present, the following dynamic equation of motion is satisfied at the time station t_k:

$$\mathbf{M}\ddot{\mathbf{x}}_k + \mathbf{C}\dot{\mathbf{x}}_k + \mathbf{K}\mathbf{x}_k = \mathbf{p}_k \tag{8.119}$$

The accelerations can still be computed using the second central difference operator. The velocities, on the other hand, are determined within time steps, not at the time stations. An average method can be used.

$$\dot{\mathbf{x}}_k = \frac{1}{2}\left(\dot{\mathbf{x}}_{k+\frac{1}{2}} + \dot{\mathbf{x}}_{k-\frac{1}{2}}\right) \tag{8.120}$$

Substituting this relation into the equation of motion we obtain the following relation for discrete dynamic equation of motion:

$$\frac{1}{\Delta t}\mathbf{M}\left(\dot{\mathbf{x}}_{k+\frac{1}{2}} - \dot{\mathbf{x}}_{k-\frac{1}{2}}\right) + \frac{1}{2}\mathbf{C}\left(\dot{\mathbf{x}}_{k+\frac{1}{2}} + \dot{\mathbf{x}}_{k-\frac{1}{2}}\right) + \mathbf{K}\mathbf{x}_k = \mathbf{p}_k \tag{8.121}$$

$$\begin{cases} \left(\dfrac{1}{\Delta t}\mathbf{M} + \dfrac{1}{2}\mathbf{C}\right)\dot{\mathbf{x}}_{k+\frac{1}{2}} = \mathbf{p}_k - \mathbf{K}\mathbf{x}_k + \left(\dfrac{1}{\Delta t}\mathbf{M} - \dfrac{1}{2}\mathbf{C}\right)\dot{\mathbf{x}}_{k-\frac{1}{2}} \\ \mathbf{x}_{k+1} = \mathbf{x}_k + \Delta t\, \dot{\mathbf{x}}_{k+\frac{1}{2}} \end{cases} \tag{8.122}$$

It can be seen that the explicit nature of the method can only be preserved if the damping matrix **C** is a diagonal matrix. Then, we only need to solve a system of equations with a diagonal coefficient matrix, which is trivial. Otherwise, the combined coefficient matrix on the left-hand side of equation is nondiagonal, which requires triangular decomposition of the matrix in the solution of a system of equations. As we will see in Section 8.3.8 on dynamic overrelaxation, often an approximate diagonal damping matrix is used in the computation.

8.3.5 Computer implementation procedure for explicit second central difference method

The steps involved in the computer implementation of the second central difference method directly follow from the equations derived earlier. The time axis is divided into equal time steps Δt. The choice of Δt depends on considerations of accuracy and stability. These aspects will be discussed in the next section. It is assumed that the values of the load vector are available at the discrete time stations. The method allows the computation of the displacement, velocity, and acceleration vectors at the discrete time stations.

Special attention must be paid in starting of the integration process. At the first step, the dynamic equation of motion is satisfied at t_1. The values of displacement and velocity at the previous step are needed. If the system starts from at rest conditions, then these initial conditions are zero. Otherwise, they are determined from other considerations, depending on the initial state of the system.

The steps in the computer implementation of the second central difference method are as follows:

Initialization

Determine Δt, consider stability and accuracy

Set initial values of $\dot{\mathbf{x}}_{1-\frac{1}{2}}$ and $\mathbf{x}_1$

If $\mathbf{x}_0$ and $\mathbf{x}_1$ are known then determine $\dot{\mathbf{x}}_{1-\frac{1}{2}}$

$$\dot{\mathbf{x}}_{1-\frac{1}{2}} = \frac{1}{\Delta t}(\mathbf{x}_1 - \mathbf{x}_0)$$

Time integration when matrices **M** and **C** diagonal

for k = 1, 2, …

$$t_{k+1} = t_k + \Delta t$$

$\mathbf{I}_k = \mathbf{K}\mathbf{x}_k$ Determined by direct assembly of element contributions

or $\mathbf{I}_k = \mathbf{I}_{k-1} + \mathbf{K}\Delta\mathbf{x}_k$ For nonlinear problems

$$\dot{\mathbf{x}}_{k+\frac{1}{2}} = \left(\frac{1}{\Delta t}\mathbf{M} + \frac{1}{2}\mathbf{C}\right)^{-1}\left[\mathbf{p}_k - \mathbf{I}_k + \left(\frac{1}{\Delta t}\mathbf{M} - \frac{1}{2}\mathbf{C}\right)\dot{\mathbf{x}}_{k-\frac{1}{2}}\right]$$

$$\mathbf{x}_{k+1} = \mathbf{x}_k + \Delta t\, \dot{\mathbf{x}}_{k+\frac{1}{2}}$$

$$\ddot{\mathbf{x}}_k = \frac{1}{\Delta t}\left(\dot{\mathbf{x}}_{k+\frac{1}{2}} - \dot{\mathbf{x}}_{k-\frac{1}{2}}\right) \quad \text{if needed}$$

Time integration when matrix $\left(\frac{1}{\Delta t}\mathbf{M} + \frac{1}{2}\mathbf{C}\right)$ is banded

Form $\bar{\mathbf{M}} = \left(\frac{1}{\Delta t}\mathbf{M} + \frac{1}{2}\mathbf{C}\right)$

Perform triangular decomposition on $\bar{\mathbf{M}}$

for k = 1, 2, …

$$t_{k+1} = t_k + \Delta t$$

$\mathbf{I}_k = \mathbf{K}\mathbf{x}_k$ Determined by direct assembly of element contributions

or $\mathbf{I}_k = \mathbf{I}_{k-1} + \mathbf{K}\Delta\mathbf{x}_k$ For nonlinear problems

$$\bar{\mathbf{p}}_k = \mathbf{p}_k - \mathbf{I}_k + \left(\frac{1}{\Delta t}\mathbf{M} - \frac{1}{2}\mathbf{C}\right)\dot{\mathbf{x}}_{k-\frac{1}{2}}$$

$\bar{\mathbf{M}}\dot{\mathbf{x}}_{k+\frac{1}{2}} = \bar{\mathbf{p}}_k$ Solve for $\dot{\mathbf{x}}_{k+\frac{1}{2}}$ by performing a vector reduction and backsubstitution

$$\mathbf{x}_{k+1} = \mathbf{x}_k + \Delta t\, \dot{\mathbf{x}}_{k+\frac{1}{2}}$$

$$\ddot{\mathbf{x}}_k = \frac{1}{\Delta t}\left(\dot{\mathbf{x}}_{k+\frac{1}{2}} - \dot{\mathbf{x}}_{k-\frac{1}{2}}\right) \quad \text{if needed}$$

8.3.6 Stability analysis of explicit second central difference method

In order to carry out a stability analysis of any direct integration method, we must first develop the integration operator matrix, which relates the vector of *state variables* at two consecutive time stations. The dimension of the integration operator matrix is mN, where m is the order of the differential equation being integrated, and N is the number of degrees of freedom in the discrete system. In this case, we are integrating a system of second-order differential equations, so the integration operator matrix, A, is 2N × 2N, and the vector of state variables has a dimension of 2N.

$$\bar{\mathbf{x}}_{k+1} = \mathbf{A}\bar{\mathbf{x}}_k \tag{8.123}$$

$$\bar{\mathbf{x}} = \begin{Bmatrix} \dot{\mathbf{x}} \\ \mathbf{x} \end{Bmatrix} \tag{8.124}$$

In order to develop this relationship for the second central difference method, we start by considering the equations of the algorithm for the case of undamped free vibration (damping and load vector normally have no influence on the stability of the underlying direct integration method).

$$\begin{aligned} \dot{\mathbf{x}}_{k+\frac{1}{2}} &= \dot{\mathbf{x}}_{k-\frac{1}{2}} - \Delta t \mathbf{M}^{-1}\mathbf{K}\mathbf{x}_k \\ \mathbf{x}_{k+1} &= \mathbf{x}_k + \Delta t\, \dot{\mathbf{x}}_{k+\frac{1}{2}} \end{aligned} \tag{8.125}$$

In order to put the right-hand side of these two equations in the correct format, we need to substitute the first equation for into the second equation.

$$\begin{aligned} \dot{\mathbf{x}}_{k+\frac{1}{2}} &= \dot{\mathbf{x}}_{k-\frac{1}{2}} - \Delta t\, \mathbf{M}^{-1}\mathbf{K}\mathbf{x}_k \\ \mathbf{x}_{k+1} &= \Delta t\, \dot{\mathbf{x}}_{k-\frac{1}{2}} + \left(\mathbf{I} - \Delta t^2\mathbf{M}^{-1}\mathbf{K}\right)\mathbf{x}_k \end{aligned} \tag{8.126}$$

$$\begin{Bmatrix} \dot{\mathbf{x}}_{k+\frac{1}{2}} \\ \mathbf{x}_{k+1} \end{Bmatrix} = \begin{bmatrix} \mathbf{I} & -\Delta t\, \mathbf{M}^{-1}\mathbf{K} \\ \Delta t\, \mathbf{I} & \left(\mathbf{I} - \Delta t^2\mathbf{M}^{-1}\mathbf{K}\right) \end{bmatrix} \begin{Bmatrix} \dot{\mathbf{x}}_{k-\frac{1}{2}} \\ \mathbf{x}_k \end{Bmatrix} \tag{8.127}$$

The matrix in the above equation is the 2N×2N integration operator matrix **A** and the state vector is at t_{k+1}.

$$\bar{\mathbf{x}} = \begin{Bmatrix} \dot{\mathbf{x}}_{k+\frac{1}{2}} \\ \mathbf{x}_{k+1} \end{Bmatrix}_{2N\times 1} \tag{8.128}$$

We note that the integration operator matrix is a function of the time step Δt and the mechanical properties of the system, which are represented by the natural frequencies, if the system is a structure.

Similar equations can be written for all the previous steps, up to the first step, which starts with initial condition.

$$\begin{aligned} \bar{\mathbf{x}}_k &= \mathbf{A}\bar{\mathbf{x}}_{k-1} \\ &= \mathbf{A}^2\bar{\mathbf{x}}_{k-2} \\ &\vdots \\ &= \mathbf{A}^k\bar{\mathbf{x}}_0 \end{aligned} \tag{8.129}$$

At this point, we can perform a spectral decomposition on **A**, such that

$$\begin{aligned} \mathbf{A}\,\psi &= \psi\,\bar{\Lambda} \\ \mathbf{A} &= \psi\,\bar{\Lambda}\,\psi^{-1} \\ \mathbf{A}^k &= \psi\,\bar{\Lambda}^k\,\psi^{-1} \end{aligned} \tag{8.130}$$

In the above relations, the diagonal terms of $\bar{\Lambda}$ are the eigenvalues of **A** and the columns of ψ are its eigenvectors. The eigenvalues and eigenvectors of **A** can in general be *complex*, since matrix **A** is unsymmetric. Moreover, the eigenvalue matrix may have nonzero terms above the diagonal, but all the terms below the diagonal are zero. This is because any matrix can be reduced to an upper triangular matrix through similarity transformation. Now we can substitute the spectral decomposition of **A** into Equation 8.129

$$\bar{\mathbf{x}}_k = \psi\,\bar{\Lambda}^k\,\psi^{-1}\bar{\mathbf{x}}_0 \tag{8.131}$$

The matrix $\bar{\Lambda}^k$ is either diagonal or upper triangular. In either case, its diagonal terms are the eigenvalues of **A** raised to the power of k. As the step number k increases, the eigenvalues are raised to higher and higher power, and this process can continue without any limit. It is easy to see that if the modulus of the eigenvalues of **A** is larger than one, the values of the state vector (displacements and velocities) will increase without bound as the time step increases. Of course, this is physically impossible in the case of the undamped free vibration. As a matter of fact, unbounded growth of the state variables is the definition of the *instability* of the solution process.

This shows that for stability, in the limit, the state variables must be bounded. That is, the integration operator must be a convergent matrix. Let $\rho = \left|\bar{\lambda}_{max}(\mathbf{A})\right|$, which is the spectral radius of matrix **A**, then the stability criterion is as follows:

$$\rho(\mathbf{A}) \le 1 \tag{8.132}$$

This is a very interesting condition. If physical damping and external loading are excluded from the system, then the eigenvalue property of **A** would expose certain aspect of the numerical integration method. For example, if, $\bar{\lambda}_k < 1$ for $k = 1,\ldots, 2n$, then $\|\bar{\mathbf{x}}_k\| < \|\bar{\mathbf{x}}_0\|$, which indicates the method is stable but it generates some amplitude decays as the integration process progresses. Therefore, the method is introducing numerical damping to the original undamped structural system. On the other hand, if $\bar{\lambda}_k = 1$ for $k = 1,\ldots, 2n$ then $\|\bar{\mathbf{x}}_k\| = \|\bar{\mathbf{x}}_0\|$. This method would have no amplitude decay and does not generate any spurious numerical damping to the structural system during the integration process. This appears to be the ideal condition for a numerical integration method.

Now the question is: what is the eigenvalue property of integration operator in the second central difference method. Instead of directly computing the eigenvalues of the integration operator matrix **A** which has a size of 2N × 2N, we can perform a coordinate transformation which transforms **A** into N uncoupled 2 × 2 matrices, which is a much simpler task. We start by performing an eigenvalue analysis of undamped free vibration of the structure.

$$\mathbf{K\Phi} = \mathbf{M\Phi\Lambda}$$

$$\Lambda = \begin{bmatrix} \lambda_1 & & \\ & \ddots & \\ & & \lambda_n \end{bmatrix} = \begin{bmatrix} \omega_1^2 & & \\ & \ddots & \\ & & \omega_n^2 \end{bmatrix} \tag{8.133}$$

$$\Phi = \left[\phi_1 | \cdots | \phi_n\right]$$

The above equation can be written as follows:

$$\mathbf{M}^{-1}\mathbf{K} = \mathbf{\Phi\Lambda\Phi}^{-1} \tag{8.134}$$

Now, we can substitute in Equation 8.127

$$\begin{Bmatrix} \dot{\mathbf{x}}_{k+\frac{1}{2}} \\ \mathbf{x}_{k+1} \end{Bmatrix} = \begin{bmatrix} \mathbf{I} & -\Delta t\, \mathbf{\Phi\Lambda\Phi}^{-1} \\ \Delta t\, \mathbf{I} & \mathbf{\Phi}\left(\mathbf{I} - \Delta t^2 \mathbf{\Lambda}\right)\mathbf{\Phi}^{-1} \end{bmatrix} \begin{Bmatrix} \dot{\mathbf{x}}_{k-\frac{1}{2}} \\ \mathbf{x}_k \end{Bmatrix} \tag{8.135}$$

Now, using the mode shape matrix, we perform coordinate transformation on the state variable as follows:

$$\begin{Bmatrix} \dot{\mathbf{x}}_{k+\frac{1}{2}} \\ \mathbf{x}_{k+1} \end{Bmatrix} = \begin{bmatrix} \mathbf{\Phi} & \\ & \mathbf{\Phi} \end{bmatrix} \begin{Bmatrix} \dot{\mathbf{y}}_{k+\frac{1}{2}} \\ \mathbf{y}_{k+1} \end{Bmatrix} \Rightarrow \begin{Bmatrix} \dot{\mathbf{y}}_{k+\frac{1}{2}} \\ \mathbf{y}_{k+1} \end{Bmatrix} = \begin{bmatrix} \mathbf{\Phi}^{-1} & \\ & \mathbf{\Phi}^{-1} \end{bmatrix} \begin{Bmatrix} \dot{\mathbf{x}}_{k+\frac{1}{2}} \\ \mathbf{x}_{k+1} \end{Bmatrix} \tag{8.136}$$

$$\begin{Bmatrix} \dot{\mathbf{x}}_{k-\frac{1}{2}} \\ \mathbf{x}_k \end{Bmatrix} = \begin{bmatrix} \mathbf{\Phi} & \\ & \mathbf{\Phi} \end{bmatrix} \begin{Bmatrix} \dot{\mathbf{y}}_{k-\frac{1}{2}} \\ \mathbf{y}_k \end{Bmatrix} \tag{8.137}$$

After substitutions of above relations into Equation 8.135, we have the following:

$$\begin{Bmatrix} \dot{\mathbf{y}}_{k+\frac{1}{2}} \\ \mathbf{y}_{k+1} \end{Bmatrix} = \begin{bmatrix} \mathbf{\Phi}^{-1} & \\ & \mathbf{\Phi}^{-1} \end{bmatrix} \begin{bmatrix} \mathbf{I} & -\Delta t\, \mathbf{\Phi\Lambda\Phi}^{-1} \\ \Delta t\, \mathbf{I} & \mathbf{\Phi}\left(\mathbf{I} - \Delta t^2 \mathbf{\Lambda}\right)\mathbf{\Phi}^{-1} \end{bmatrix} \begin{bmatrix} \mathbf{\Phi} & \\ & \mathbf{\Phi} \end{bmatrix} \begin{Bmatrix} \dot{\mathbf{y}}_{k-\frac{1}{2}} \\ \mathbf{y}_k \end{Bmatrix} \tag{8.138}$$

$$\begin{Bmatrix} \dot{\mathbf{y}}_{k+\frac{1}{2}} \\ \mathbf{y}_{k+1} \end{Bmatrix} = \begin{bmatrix} \mathbf{I} & -\Delta t\, \mathbf{\Lambda} \\ \Delta t\, \mathbf{I} & \left(\mathbf{I} - \Delta t^2 \mathbf{\Lambda}\right) \end{bmatrix} \begin{Bmatrix} \dot{\mathbf{y}}_{k-\frac{1}{2}} \\ \mathbf{y}_k \end{Bmatrix} \tag{8.139}$$

This can be written in the following compact form:

$$\bar{y}_{k+1} = \mathbf{B}\ \bar{y}_k \tag{8.140}$$

Because of the diagonal structure of matrices **I** and **Λ**, the matrix **B** is made up of four diagonal submatrices. Moreover, matrices **B** and **A** are similar and their eigenvalues are the same. We also note that interchanging rows and columns of a matrix does not affect its eigenvalues. We can perform row and column interchange operations to transform matrix **B** into a form with 2×2 submatrices along its diagonal.

$$\bar{y}_{k+1} = \bar{\mathbf{B}}\ \bar{y}_k \tag{8.141}$$

$$\begin{Bmatrix} \bar{y}_{k+1,1} \\ \vdots \\ \bar{y}_{k+1,n} \end{Bmatrix} = \begin{bmatrix} \bar{\mathbf{B}}_{11} & & \\ & \ddots & \\ & & \bar{\mathbf{B}}_{nn} \end{bmatrix} \begin{Bmatrix} \bar{y}_{k,1} \\ \vdots \\ \bar{y}_{k,n} \end{Bmatrix} \tag{8.142}$$

We can express the above matrix equation in its equivalent 2×2 decoupled form.

$$\bar{y}_{k+1,j} = \bar{\mathbf{B}}_{jj}\ \bar{y}_{k,j},\ \text{for } j = 1,\ \ldots,\ n \tag{8.143}$$

$$\begin{Bmatrix} \dot{y}_{k+\frac{1}{2}} \\ y_{k+1} \end{Bmatrix}_{,j} = \begin{bmatrix} 1 & -\Delta t\ \lambda_j \\ \Delta t & 1 - \Delta t^2 \lambda_j \end{bmatrix} \begin{Bmatrix} \dot{y}_{k-\frac{1}{2}} \\ y_k \end{Bmatrix}_{,j} \tag{8.144}$$

This shows that each one of these submatrices contains two eigenvalues of the original **B** matrix that are the same as the eigenvalues of the **A** matrix due to similarity transformation. In this case, the stability criterion can be expressed in terms of the maximum eigenvalues.

$$\max_j \rho\left(\bar{\mathbf{B}}_{jj}\right) \leq 1 \tag{8.145}$$

We can write the 2×2 submatrices in terms of the natural frequencies of the structural system.

$$\bar{\mathbf{B}}_{jj} = \begin{bmatrix} 1 & -\Delta t\ \omega_j^2 \\ \Delta t & 1 - \Delta t^2 \omega_j^2 \end{bmatrix} \tag{8.146}$$

The eigenvalue of the submatrix can be directly computed as the roots of the following characteristic equation:

$$\det\left(\begin{bmatrix} 1 & -\Delta t\ \omega_j^2 \\ \Delta t & 1 - \Delta t^2 \omega_j^2 \end{bmatrix} - \rho \mathbf{I}\right) = 0 \tag{8.147}$$

$$(1-\rho)^2 - \Delta t^2 \omega_j^2 (1-\rho) - \Delta t^2 \omega_j^2 = 0 \tag{8.148}$$

$$\begin{aligned} \rho &= 1 - \frac{\gamma}{2}\left(1 \pm \sqrt{1 - \frac{4}{\gamma}}\right) \\ \gamma &= \Delta t^2 \omega_j^2 \end{aligned} \tag{8.149}$$

Therefore, the stability criterion is follows:

$$\max|\rho| = \max\left|1 - \frac{\gamma}{2}\left(1 \pm \sqrt{1 - \frac{4}{\gamma}}\right)\right| \le 1 \tag{8.150}$$

From this relation, we can observe that if $\gamma > 4$, then ρ is a real quantity and we always have $|\rho| > 1$. This means that the method is unstable when $\gamma > 4$. For $\gamma < 4$, we have the following:

$$\rho = \left(1 - \frac{\gamma}{2}\right) \pm i\frac{\gamma}{2}\sqrt{\frac{4}{\gamma} - 1} \tag{8.151}$$

$$|\rho| = \left[\left(1 - \frac{\gamma}{2}\right)^2 + \left(\frac{\gamma}{2}\right)^2\left(\frac{4}{\gamma} - 1\right)\right]^{1/2} = 1 \tag{8.152}$$

This shows that the second central difference method is conditionally stable, and the stability criterion limits the time step.

$$\gamma_{max} = \Delta t^2 \omega_n^2 \le 4 \tag{8.153}$$

$$\Delta t \le \Delta t_{cr} = \frac{2}{\omega_n} \tag{8.154}$$

Δt_{cr} = critical time step

From this analysis, we can conclude that the second central difference method is *conditionally stable*. The stability condition imposes a limit on the size of time step, Δt, which must be smaller than the critical time step. Moreover, we note that the limiting critical time step is a function of the highest natural frequency of the system.

8.3.7 Courant stability criterion

In finite elements, the highest frequency of a system is equal to the highest frequency of any of the elements in the finite element discretization of the structural system. For a one-dimensional finite difference grid with grid size Δx or finite element mesh with equal size elements of Δx, the highest frequency is given by the following equation:

$$\omega_n = \frac{2c}{\Delta x} \tag{8.155}$$

where:

c is the wave velocity
Δx is the length of the shortest element

For one-dimensional linear elastic problems, the wave velocity can be obtained as $c = (E/\rho)^{1/2}$, in which E is the Young's modulus and ρ the mass density of the material. Then the critical time step is the time for wave signals to propagate across one element. This is the *Courant stability criterion.*

$$\Delta t \le \Delta t_{cr} = \frac{\Delta x}{c} \tag{8.156}$$

We can also write the Courant stability criterion in terms of a parameter α.

$$\alpha = \frac{c\Delta t}{\Delta x} \le 1 \tag{8.157}$$

In the special case of one-dimensional wave propagation, when we use a uniform mesh, the critical time step gives exact solution. In this case the signals propagate one element per time step. In nonuniform finite element mesh in 2D and 3D problems, the above stability criterion is approximate and is based on the smallest element in the finite element system.

8.3.8 Dynamic relaxation

The basic concept of dynamic relaxation method is to predict the static response of a structural system by performing dynamic analysis of an overdamped system subjected to a step function load ($\mathbf{p}$ = constant). We start with the dynamic equation of motion of damped structural system.

$$\mathbf{M}\ddot{\mathbf{x}}(t) + \mathbf{C}\dot{\mathbf{x}}(t) + \mathbf{K}\mathbf{x}(t) = \mathbf{p}(t) \tag{8.158}$$

For a step function load, the velocities and accelerations of the response of the structural system approach zero with time, as illustrated in Figure 8.10. Then the above system of dynamic equations of motion approaches the static system of equilibrium equations.

$$\mathbf{K}\mathbf{x}(t) = \mathbf{p}(t) \tag{8.159}$$

This is self-evident if we consider the dynamic response of an overdamped structure. Its steady-state dynamic response will approach the static response of the structure after a certain period of time. For highly damped systems, the convergence to static solution is very fast.

As discussed earlier, the damping matrix of the structure can be of many different forms. For simplicity, we can assume a mass-proportional damping, that is, $\mathbf{C} = \alpha\,\mathbf{M}$. This will facilitate the computations because the mass matrix is usually diagonal. We can substitute into the dynamic equation of motion at time station k.

$$\mathbf{M}\ddot{\mathbf{x}}_k + \alpha\mathbf{M}\dot{\mathbf{x}}_k + \mathbf{K}\mathbf{x}_k = \mathbf{p}_k \tag{8.160}$$

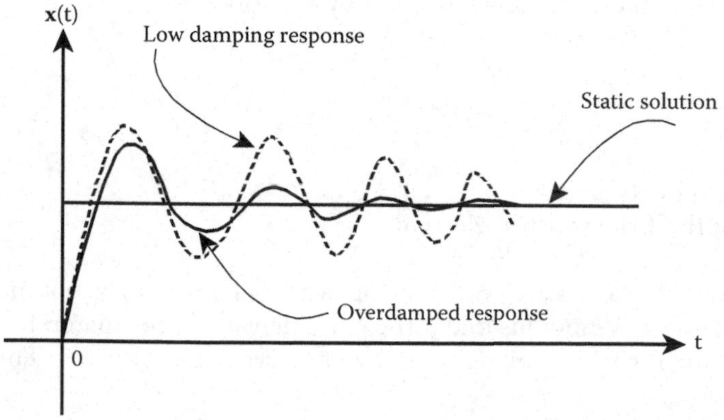

Figure 8.10 Response of overdamped structure subjected to a step function load under dynamic analysis and static analysis.

We can use the second central difference operator for acceleration vector and average velocity vector to arrive at the following relations that describe the computational process:

$$\frac{1}{\Delta t}\mathbf{M}\left(\dot{\mathbf{x}}_{k+\frac{1}{2}}-\dot{\mathbf{x}}_{k-\frac{1}{2}}\right)+\frac{1}{2}\alpha\mathbf{M}\left(\dot{\mathbf{x}}_{k+\frac{1}{2}}+\dot{\mathbf{x}}_{k-\frac{1}{2}}\right)+\mathbf{K}\mathbf{x}_k=\mathbf{p}_k \tag{8.161}$$

$$\mathbf{M}\dot{\mathbf{x}}_{k+\frac{1}{2}}=\frac{2-\alpha\Delta t}{2+\alpha\Delta t}\mathbf{M}\dot{\mathbf{x}}_{k-\frac{1}{2}}+\frac{2\Delta t}{2+\alpha\Delta t}\left(\mathbf{p}_k-\mathbf{K}\mathbf{x}_k\right) \tag{8.162}$$

$$\begin{cases}\mathbf{M}\dot{\mathbf{x}}_{k+\frac{1}{2}}=a_1\mathbf{M}\dot{\mathbf{x}}_{k-\frac{1}{2}}+a_2\left(\mathbf{p}_k-\mathbf{K}\mathbf{x}_k\right)\\[2ex] \mathbf{x}_{k+1}=\mathbf{x}_k+\Delta t\,\dot{\mathbf{x}}_{k+\frac{1}{2}}\end{cases} \tag{8.163}$$

$$\begin{cases}a_1=\dfrac{2-\alpha\Delta t}{2+\alpha\Delta t}\\[2ex] a_2=\dfrac{2\Delta t}{2+\alpha\Delta t}\end{cases} \tag{8.164}$$

8.4 IMPLICIT DIRECT INTEGRATION METHODS FOR DYNAMIC RESPONSE OF STRUCTURAL SYSTEMS

From our discussion earlier, we know that explicit direct integration methods are obtained by satisfying the equation of motion at the beginning of the time step, at time station t_k in the time step $[t_k, t_{k+1}]$. For implicit methods, the equation of motion is satisfied at the end of the time step, at time station t_{k+1}.

$$\mathbf{M}\ddot{\mathbf{x}}_{k+1}+\mathbf{C}\dot{\mathbf{x}}_{k+1}+\mathbf{K}\mathbf{x}_{k+1}=\mathbf{p}_{k+1} \tag{8.165}$$

In a general case, the displacement, velocity, and acceleration vectors are known at the beginning of the time step. Our objective is to compute them at the end of the time step. There are three unknown vectors. Therefore, we need two more equations in order to solve the problem. These two system of equations can only be obtained from temporal discretization method. In general, the temporal discretization is of the following form:

$$\begin{cases}\dot{\mathbf{x}}_{k+1}=a_0\mathbf{x}_{k+1}+a_1\mathbf{x}_k+a_2\dot{\mathbf{x}}_k+a_3\ddot{\mathbf{x}}_k\\ \ddot{\mathbf{x}}_{k+1}=b_0\mathbf{x}_{k+1}+b_1\mathbf{x}_k+b_2\dot{\mathbf{x}}_k+b_3\ddot{\mathbf{x}}_k\end{cases} \tag{8.166}$$

Integration constants are $a_0, \ldots, a_3, b_0, \ldots, b_3$, which are functions of the time step and *integration parameters*. The above equations can also be written as follows:

$$\begin{cases}\dot{\mathbf{x}}_{k+1}=a_0\mathbf{x}_{k+1}+\mathbf{g}_k\\[1ex] \ddot{\mathbf{x}}_{k+1}=b_0\mathbf{x}_{k+1}+\mathbf{h}_k\end{cases} \tag{8.167}$$

$$\begin{cases}\mathbf{g}_k=a_1\mathbf{x}_k+a_2\dot{\mathbf{x}}_k+a_3\ddot{\mathbf{x}}_k\\[1ex] \mathbf{h}_k=b_1\mathbf{x}_k+b_2\dot{\mathbf{x}}_k+b_3\ddot{\mathbf{x}}_k\end{cases} \tag{8.168}$$

The above equations lead to the following discrete equation of motion:

$$(b_0\mathbf{M} + a_0\mathbf{C} + \mathbf{K})\mathbf{x}_{k+1} = \mathbf{p}_{k+1} - \mathbf{M}\mathbf{h}_k - \mathbf{C}\mathbf{g}_k \tag{8.169}$$

$$\bar{\mathbf{K}}\mathbf{x}_{k+1} = \bar{\mathbf{p}}_{k+1}$$

$$\begin{cases} \bar{\mathbf{K}} = b_0\mathbf{M} + a_0\mathbf{C} + \mathbf{K} \\ \bar{\mathbf{p}}_{k+1} = \mathbf{p}_{k+1} - \mathbf{M}\mathbf{h}_k - \mathbf{C}\mathbf{g}_k \end{cases} \tag{8.170}$$

In the above equation we have the equivalent stiffness matrix and equivalent load vector. The above relation is called the equivalent static system of equations, which is linear at every time step. This shows that computing $\mathbf{x}_{k+1}$ requires the solution of a system of equations with $\bar{\mathbf{K}}$, and involves the use of equilibrium equations at time station t_{k+1}. For this reason, this class of direct integration methods are called *implicit* integration methods.

It is observed that if the structural system is linear and the time step Δt is constant during the time integration process, the equivalent stiffness matrix needs to be formed only once at the beginning. We can then perform LU decomposition of the matrix and store them to facilitate the solution of the system of equations at each step.

So far we have presented the general framework for implicit direct integration methods of the solution of dynamic equations of motion. In structural dynamics, the most well-known practical methods in direct integration are the Newmark family of methods, the Wilson-θ method, as well as some other multistep methods. This is a unique class of methods developed specifically for the solution of second-order ordinary differential equations arising from finite element spatial discretization in structural dynamics. In the following sections, we will discuss some of these methods.

8.4.1 Linear acceleration method

In the temporal discretization of dynamic equations of motion, the accelerations, velocities, and displacements are related through either differentiation or integration operations.

$$\dot{\mathbf{x}} = \frac{d}{dt}\mathbf{x}, \; \mathbf{x} = \int \dot{\mathbf{x}}\, dt \tag{8.171}$$

It is observed that with numerical differentiation, the original error present in $\mathbf{x}$ will get expanded after the differential operation. However, with numerical integration, the original error presented in the integrand will get smoothed out after the operation. This means that performing numerical integration operation is potentially of advantage in reducing or smoothing out the original error present within the integrand. This observation serves as the basis of many engineering based temporal discretization methods. That is, instead of starting with displacements, we make assumptions on the variation of accelerations within a time step, and obtain relations for velocities and displacements within the time step through numerical integration. By so doing, we will obtain numerical integration method with higher degree of accuracy.

As the name implies, with the linear acceleration method, we assume linear variation of accelerations within the time step, as shown in Figure 8.11.

$$\begin{aligned} \ddot{\mathbf{x}}(t+\tau) &= \ddot{\mathbf{x}}_k + \frac{1}{\Delta t}(\ddot{\mathbf{x}}_{k+1} - \ddot{\mathbf{x}}_k)\tau \\ \tau &= t - t_k \end{aligned} \tag{8.172}$$

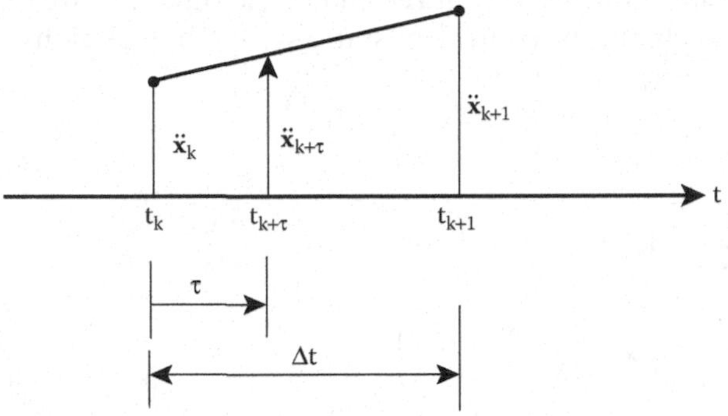

Figure 8.11 Assumption of linear variation of acceleration within a time step in the linear acceleration method.

The velocity and displacement can be obtained through integration of the equation above.

$$\begin{aligned}\dot{x}(t+\tau) &= \dot{x}_k + \int_0^{\tau} \ddot{x}(t+\tau)\, d\tau \\ &= \dot{x}_k + \ddot{x}_k \tau + \frac{1}{2\Delta t}\left(\ddot{x}_{k+1} - \ddot{x}_k\right)\tau^2\end{aligned} \tag{8.173}$$

$$\begin{aligned}x(t+\tau) &= x_k + \int_0^{\tau} \dot{x}(t+\tau)\, dt \\ &= x_k + \dot{x}_k \tau + \frac{1}{2}\ddot{x}_k \tau^2 + \frac{1}{6\Delta t}\left(\ddot{x}_{k+1} - \ddot{x}_k\right)\tau^3\end{aligned} \tag{8.174}$$

Using these equations, we can determine the velocity and displacement at the end of the time step, $\tau = \Delta t$.

$$\begin{cases} \dot{x}_{k+1} = \dot{x}_k + \dfrac{1}{2}\Delta t\, \ddot{x}_k + \dfrac{1}{2}\Delta t\, \ddot{x}_{k+1} \\ x_{k+1} = x_k + \Delta t\, \dot{x}_k + \dfrac{1}{3}\Delta t^2\, \ddot{x}_k + \dfrac{1}{6}\Delta t^2\, \ddot{x}_{k+1} \end{cases} \tag{8.175}$$

These two equations and the equation of motion at the end of the time step are the three equations needed to determine the three variables at the end of the time step. In the numerical computational process, we normally arrive at an equivalent static equilibrium equation, Equation 8.170. To achieve this, we determine the acceleration at the end of the time step from the second equation above, and substitute in the first equation.

$$\ddot{x}_{k+1} = \frac{6}{\Delta t^2}\left(x_{k+1} - x_k\right) - \frac{6}{\Delta t}\dot{x}_k - 2\ddot{x}_k \tag{8.176}$$

$$\dot{x}_{k+1} = \frac{3}{\Delta t}\left(x_{k+1} - x_k\right) - 2\dot{x}_k - \frac{\Delta t}{2}\ddot{x}_k \tag{8.177}$$

The resulting acceleration and velocity at the end of the time step are then substituted in the equation of motion to obtain the equivalent static system of equations.

$$\bar{\mathbf{K}}\mathbf{x}_{k+1} = \bar{\mathbf{p}}_{k+1} \tag{8.178}$$

$$\begin{cases} \bar{\mathbf{K}} = \dfrac{6}{\Delta t^2}\mathbf{M} + \dfrac{3}{\Delta t}\mathbf{C} + \mathbf{K} \\ \bar{\mathbf{p}}_{k+1} = \mathbf{p}_{k+1} + \mathbf{M}\left(\dfrac{6}{\Delta t^2}\mathbf{x}_k + \dfrac{6}{\Delta t}\dot{\mathbf{x}}_k + 2\ddot{\mathbf{x}}_k\right) \\ \qquad + \mathbf{C}\left(\dfrac{3}{\Delta t}\mathbf{x}_k + 2\dot{\mathbf{x}}_k + \dfrac{\Delta t}{2}\ddot{\mathbf{x}}_k\right) \end{cases} \tag{8.179}$$

Once the displacement is determined, we can then compute the acceleration and velocity and carry out the time stepping direct integration process. We will see in Section 8.4.4 that this method is conditionally stable.

8.4.2 Newmark family of methods

The Newmark family of methods was developed in 1959 (Newmark, 1959). It is an implicit method similar to linear acceleration method. It may be considered a generalization of the linear acceleration method. Newmark method differs from linear acceleration method by introducing two parameters, γ and β, into the temporal discretization method within the time step.

$$\begin{cases} \dot{\mathbf{x}}_{k+1} = \dot{\mathbf{x}}_k + \Delta t\left[(1-\gamma)\,\ddot{\mathbf{x}}_k + \gamma\,\ddot{\mathbf{x}}_{k+1}\right] \\ \mathbf{x}_{k+1} = \mathbf{x}_k + \Delta t\,\dot{\mathbf{x}}_k + \Delta t^2\left[\left(\dfrac{1}{2} - \beta\right)\ddot{\mathbf{x}}_k + \beta\,\ddot{\mathbf{x}}_{k+1}\right] \end{cases} \tag{8.180}$$

Computational steps and implementation of Newmark family of methods is similar to other implicit methods described in Equations 8.165 through 8.170.

We can see that the linear acceleration method corresponds to setting $\gamma = 1/2$ and $\beta = 1/6$. Newmark method can also be thought of as obtained from expressing the solution at t_{k+1} as a Taylor series with the remainder approximated by quadrature formulas. There are other ways to derive direct integration methods including the Newmark method, which will be discussed in later sections and in Chapter 9.

We can show later that by giving different values to the Newmark integration parameters γ and β, we can obtain a whole spectrum of direct integration methods. For this reason, the Newmark-β method is usually referred to as the Newmark family of methods. However, for stability and accuracy consideration, the integration parameters are usually chosen in the following intervals:

$$\begin{cases} 0 \le \beta \le \dfrac{1}{2} \\ \dfrac{1}{2} \le \gamma \le 1 \end{cases} \tag{8.181}$$

Usually, we fix the parameter $\gamma = 1/2$. It can be shown that if $\gamma < 1/2$, a *negative* numerical damping is introduced by the algorithm which may result in unreasonable dynamic response, even unbounded responses. For $\gamma > 1/2$, a *positive* numerical damping is introduced by the algorithm. When $\gamma = 1/2$, *no* numerical damping is introduced by the algorithm. However, there is no optimal value for the integration parameter β. Usually, the method corresponding to $\gamma = 1/2$ and $\beta = 1/4$ is used in practice. This method is referred to as the *constant acceleration method* in structural dynamics.

We have mentioned that the Newmark method is a general method that encompasses a spectrum of well-known direct integration methods. It has been shown that by setting $\gamma = 1/2$ and $\beta = 1/6$, we obtain the linear acceleration method which is conditionally stable. The stability criterion of this method is $\Delta t < 0.551$ T, where T is the system's smallest natural period, corresponding to the highest eigenvalue. The method becomes the constant average acceleration method by setting $\gamma = 1/2$ and $\beta = 1/4$, which is unconditionally stable. In addition, Fox–Goodwin method (Fox and Goodwin, 1949) corresponds to $\gamma = 1/2$ and $\beta = 1/12$. The Fox–Goodwin method is conditionally stable, with the stability criterion $\Delta t < 0.389$ T. All these methods are implicit methods and only the average acceleration method is unconditionally stable.

We can also obtain an explicit method by setting $\gamma = 1/2$ and $\beta = 0$; this is equivalent to the second central difference method, whose stability criterion has been obtained as $\Delta t < T/\pi = 0.318$ T. When $\gamma \neq 1/2$, we will obtain some methods with *positive* numerical damping. The Galerkin method corresponds to $\gamma = 3/2$ and $\beta = 4/5$. When $\gamma = 3/2$ and $\beta = 1$, the method is the backward difference method. And when $\gamma = -1/2$ and $\beta = 0$, we obtain an unstable method equivalent to the forward difference method.

8.4.3 Constant average acceleration method

This method was proposed by Newmark by assuming a constant average acceleration within the time step as shown in Figure 8.12.

$$\ddot{x}(t+\tau) = \frac{1}{2}\left(\ddot{x}_{k+1} + \ddot{x}_k\right) \tag{8.182}$$

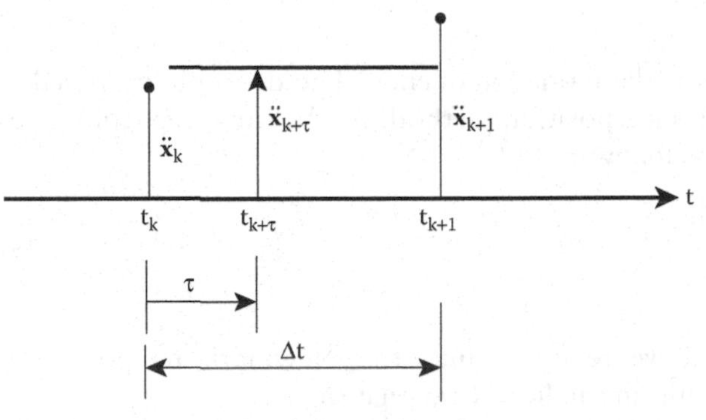

Figure 8.12 Assumption of constant average acceleration within a time step in the constant average acceleration method.

Through direct integration, we obtain the velocity and displacement vector.

$$\begin{aligned}\dot{\mathbf{x}}(t+\tau) &= \dot{\mathbf{x}}_k + \int_0^{\tau} \ddot{\mathbf{x}}(t+\tau)\, d\tau \\ &= \dot{\mathbf{x}}_k + \frac{1}{2}\left(\ddot{\mathbf{x}}_{k+1} + \ddot{\mathbf{x}}_k\right)\tau\end{aligned} \tag{8.183}$$

$$\begin{aligned}\mathbf{x}(t+\tau) &= \mathbf{x}_k + \int_0^{\tau} \dot{\mathbf{x}}(t+\tau) dt \\ &= \mathbf{x}_k + \dot{\mathbf{x}}_k \tau + \frac{1}{4}\left(\ddot{\mathbf{x}}_{k+1} + \ddot{\mathbf{x}}_k\right)\tau^2\end{aligned} \tag{8.184}$$

At $\tau = \Delta t$, we obtain the velocity and displacement vectors at the end of the time step.

$$\begin{cases}\dot{\mathbf{x}}_{k+1} = \dot{\mathbf{x}}_k + \dfrac{1}{2}\Delta t\,\left(\ddot{\mathbf{x}}_k + \ddot{\mathbf{x}}_{k+1}\right) \\ \mathbf{x}_{k+1} = \mathbf{x}_k + \Delta t\ \dot{\mathbf{x}}_k + \dfrac{1}{4}\Delta t^2\left(\ddot{\mathbf{x}}_k + \ddot{\mathbf{x}}_{k+1}\right)\end{cases} \tag{8.185}$$

We can easily verify that this method corresponds to taking $\gamma = 1/2$ and $\beta = 1/4$ in the original Newmark family of methods. We will show later that this method is unconditionally stable and does not introduce any numerical damping to the system. The only numerical error introduced by this algorithm is in terms of period elongation, as will be discussed in Section 8.4.5.

8.4.4 Stability analysis of Newmark method

As has been discussed earlier, the stability analysis of a direct integration method can be carried out by considering a single-mode response. By ignoring physical damping and the external loading acting on the system, we can decouple the following system of dynamic equations of motion into its modal equation of motion:

$$\mathbf{M}\ddot{\mathbf{x}}(t) + \mathbf{K}\mathbf{x}(t) = 0 \tag{8.186}$$

$$\ddot{x}(t) + \omega^2 x(t) = 0 \tag{8.187}$$

In this equation ω is the modal frequency. The decoupling procedure has been discussed earlier in the mode superposition method. At the time step from t_k to t_{k+1}, we have the discretized relations as follows:

$$\begin{cases}\ddot{x}_k = -\omega^2 x_k \\ \ddot{x}_{k+1} = -\omega^2 x_{k+1}\end{cases} \tag{8.188}$$

Substituting the above relations into the Newmark temporal discretization formulae Equation 8.180 results in the following relations:

$$\begin{aligned}\Delta t\dot{x}_{k+1} &= \Delta t\dot{x}_k + \Delta t^2\left[(1-\gamma)\ddot{x}_k + \gamma\ddot{x}_{k+1}\right] \\ &= \Delta t\dot{x}_k - \left[\omega^2\Delta t^2(1-\gamma)x_k + \omega^2\Delta t^2\gamma x_{k+1}\right]\end{aligned} \tag{8.189}$$

$$\begin{aligned} x_{k+1} &= x_k + \Delta t\ \dot{x}_k + \Delta t^2 \left[\left(\frac{1}{2} - \beta \right) \ddot{x}_k + \beta\ \ddot{x}_{k+1} \right] \\ &= x_k + \Delta t\ \dot{x}_k - \left[\omega^2 \Delta t^2 \left(\frac{1}{2} - \beta \right) x_k + \omega^2 \Delta t^2 \beta\ x_{k+1} \right] \end{aligned} \tag{8.190}$$

Let $\delta = \omega\Delta t$, then the above two equations can be rearranged in the following form:

$$\begin{cases} \left(1+\delta^2\beta\right) x_{k+1} = \left[1 - \delta^2\left(\frac{1}{2} - \beta\right)\right] x_k + \Delta t\ \dot{x}_k \\ \delta^2 \gamma x_{k+1} + \Delta t \dot{x}_{k+1} = -\delta^2 (1-\gamma) x_k + \Delta t \dot{x}_k \end{cases} \tag{8.191}$$

$$\begin{bmatrix} 1+\delta^2\beta & 0 \\ \delta^2\gamma & 1 \end{bmatrix} \begin{Bmatrix} x_{k+1} \\ \Delta t \dot{x}_{k+1} \end{Bmatrix} = \begin{bmatrix} 1-\delta^2(1/2-\beta) & 1 \\ -\delta^2(1-\gamma) & 1 \end{bmatrix} \begin{Bmatrix} x_k \\ \Delta t \dot{x}_k \end{Bmatrix} \tag{8.192}$$

The undamped free vibration response of this single degree of freedom system is as follows.

$$\begin{aligned} x_k &= Ae^{\omega t_k} = Ae^{\omega(k\Delta t)} \\ &= A\left(e^{\omega\Delta t}\right)^k = A\lambda^k \end{aligned} \tag{8.193}$$

$$\Delta t\ \dot{x}_k = B\lambda^k \tag{8.194}$$

$$\lambda = e^{\omega\Delta t} \tag{8.195}$$

A and B are constants depending on the initial conditions.

$$\begin{Bmatrix} x_k \\ \Delta t\ \dot{x}_k \end{Bmatrix} = \lambda^k \begin{Bmatrix} A \\ B \end{Bmatrix}; \begin{Bmatrix} x_{k+1} \\ \Delta t\ \dot{x}_{k+1} \end{Bmatrix} = \lambda^{k+1} \begin{Bmatrix} A \\ B \end{Bmatrix} \tag{8.196}$$

After substitutions, we obtain the following eigenvalue problem:

$$\lambda \begin{bmatrix} 1+\delta^2\beta & 0 \\ \delta^2\gamma & 1 \end{bmatrix} \begin{Bmatrix} A \\ B \end{Bmatrix} = \begin{bmatrix} 1-\delta^2(1/2-\beta) & 1 \\ -\delta^2(1-\gamma) & 1 \end{bmatrix} \begin{Bmatrix} A \\ B \end{Bmatrix} \tag{8.197}$$

$$\begin{bmatrix} 1-(1/2-\beta)\delta^2 - \lambda\left(1+\delta^2\beta\right) & 1 \\ -(1-\gamma)\delta^2 - \lambda\gamma\delta^2 & 1-\lambda \end{bmatrix} \begin{Bmatrix} A \\ B \end{Bmatrix} = \begin{Bmatrix} 0 \\ 0 \end{Bmatrix} \tag{8.198}$$

This leads to the following characteristic equation:

$$\left(1+\delta^2\beta\right)\lambda^2 - \left[2\left(1+\beta\delta^2\right) - \left(\frac{1}{2}+\gamma\right)\delta^2\right]\lambda + \left[\left(1+\beta\delta^2\right) + \left(\frac{1}{2}-\gamma\right)\delta^2\right] = 0 \tag{8.199}$$

Since β is a nonnegative constant, $1 + \beta\delta^2 > 0$, then the above quadratic equation can be simplified as follows.

$$\lambda^2 - c_1\lambda + c_2 = 0 \tag{8.200}$$

$$\begin{cases} c_1 = 2 - \dfrac{\left(\dfrac{1}{2} + \gamma\right)\delta^2}{1 + \delta^2\beta} \\ c_2 = 1 + \dfrac{\left(\dfrac{1}{2} - \gamma\right)\delta^2}{1 + \delta^2\beta} \end{cases} \tag{8.201}$$

The roots of this quadratic equations are λ_1 and λ_2.

$$\begin{cases} \lambda_1 = a\ e^{i\phi} = a\left(\cos\phi + i\ \sin\phi\right) \\ \lambda_2 = \bar{\lambda}_1 = a\ e^{-i\phi} = a\left(\cos\phi - i\ \sin\phi\right) \end{cases} \tag{8.202}$$

$$\left(i = \sqrt{-1}\right)$$

$$a^2 = c_2;\ \phi = \cos^{-1}\left(\frac{c_1}{2\sqrt{c_2}}\right) \tag{8.203}$$

$$\left(\lambda - \lambda_1\right)\left(\lambda - \lambda_2\right) = 0 \tag{8.204}$$

$$\lambda^2 - \left(\lambda_1 + \lambda_2\right)\lambda + \lambda_1\lambda_2 = 0 \tag{8.205}$$

$$\begin{cases} \lambda_1 + \lambda_2 = c_1 \\ \lambda_1\lambda_2 = c_2 \end{cases} \tag{8.206}$$

$$x_k = A_1\lambda_1^k + A_2\lambda_2^k \tag{8.207}$$

A special case is represented by $c_2 = 1$, which in turn means that there should be $\gamma = 1/2$. We have discussed earlier this corresponds to no numerical damping. If $\gamma > 1/2$ which means that $c_2 < 1$, then *positive* numerical damping is introduced to the system, the amplitude of the resulting vibration will be decreasing with time. However, if $\gamma < 1/2$ which means that $c_2 > 1$, then *negative* numerical damping is introduced to the system and the amplitude of the resulting vibration will be increasing with time. In the case of $c_2 = 1$ we have the following equation that leads to the stability criterion:

$$\lambda^2 - \left(2 - \frac{\delta^2}{1 + \delta^2\beta}\right)\lambda + 1 = 0 \tag{8.208}$$

$$c_1 = \lambda_1 + \lambda_2 = 2 - \frac{\delta^2}{1 + \delta^2\beta} = 2\cos\phi \tag{8.209}$$

$$\cos\phi = 1 - \frac{\delta^2}{2\left(1+\delta^2\beta\right)}$$

$$\left|1 - \frac{\delta^2}{2\left(1+\delta^2\beta\right)}\right| \leq 1 \tag{8.210}$$

$$\delta^2\left(1-4\beta\right) \leq 4 \tag{8.211}$$

Hence, the stability criterion of Newmark method is obtained as follows:

$$\Delta t \leq \frac{2}{\omega\sqrt{1-4\beta}} \tag{8.212}$$

Obviously, for $\beta = 1/4$, the stability criterion is always satisfied, which indicates that the constant average acceleration method corresponding to $(\gamma, \beta) = (1/2, 1/4)$ is unconditionally stable. If $\beta \neq 1/4$, then the resulting direct integration method is conditionally stable. When $\beta = 0$, the method corresponds to the second central difference method, and we get the Courant stability criterion derived earlier.

8.4.5 Numerical integration errors in Newmark method

The numerical integration errors in the dynamic responses of structural systems can be described in terms of amplitude decay and period elongation. For the case of undamped free vibration, these two measures of error are illustrated in Figure 8.13. To analyze the numerical error in computed period of response or period elongation, we start with the definition of vibration period.

$$T = \frac{2\pi}{\omega} = \frac{2\pi}{\delta}\Delta t \tag{8.213}$$

This is the actual period of the system. The pseudo-period computed from Newmark method is

$$T_{sp} = \frac{2\pi}{\phi}\Delta t \tag{8.214}$$

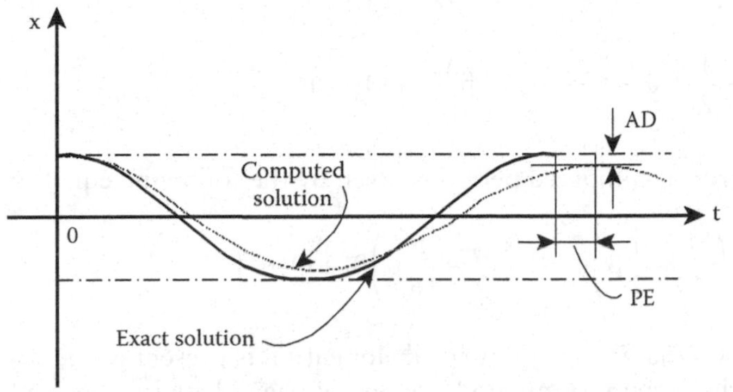

Figure 8.13 Illustration of numerical integration errors: amplitude decay (AD) and period elongation (PE) (assuming no physical damping and free vibration).

From the above relations, it can be obtained that

$$\frac{T}{T_{sp}} = \frac{\phi}{\delta} \tag{8.215}$$

This relation represents the numerical error in computed period of the structural response. We recall the following relation (Equation 8.209) and expand on that relation:

$$\cos\phi = 1 - \frac{\delta^2}{2\left(1+\delta^2\beta\right)} \tag{8.216}$$

$$2\sin^2\frac{\phi}{2} = \frac{\delta^2}{2\left(1+\delta^2\beta\right)} \tag{8.217}$$

$$\begin{aligned} \sin\frac{\phi}{2} &= \frac{\delta}{2\sqrt{1+\delta^2\beta}} \\ \phi &= 2\ \arcsin\frac{\delta}{2\sqrt{1+\delta^2\beta}} \end{aligned} \tag{8.218}$$

Next, we use the following series for determining the arcsine:

$$\arcsin\ x = x + \left(\frac{1}{2}\right)\frac{x^3}{3} + \left(\frac{1\cdot 3}{2\cdot 4}\right)\frac{x^5}{5} + \left(\frac{1\cdot 3\cdot 5}{2\cdot 4\cdot 6}\right)\frac{x^7}{7} + \cdots \quad \text{for } |x| < 1 \tag{8.219}$$

We will use the first two terms from this series for arcsine.

$$\begin{aligned} \phi &= 2\left[\frac{\delta}{2\sqrt{1+\delta^2\beta}} + \frac{1}{6}\frac{\delta^3}{8\left(1+\delta^2\beta\right)^{3/2}} + \cdots\right] \\ &= 2\left[\frac{\delta}{2}\left(1 - \frac{1}{2}\delta^2\beta + \frac{3}{8}\delta^4\beta^2 + \cdots\right) + \frac{\delta^3}{48}\left(1 - \frac{3}{2}\delta^2\beta + \cdots\right)\right] \\ &= \delta + \left(\frac{1}{24} - \frac{1}{2}\beta\right)\delta^3 + \left(\frac{3}{8}\beta^2 - \frac{1}{16}\beta\right)\delta^5 + O\left(\delta^7\right) \end{aligned} \tag{8.220}$$

Therefore, the error in computed period is given by the following equation:

$$\frac{T}{T_{sp}} = \frac{\phi}{\delta} = 1 + \left(\frac{1}{24} - \frac{1}{2}\beta\right)\delta^2 + \left(\frac{3}{8}\beta^2 - \frac{1}{16}\beta\right)\delta^4 + O\left(\delta^6\right) \tag{8.221}$$

This relation shows that in general, period elongation is present in the computed responses obtained using the Newmark method. Below, we have listed the period elongation error for three direct integration methods. We can see that the error is largest for the Newmark constant average acceleration method and is smallest for Fox–Goodwin method.

Constant average acceleration method

$$\beta = \frac{1}{4}, \quad \gamma = \frac{1}{2}$$

$$\frac{T}{T_{sp}} = 1 - \frac{1}{12}\delta^2 + O\left(\delta^4\right)$$

Linear acceleration method

$$\beta = \frac{1}{6}, \quad \gamma = \frac{1}{2}$$

$$\frac{T}{T_{sp}} = 1 - \frac{1}{24}\delta^2 + O\left(\delta^4\right)$$

Fox–Goodwin method

$$\beta = \frac{1}{12}, \quad \gamma = \frac{1}{2}$$

$$\frac{T}{T_{sp}} = 1 - \frac{1}{384}\delta^4 + O\left(\delta^6\right)$$

To study the amplitude decay error, we need to consider the eigenvectors of the system, which represents the mode shapes. We start with the following derivations from previous section, for $\gamma = 1/2$, and the eigenvalue equation:

$$\begin{cases} \lambda_1 = e^{i\phi} = \cos\phi + i\ \sin\phi \\ \lambda_2 = e^{-i\phi} = \cos\phi - i\ \sin\phi \end{cases}$$

$$\cos\phi = 1 - \frac{\delta^2}{2\left(1 + \delta^2\beta\right)} \tag{8.222}$$

$$\begin{bmatrix} 1 - \left(\frac{1}{2} - \beta\right)\delta^2 - \lambda\left(1 + \delta^2\beta\right) & 1 \\ -(1-\gamma)\delta^2 - \lambda\gamma\delta^2 & 1 - \lambda \end{bmatrix} \begin{Bmatrix} A \\ B \end{Bmatrix} = \begin{Bmatrix} 0 \\ 0 \end{Bmatrix} \tag{8.223}$$

To facilitate the computation of the eigenvectors, we first substitute λ_1 into the first equation and determine the ratio of B/A.

$$\begin{aligned} \frac{B}{A} &= -1 + \left(\frac{1}{2} - \beta\right)\delta^2 + \lambda_1\left(1 + \delta^2\beta\right) \\ &= -1 + \left(\frac{1}{2} - \beta\right)\delta^2 + \left(\cos\phi + i\ \sin\phi\right)\left(1 + \delta^2\beta\right) \\ &= -1 + \left(\frac{1}{2} - \beta\right)\delta^2 + \frac{1}{2}\left(2 + 2\delta^2\beta - \delta^2\right) + i\ \left(1 + \delta^2\beta\right)\sin\phi \\ &= i\ \left(1 + \delta^2\beta\right)\sin\phi \end{aligned} \tag{8.224}$$

We can also determine sin ϕ directly from the expression for the cos ϕ in Equation 8.222.

$$\cos^2\phi = \frac{4\left(1+\delta^2\beta\right)^2 - 4\delta^2\left(1+\delta^2\beta\right)+\delta^4}{4\left(1+\delta^2\beta\right)^2} \tag{8.225}$$

$$\sin^2\phi = 1-\cos^2\phi = \frac{4\delta^2\left(1+\delta^2\beta\right)-\delta^4}{4\left(1+\delta^2\beta\right)^2} \tag{8.226}$$

$$\sin\phi = \frac{\delta\sqrt{1+\delta^2\left(\beta-\dfrac{1}{4}\right)}}{\left(1+\delta^2\beta\right)} \tag{8.227}$$

Now, we can substitute in Equation 8.224

$$\frac{B}{A} = i\delta\sqrt{1+\delta^2\left(\beta-\frac{1}{4}\right)} \tag{8.228}$$

We can arrive at a similar relation for λ_2.

$$\frac{B}{A} = -i\delta\sqrt{1+\delta^2\left(\beta-\frac{1}{4}\right)} \tag{8.229}$$

From Equation 8.196 we can obtain the solution at time t_k.

$$\begin{Bmatrix} x_k \\ \Delta t\ \dot{x}_k \end{Bmatrix} = \lambda^k \begin{Bmatrix} A \\ B \end{Bmatrix} \tag{8.230}$$

$$x_k = A_1 e^{ik\phi} + A_2 e^{-ik\phi} \tag{8.231}$$

$$\begin{aligned} \Delta t\ \dot{x}_k &= B_1 e^{ik\phi} + B_2 e^{-ik\phi} \\ &= A_1 i\delta\sqrt{1+\delta^2\left(\beta-\frac{1}{4}\right)}e^{ik\phi} - A_2 i\delta\sqrt{1+\delta^2\left(\beta-\frac{1}{4}\right)}e^{-ik\phi} \end{aligned} \tag{8.232}$$

The constants A_1 and A_2 can be determined from initial condition and we assume them to be as follows.

$$\begin{cases} A_1 = c_0 e^{i\phi_o} \\ A_2 = c_0 e^{-i\phi_o} \end{cases} \tag{8.233}$$

$$\begin{aligned} x_k &= c_0 e^{i(\phi_o+k\phi)} + c_0 e^{-i(\phi_o+k\phi)} \\ &= 2c_0\cos\left(\phi_o + k\phi\right) \\ &= 2c_0\cos\left(\phi_o + \omega k\Delta t\right) \\ &= 2c_0\cos\left(\phi_o + \omega t_k\right) \end{aligned} \tag{8.234}$$

$$\Delta t\ \dot{x}_k = i\delta\sqrt{1+\delta^2\left(\beta-\frac{1}{4}\right)}\left[2ic_0\sin(\phi_o+k\phi)\right]$$
$$= -\sqrt{1+\delta^2\left(\beta-\frac{1}{4}\right)}\left[2c_0\delta\sin(\phi_o+\omega t_k)\right] \tag{8.235}$$

However, by direct calculation we can have the exact relation from x_k.

$$\Delta t\ \dot{x}_k = -2c_0\delta\sin(\phi_o+\omega t_k) \tag{8.236}$$

This is consistent with the displacement response of the system. This shows that in order to conserve the energy of the system, there must be $\beta = 1/4$. Otherwise, when $\beta < 1/4$ for stability considerations, extra energy has to be dissipated from the system because the kinetic energy is related to the velocity of the system. This indicates that the constant average acceleration method possesses the best numerical property in not having any amplitude decay, but it still generates period elongation error. Whereas, all the other methods in the Newmark family have errors in both amplitude decay and period elongation.

The stability characteristics of Newmark method can also be investigated by using the same approach in analyzing the second central difference method. Ignoring external loading, we can write the Newmark time stepping method in the following recursive form (Bathe and Wilson, 1973):

$$\begin{Bmatrix} \ddot{x}_{k+1} \\ \dot{x}_{k+1} \\ x_{k+1} \end{Bmatrix} = [\mathbf{A}]\begin{Bmatrix} \ddot{x}_k \\ \dot{x}_k \\ x_k \end{Bmatrix} \tag{8.237}$$

The matrix [**A**] is the approximation operator.

$$[\mathbf{A}] = \begin{bmatrix} -\left(\frac{1}{2}-\beta\right)\eta - 2(1-\gamma)\kappa & \frac{1}{\Delta t}(-\eta-2\kappa) & \frac{1}{\Delta t^2}(-\eta) \\ \Delta t\left[(1-\gamma)-\left(\frac{1}{2}-\beta\right)\gamma\eta - 2(1-\gamma)\gamma\kappa\right] & (1-\gamma\eta-2\eta\kappa) & \frac{1}{\Delta t}(-\eta\gamma) \\ \Delta t^2\left[\left(\frac{1}{2}-\beta\right)-\left(\frac{1}{2}-\beta\right)\beta\eta - 2(1-\gamma)\beta\kappa\right] & \Delta t(1-\beta\eta-2\beta\kappa) & (1-\beta\eta) \end{bmatrix} \tag{8.238}$$

$$\eta = \left(\frac{1}{\omega^2\Delta t^2} + \frac{2\xi\gamma}{\omega\Delta t} + \beta\right)^{-1}, \quad \kappa = \frac{\xi\eta}{\omega\Delta t} \tag{8.239}$$

ξ is the modal damping ratio. The stability of the method requires that the following condition be satisfied:

$$\rho(\mathbf{A}) = \max|\lambda_j| < 1 \tag{8.240}$$

It is obvious that the spectral radii of the Newmark methods depend on $\Delta t/T$, ξ, γ, and β. Given values to the integration parameters $(\gamma, \beta) = (1/2, \beta)$ and modal damping ratio ξ, we can then compute the largest eigenvalues for the corresponding matrix **A** as a function of $\Delta t/T$. The stability conditions can be determined from the curve $\rho(\mathbf{A}) \sim \Delta t/T$ computed numerically.

8.4.6 Wilson's θ-method

Wilson's θ-method (Wilson et al., 1972) is an implicit method based on some observations made on the solution of diffusion type of first-order ordinary differential equations or quasi-static problems. It was observed that if solutions at discrete points which are apart by Δt are obtained through some direct integration method, then a curve that joins the midpoints of the computed solutions x_k and x_{k+1} at t_k and t_{k+1} would give a better and closer approximation to the actual solution than the computed results. This phenomenon is schematically shown in Figure 8.14.

This simple observation serves as the basis for the Wilson's θ-method, which makes the assertion that the acceleration computed by the linear acceleration method is likely to be more accurate at the midpoint of the time interval. We start with the solution already determined at time t_k. The equation of motion is satisfied and the solution is determined at time $t_{k+\theta} = (t_k + \theta\Delta t)$, where $\theta > 1.0$. The accelerations at time $t_{k+1} = (t_k + \Delta t)$ are determined by linear interpolation between t_k and $(t_k + \theta\Delta t)$. This method is an extension of the linear acceleration method such that the variation of acceleration is assumed to be linear between the time step from t_k to $(t_k + \theta\Delta t)$, as shown in Figure 8.15. The dynamic equation of motion is satisfied at $t_{k+\theta} = (t_k + \theta\Delta t)$.

$$\mathbf{M}\ddot{\mathbf{x}}_{k+\theta} + \mathbf{C}\dot{\mathbf{x}}_{k+\theta} + \mathbf{K}\mathbf{x}_{k+\theta} = \mathbf{p}_{k+\theta} \tag{8.241}$$

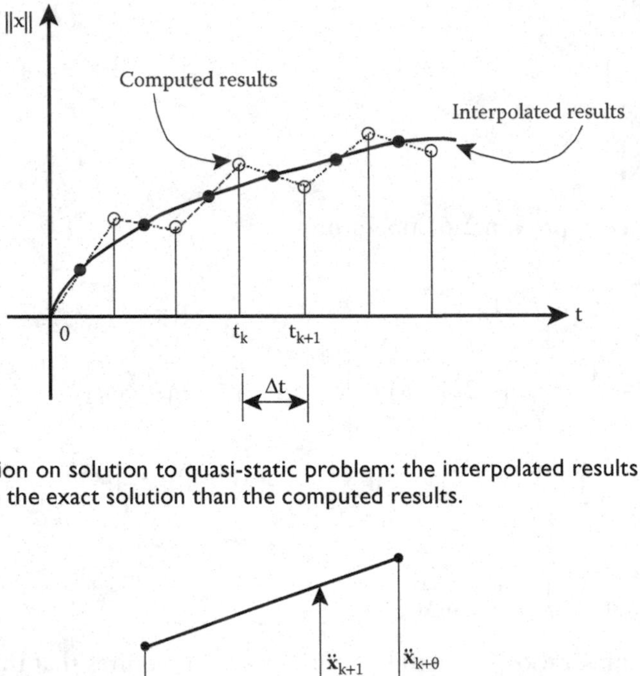

Figure 8.14 Observation on solution to quasi-static problem: the interpolated results give better approximation to the exact solution than the computed results.

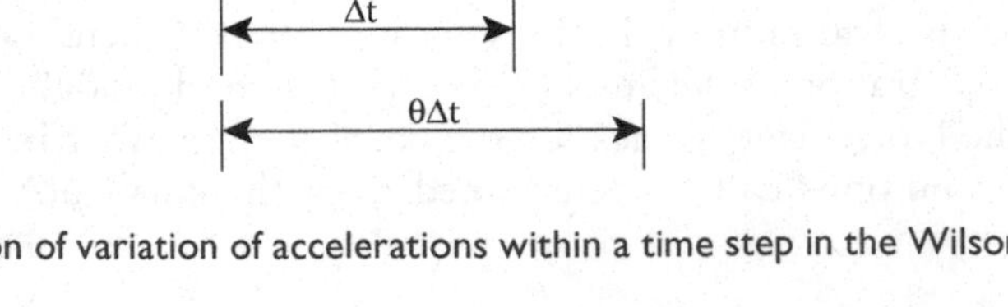

Figure 8.15 Assumption of variation of accelerations within a time step in the Wilson's θ-method.

The load vector is normally known at discrete times Δt apart; in this cases they are known at t_k and t_{k+1}. Since the equation of motion is satisfied at time $t_k + \theta\Delta t$ the load vector is needed at that time. It can be determined either by linear projection from t_k to t_{k+1} or by linear interpolation between t_{k+1} to t_{k+2} as shown in the equations below.

$$\mathbf{p}_{k+\theta} = \mathbf{p}_k + \theta(\mathbf{p}_{k+1} - \mathbf{p}_k) \tag{8.242}$$

$$\mathbf{p}_{k+\theta} = \mathbf{p}_{k+1} + (\theta - 1)(\mathbf{p}_{k+2} - \mathbf{p}_{k+1}) \tag{8.243}$$

The temporal discretization formulae for Wilson's θ-method can be readily obtained by substituting with $\theta\Delta t$ in the equations for linear acceleration method developed earlier.

$$\tau = t - t_k;\ 0 \le \tau \le \theta\Delta t \tag{8.244}$$

$$\ddot{\mathbf{x}}(t+\tau) = \ddot{\mathbf{x}}_k + \frac{\tau}{\theta\Delta t}(\ddot{\mathbf{x}}_{k+\theta} - \ddot{\mathbf{x}}_k) \tag{8.245}$$

$$\begin{cases} \dot{\mathbf{x}}(t+\tau) = \dot{\mathbf{x}}_k + \tau\ddot{\mathbf{x}}_k + \dfrac{\tau^2}{2\theta\Delta t}(\ddot{\mathbf{x}}_{k+\theta} - \ddot{\mathbf{x}}_k) \\ \mathbf{x}(t+\tau) = \mathbf{x}_k + \tau\dot{\mathbf{x}}_k + \dfrac{\tau^2}{2}\ddot{\mathbf{x}}_k + \dfrac{\tau^3}{6\theta\Delta t}(\ddot{\mathbf{x}}_{k+\theta} - \ddot{\mathbf{x}}_k) \end{cases} \tag{8.246}$$

At $\tau = \theta\Delta t$, we have the following:

$$\begin{cases} \dot{\mathbf{x}}_{k+\theta} = \dot{\mathbf{x}}_k + \dfrac{\theta\Delta t}{2}(\ddot{\mathbf{x}}_k + \ddot{\mathbf{x}}_{k+\theta}) \\ \mathbf{x}_{k+\theta} = \mathbf{x}_k + \theta\Delta t\dot{\mathbf{x}}_k + \dfrac{\theta^2\Delta t^2}{6}(2\ddot{\mathbf{x}}_k + \ddot{\mathbf{x}}_{k+\theta}) \end{cases} \tag{8.247}$$

Therefore, the temporal discretization formulae become as follows.

$$\begin{cases} \ddot{\mathbf{x}}_{k+\theta} = \dfrac{6}{\theta^2\Delta t^2}(\mathbf{x}_{k+\theta} - \mathbf{x}_k) - \dfrac{6}{\theta\Delta t}\dot{\mathbf{x}}_k - 2\ddot{\mathbf{x}}_k \\ \dot{\mathbf{x}}_{k+\theta} = \dfrac{3}{\theta\Delta t}(\mathbf{x}_{k+\theta} - \mathbf{x}_k) - 2\dot{\mathbf{x}}_k - \dfrac{\theta\Delta t}{2}\ddot{\mathbf{x}}_k \end{cases} \tag{8.248}$$

Substituting the above relations into the system of equations of motion, we obtain the equivalent static system of equations.

$$\bar{\mathbf{K}}\mathbf{x}_{k+\theta} = \bar{\mathbf{p}}_{k+\theta} \tag{8.249}$$

$$\bar{\mathbf{K}} = \mathbf{K} + \frac{6}{\theta^2\Delta t^2}\mathbf{M} + \frac{3}{\theta\Delta t}\mathbf{C} \tag{8.250}$$

$$\begin{aligned} \bar{\mathbf{p}}_{k+\theta} = {} & \mathbf{p}_k + \theta(\mathbf{p}_{k+1} - \mathbf{p}_k) + \mathbf{M}\left(\frac{6}{\theta^2\Delta t^2}\mathbf{x}_k + \frac{6}{\theta\Delta t}\dot{\mathbf{x}}_k + 2\ddot{\mathbf{x}}_k\right) \\ & + \mathbf{C}\left(\frac{3}{\theta\Delta t}\mathbf{x}_k + 2\dot{\mathbf{x}}_k + \frac{\theta\Delta t}{2}\ddot{\mathbf{x}}_k\right) \end{aligned} \tag{8.251}$$

After computing $\mathbf{x}_{k+\theta}$ from solving the equivalent static system of equations, we can then compute the accelerations at $t_{k+\theta}$ and interpolate to determine the acceleration at t_{k+1}. Displacement and velocity vectors at t_{k+1} are determined using the linear acceleration method.

$$\ddot{\mathbf{x}}_{k+1} = \frac{\theta-1}{\theta}\ddot{\mathbf{x}}_k + \frac{1}{\theta}\ddot{\mathbf{x}}_{k+\theta} \tag{8.252}$$

$$\begin{cases} \dot{\mathbf{x}}_{k+1} = \dot{\mathbf{x}}_k + \dfrac{\Delta t}{2}\left(\ddot{\mathbf{x}}_k + \ddot{\mathbf{x}}_{k+1}\right) \\ \mathbf{x}_{k+1} = \mathbf{x}_k + \Delta t\dot{\mathbf{x}}_k + \dfrac{\Delta t^2}{6}\left(2\ddot{\mathbf{x}}_k + \ddot{\mathbf{x}}_{k+1}\right) \end{cases} \tag{8.253}$$

Acceleration, velocity, and displacement vectors at t_{k+1} can also be directly computed from the displacements at $t_{k+\theta}$. The following equations can be used in practical computational implementation:

$$\begin{cases} \ddot{\mathbf{x}}_{k+1} = \dfrac{6}{\theta^3\Delta t^2}\left(\mathbf{x}_{k+\theta} - \mathbf{x}_k\right) - \dfrac{6}{\theta^2\Delta t}\dot{\mathbf{x}}_k + \dfrac{\theta-3}{\theta}\ddot{\mathbf{x}}_k \\ \dot{\mathbf{x}}_{k+1} = \dfrac{3}{\theta^3\Delta t}\left(\mathbf{x}_{k+\theta} - \mathbf{x}_k\right) + \dfrac{\theta^2-3}{\theta^2}\dot{\mathbf{x}}_k + \dfrac{(2\theta-3)\Delta t}{2\theta}\ddot{\mathbf{x}}_k \\ \mathbf{x}_{k+1} = \dfrac{1}{\theta^3}\mathbf{x}_{k+\theta} + \dfrac{\theta^3-1}{\theta^3}\mathbf{x}_k + \dfrac{(\theta^2-1)\Delta t}{\theta^2}\dot{\mathbf{x}}_k + \dfrac{(\theta-1)\Delta t^2}{2\theta}\ddot{\mathbf{x}}_k \end{cases} \tag{8.254}$$

It can be shown that the Wilson's θ-method is unconditionally stable for $\theta \geq 1.37$. Normally $\theta = 1.40$ is used in practice for convenience. For $\theta < 1.37$, the method is conditionally stable, and the stability criterion on time step limit depends on the physical damping in the system. In general, higher value of θ (greater than 1.4) introduces spurious numerical damping to the system. Because of this property, in linear dynamic analysis, this method can compensate for the spurious oscillations inherent in spatial discretization using finite element or finite difference method. In other words, some responses from the higher modes get damped out during the solution process.

8.4.7 Stability analysis of Wilson's θ-method

The stability analysis can be carried out by studying the spectral radii of the Wilson's approximation operator, which can be determined as follows (Bathe and Wilson, 1973):

$$[\mathbf{A}] = \begin{bmatrix} \left(1 - \dfrac{1}{\theta} - \dfrac{\eta\theta^2}{3} - \kappa\theta\right) & \dfrac{1}{\Delta t}(-\eta\theta - 2\kappa) & \dfrac{1}{\Delta t^2}(-\eta) \\ \Delta t\left(1 - \dfrac{1}{2\theta} - \dfrac{\eta\theta^2}{6} - \dfrac{\kappa\theta}{2}\right) & \left(1 - \dfrac{\eta\theta}{2} - \kappa\right) & \dfrac{1}{\Delta t}\left(-\dfrac{\eta}{2}\right) \\ \Delta t^2\left(\dfrac{1}{2} - \dfrac{1}{6\theta} - \dfrac{\eta\theta^2}{18} - \dfrac{\kappa\theta}{6}\right) & \Delta t\left(1 - \dfrac{\eta\theta}{6} - \dfrac{\kappa}{3}\right) & \left(1 - \dfrac{\eta}{6}\right) \end{bmatrix} \tag{8.255}$$

$$\eta = \left(\frac{1}{\omega^2\Delta t^2} + \frac{2\xi}{\omega\Delta t} + \frac{1}{6}\right)^{-1}, \quad \kappa = \frac{\xi\eta}{\omega\Delta t} \tag{8.256}$$

Obviously, the spectral radii of the Wilson's approximation operator is a function of θ, $\Delta t/T$, and ξ. By giving different values to $\Delta t/T$ and θ, we can determine the relationship of $\rho(\mathbf{A}) \sim \theta$ through numerical computations. It can be observed that for $\theta \geq 1.37$, $\rho(\mathbf{A}) \leq 1$ for any $\Delta t/T$, which means that the method is unconditionally stable when $\theta \geq 1.37$. For $\theta < 1.37$, the method is conditionally stable.

8.4.8 Newmark's original iterative method

The original Newmark method (Newmark, 1959) was presented as an iterative method for the solution of the dynamic response of structures. In the absence of physical damping in the system two equations are used in iterations in each time step. The first equation is the equation of motion where internal resisting force vector $\mathbf{I}$ is used and the stiffness matrix is not explicitly formed.

$$\mathbf{M}\ddot{\mathbf{x}}_{k+1} + \mathbf{I}_{k+1} = \mathbf{p}_{k+1} \tag{8.257}$$

$$\mathbf{I}_{k+1} = \mathbf{K}\mathbf{x}_{k+1} \tag{8.258}$$

The second equation is one of the two Newmark method discretization formulae discussed earlier, Equation 8.180.

$$\mathbf{x}_{k+1} = \mathbf{x}_k + \Delta t\,\dot{\mathbf{x}}_k + \Delta t^2\left[\left(\frac{1}{2} - \beta\right)\ddot{\mathbf{x}}_k + \beta\,\ddot{\mathbf{x}}_{k+1}\right] \tag{8.259}$$

This equation is rewritten in terms of increment of the displacement vector.

$$\Delta\mathbf{x}_{k+1} = \beta\Delta t^2\ddot{\mathbf{x}}_{k+1} + \left[\Delta t\,\dot{\mathbf{x}}_k + \left(\frac{1}{2} - \beta\right)\Delta t^2\ddot{\mathbf{x}}_k\right] \tag{8.260}$$

This method can be directly implemented as an iterative method in each time step in the following procedures, in which the superscript represents iteration number.

Initialization

$$\ddot{\mathbf{x}}_{k+1}^{(0)} = \ddot{\mathbf{x}}_k$$

for $j = 1, 2, \ldots$

$$\Delta\mathbf{x}_{k+1}^{(j)} = \beta\Delta t^2\ddot{\mathbf{x}}_{k+1}^{(j-1)} + \left[\Delta t\,\dot{\mathbf{x}}_k + \left(\frac{1}{2} - \beta\right)\Delta t^2\ddot{\mathbf{x}}_k\right]$$

$$\mathbf{x}_{k+1}^{(j)} \leftarrow \mathbf{x}_k + \Delta\mathbf{x}_{k+1}^{(j)}$$

$$\mathbf{I}_{k+1}^{(j)} \leftarrow \mathbf{I}\left(\mathbf{x}_{k+1}^{(j)}\right)$$

$$\ddot{\mathbf{x}}_{k+1}^{(j)} \leftarrow \mathbf{M}^{-1}\left(\mathbf{p}_{k+1} - \mathbf{I}_{k+1}^{(j)}\right)$$

It can be seen that this scheme is well suited for solving very large problems in finite element analysis because there is no need to explicitly formulate the structural stiffness matrix. The computation of the internal resisting force vector can be carried out on the element level. Now the question is what is the convergence property of this iterative method?

8.4.9 Analysis of iterative Newmark method

Newmark iterative method is well suited for solving problems in nonlinear dynamic analysis. Excluding physical damping in the system, under incremental formulation for nonlinear analysis, the above algorithm can be modified as follows:

$$\ddot{\mathbf{x}}_{k+1}^{(j)} = \mathbf{M}^{-1}\left[\mathbf{p}_k + \Delta\mathbf{p}_{k+1} - \mathbf{I}_k - \mathbf{I}\left(\Delta\mathbf{x}_{k+1}^{(j)}\right)\right] \tag{8.261}$$

We can use the dynamic equilibrium equation at t_k, to arrive at the incremental form of the equation of motion.

$$\mathbf{M}\ddot{\mathbf{x}}_k = \mathbf{p}_k - \mathbf{I}_k \tag{8.262}$$

$$\ddot{\mathbf{x}}_{k+1}^{(j)} = \ddot{\mathbf{x}}_k + \mathbf{M}^{-1}\left[\Delta\mathbf{p}_{k+1} - \mathbf{I}\left(\Delta\mathbf{x}_{k+1}^{(j)}\right)\right] \tag{8.263}$$

Substituting this relation into the original Newmark formulas yields the following:

$$\begin{aligned}\Delta\mathbf{x}_{k+1}^{(j)} &= \beta\Delta t^2\mathbf{M}^{-1}\left[\Delta\mathbf{p}_{k+1} - \mathbf{I}\left(\Delta\mathbf{x}_{k+1}^{(j-1)}\right)\right] \\ &\quad + \beta\Delta t^2\ddot{\mathbf{x}}_k + \left[\Delta t\,\dot{\mathbf{x}}_k + \left(\frac{1}{2} - \beta\right)\Delta t^2\ddot{\mathbf{x}}_k\right]\end{aligned} \tag{8.264}$$

$$\Delta\mathbf{x}_{k+1}^{(1)} = \beta\Delta t^2\ddot{\mathbf{x}}_k + \left[\Delta t\,\dot{\mathbf{x}}_k + \left(\frac{1}{2} - \beta\right)\Delta t^2\ddot{\mathbf{x}}_k\right] \tag{8.265}$$

$$\Delta\mathbf{x}_{k+1}^{(j)} = \beta\Delta t^2\mathbf{M}^{-1}\left[\Delta\mathbf{p}_{k+1} - \mathbf{I}\left(\Delta\mathbf{x}_{k+1}^{(j-1)}\right)\right] + \Delta\mathbf{x}_{k+1}^{(1)} \tag{8.266}$$

By ignoring the increment of the load vector, we arrive at the basic iterative equation with tangent stiffness matrix and initial conditions.

$$\Delta\mathbf{x}_{k+1}^{(j)} = -\beta\Delta t^2\mathbf{M}^{-1}\mathbf{K}_t\Delta\mathbf{x}_{k+1}^{(j-1)} + \Delta\mathbf{x}_{k+1}^{(1)} \tag{8.267}$$

$$\mathbf{A} = -\beta\Delta t^2\mathbf{M}^{-1}\mathbf{K}_t \tag{8.268}$$

$$\begin{aligned}\Delta\mathbf{x}_{k+1}^{(j)} &= \mathbf{A}\Delta\mathbf{x}_{k+1}^{(j-1)} + \Delta\mathbf{x}_{k+1}^{(1)} \\ &= \mathbf{A}\left(\mathbf{A}\Delta\mathbf{x}_{k+1}^{(j-2)} + \Delta\mathbf{x}_{k+1}^{(1)}\right) + \Delta\mathbf{x}_{k+1}^{(1)} \\ &\;\vdots \\ &= \left(\mathbf{A}^{j+1} + \cdots + \mathbf{A} + \mathbf{I}\right)\Delta\mathbf{x}_{k+1}^{(1)}\end{aligned} \tag{8.269}$$

This series is convergent if and only if matrix **A** is convergent; its spectral radius must satisfy the following relation:

$$\rho(\mathbf{A}) < 1 \tag{8.270}$$

$$\lim_{k\to\infty}\left(\mathbf{I} + \mathbf{A} + \cdots + \mathbf{A}^k\right) \to (\mathbf{I} - \mathbf{A})^{-1} \tag{8.271}$$

Now we determine the spectral radius or the largest eigenvalue λ_n for matrix **A**.

$$\mathbf{A}\phi_n = \lambda_n\phi_n \tag{8.272}$$

$$-\beta\Delta t^2\mathbf{M}^{-1}\mathbf{K}_t\phi_n = \lambda_n\phi_n \tag{8.273}$$

$$\mathbf{K}_t\phi_n = -\frac{\lambda_n}{\beta\Delta t^2}\mathbf{M}\phi_n \tag{8.274}$$

Comparing this relation with the equation obtained in structural frequency analysis, we have the following:

$$\omega_n^2 = -\frac{\lambda_n}{\beta\Delta t^2} \tag{8.275}$$

$$\left|\lambda_{max}\right| = \beta\Delta t^2\omega_{max}^2 \tag{8.276}$$

From the condition that $\left|\lambda_{max}\right| < 1$, we obtain the criterion for convergence of the iterative Newmark method.

$$\Delta t < \frac{1}{\omega_{max}\sqrt{\beta}} \tag{8.277}$$

For constant average acceleration method, where $\beta = 1/4$, the convergence criterion becomes

$$\Delta t < \frac{2}{\omega_{max}} \tag{8.278}$$

which is the same as the Courant stability criterion.

At convergence of the iterative Newmark method, we have the following:

$$\Delta\mathbf{x}_{k+1} = \mathbf{A}\Delta\mathbf{x}_{k+1} + \Delta\mathbf{x}_{k+1}^{(1)} \tag{8.279}$$

$$\Delta\mathbf{x}_{k+1} = (\mathbf{I} - \mathbf{A})^{-1}\Delta\mathbf{x}_{k+1}^{(1)} \tag{8.280}$$

This shows that if the Courant stability criterion is satisfied, the iterative Newmark method converges to the exact solution.

8.5 OTHER DIRECT INTEGRATION METHODS

In this section, we will discuss some of the other implicit direct integration methods for the solution of the dynamic response of structural systems. These include the Houbolt method, Hilber–Hughes–Taylor method (α-method), SSpj method, and successive symmetric quadrature. This is not intended to provide a comprehensive summary of all the other methods proposed. Rather, our emphasis is on presenting the basic concepts and ideas embedded in each method that has some historical significance and unique numerical properties.

8.5.1 Houbolt method

The Houbolt method was proposed in 1950 for computing the dynamic response of aircraft (Houbolt, 1950), which is one of the first implicit numerical integration methods proposed for

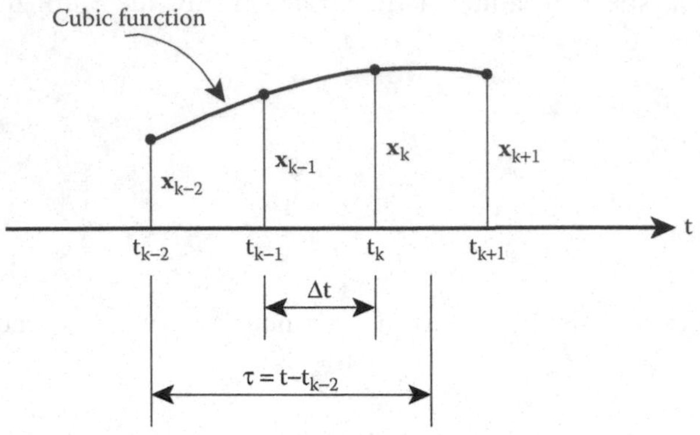

Figure 8.16 Assumption of cubic variation of displacements within three time steps in the Houbolt method.

determining the dynamic response of structural systems. It is a multistep (three-step) method by using information on previous three time stations to compute displacements, velocities, and accelerations at the current time station. Considering the four time stations, we assume that the displacement is a cubic function passing through these four time stations as shown in Figure 8.16. Using Lagrange functions, we can determine the displacement as a function of $\tau = t - t_{k-2}$.

$$\begin{aligned}\mathbf{x}(\tau) &= \frac{1}{6}\left(1-\frac{\tau}{\Delta t}\right)\left(\frac{\tau}{\Delta t}-2\right)\left(\frac{\tau}{\Delta t}-3\right)\mathbf{x}_{k-2} + \frac{1}{2}\frac{\tau}{\Delta t}\left(\frac{\tau}{\Delta t}-2\right)\left(\frac{\tau}{\Delta t}-3\right)\mathbf{x}_{k-1} \\ &+ \frac{1}{2}\frac{\tau}{\Delta t}\left(1-\frac{\tau}{\Delta t}\right)\left(\frac{\tau}{\Delta t}-3\right)\mathbf{x}_k + \frac{1}{6}\frac{\tau}{\Delta t}\left(\frac{\tau}{\Delta t}-1\right)\left(\frac{\tau}{\Delta t}-2\right)\mathbf{x}_{k+1}\end{aligned} \tag{8.281}$$

The velocity and acceleration can be directly obtained through differentiation of this displacement function.

$$\dot{\mathbf{x}}_{k+1} = \dot{\mathbf{x}}(\tau = 3\Delta t) = \frac{1}{6\Delta t}\left(-2\mathbf{x}_{k-2} + 9\mathbf{x}_{k-1} - 18\mathbf{x}_k + 11\mathbf{x}_{k+1}\right) \tag{8.282}$$

$$\ddot{\mathbf{x}}_{k+1} = \ddot{\mathbf{x}}(\tau = 3\Delta t) = \frac{1}{\Delta t^2}\left(-\mathbf{x}_{k-2} + 4\mathbf{x}_{k-1} - 5\mathbf{x}_k + 2\mathbf{x}_{k+1}\right) \tag{8.283}$$

Satisfying the system of equilibrium equations at t_{k+1}, we have the discrete dynamic equation of motion.

$$\mathbf{M}\ddot{\mathbf{x}}_{k+1} + \mathbf{C}\dot{\mathbf{x}}_{k+1} + \mathbf{K}\mathbf{x}_{k+1} = \mathbf{p}_{k+1} \tag{8.284}$$

$$\bar{\mathbf{K}}\mathbf{x}_{k+1} = \bar{\mathbf{p}}_{k+1} \tag{8.285}$$

$$\bar{\mathbf{K}} = \frac{2}{\Delta t^2}\mathbf{M} + \frac{11}{6\Delta t}\mathbf{C} + \mathbf{K}$$

$$\begin{aligned}\bar{\mathbf{p}}_{k+1} &= \mathbf{p}_{k+1} + \frac{1}{\Delta t^2}\mathbf{M}\left(5\mathbf{x}_k - 4\mathbf{x}_{k-1} + \mathbf{x}_{k-2}\right) \\ &+ \frac{1}{6\Delta t}\mathbf{C}\left(18\mathbf{x}_k - 9\mathbf{x}_{k-1} + 2\mathbf{x}_{k-2}\right)\end{aligned} \tag{8.286}$$

As a multistep method, this procedure requires computing the initial starting values for displacements a t_{k-1} and t_{k-2}, which can usually be computed by some other method.

This method can be shown to be unconditionally stable but the requirement on a special starting scheme makes it difficult to use in most practical applications, especially when the degrees of freedom of the finite element discretized structural system becomes very large. One interesting point is that if the mass and damping matrices are neglected, the method reduces to static analysis. This is not necessarily the case with some of the other direct integration methods.

To carry out stability analysis of this method, we consider the modal response of the system in free vibration without physical damping.

$$\ddot{\mathbf{x}}_{k+1} + \omega^2 \mathbf{x}_{k+1} = 0 \tag{8.287}$$

After substituting the Houbolt temporal discretization formulas into the above equation, we have the following, where $\delta = \omega\Delta t$:

$$\frac{1}{\Delta t^2}\left(2\mathbf{x}_{k+1} - 5\mathbf{x}_k + 4\mathbf{x}_{k-1} - \mathbf{x}_{k-2}\right) + \omega^2 \mathbf{x}_{k+1} = 0 \tag{8.288}$$

$$\mathbf{x}_{k+1} = \frac{5}{2+\delta^2}\mathbf{x}_k - \frac{4}{2+\delta^2}\mathbf{x}_{k-1} + \frac{1}{2+\delta^2}\mathbf{x}_{k-2} \tag{8.289}$$

Then the approximation matrix is as follows:

$$\begin{Bmatrix} \mathbf{x}_{k+1} \\ \mathbf{x}_k \\ \mathbf{x}_{k-1} \end{Bmatrix} = [A] \begin{Bmatrix} \mathbf{x}_k \\ \mathbf{x}_{k-1} \\ \mathbf{x}_{k-2} \end{Bmatrix} \tag{8.290}$$

$$[A] = \begin{bmatrix} \dfrac{5}{2+\delta^2} & -\dfrac{4}{2+\delta^2} & \dfrac{1}{2+\delta^2} \\ 1 & 0 & 0 \\ 0 & 1 & 0 \end{bmatrix} \tag{8.291}$$

The stability criterion can be obtained from $\rho(A) \le 1$. In this case, the characteristic equation is of the following form:

$$\left(2+\delta^2\right)\lambda^3 - 5\lambda^2 + 4\lambda - 1 = 0 \tag{8.292}$$

If we perform the transformation of $\lambda = (1 + z)/(1 - z)$ where z is a complex variable, then this operation transforms the condition of $|\lambda| \le 1$ which is a circled region into the domain in which the half plane $R_e(z) \le 0$. After this transformation, and multiplication by $(1 - z)^3$, the above characteristic equation becomes as follows:

$$a_0 z^3 + a_1 z^2 + a_2 z + a_3 = 0 \tag{8.293}$$

$$\begin{cases} a_0 = \delta^2 + 12 \\ a_1 = 3\delta^2 + 4 \\ a_2 = 3\delta^2 \\ a_3 = \delta^2 \end{cases} \tag{8.294}$$

According to Routh–Hurwitz criterion (Gantmacher, 1959), the necessary and sufficient conditions for the roots of the above characteristic equation to satisfy the condition $\mathrm{Re}(z) \leq 0$ are as follows:

$$\begin{cases} a_0 > 0 \\ a_1,\ a_3,\ a_4 \geq 0 \\ a_1 a_2 - a_0 a_3 \geq 0 \end{cases} \tag{8.295}$$

It can be easily verified that the Routh–Hurwitz criterion is automatically satisfied. This shows that the approximation matrix of Houbolt method always satisfies the condition $\rho(A) \leq 1$. Therefore, the Houbolt method is unconditionally stable.

This time discretization scheme was generated based on assumption on variation of displacements and through differentiation operations to obtain approximations for velocity and acceleration. From our earlier discussion, performing differentiation operations may expand the errors present in the displacements. Therefore, this is not a method of high accuracy, even though it is a multistep method. Numerical experiments indicate that with the increase of $\Delta t/T > 0.1$, the method introduces a large amount of numerical error into the system in both amplitude decay and period elongation. The presence of excessive numerical damping by the numerical method makes it undesirable for practical usage, which requires a small step size in order to preserve accuracy. Because of its poor numerical properties, this method is not of much practical use in structural dynamics. However, for historical reasons, this method serves as a benchmark in the analysis and comparative studies of different direct integration methods.

8.5.2 Hilber–Hughes–Taylor method (α-method)

The α-method was proposed to introduce controlled numerical damping into the system without degrading the numerical accuracy of the Newmark method (Hilber et al., 1977). The basic idea involves introducing a relaxation or diffusion parameter (α) to the discrete equations of motion.

$$\mathbf{M}\ddot{\mathbf{x}}_{k+1} + (1+\alpha)\mathbf{C}\dot{\mathbf{x}}_{k+1} - \alpha\mathbf{C}\dot{\mathbf{x}}_k + (1+\alpha)\mathbf{K}\mathbf{x}_{k+1} - \alpha\mathbf{K}\mathbf{x}_k = \mathbf{p}_{k+1} \tag{8.296}$$

In the case of linear undamped structural systems, usually the following equation of motion is satisfied and the temporal discretization is still the original Newmark method:

$$\mathbf{M}\ddot{\mathbf{x}}_{k+1} + (1+\alpha)\mathbf{K}\mathbf{x}_{k+1} - \alpha\mathbf{K}\mathbf{x}_k = \mathbf{p}_{k+1} \tag{8.297}$$

$$\begin{cases} \dot{\mathbf{x}}_{k+1} = \dot{\mathbf{x}}_k + \Delta t\left[(1-\gamma)\ddot{\mathbf{x}}_k + \gamma\ddot{\mathbf{x}}_{k+1}\right] \\ \mathbf{x}_{k+1} = \mathbf{x}_k + \Delta t\,\dot{\mathbf{x}}_k + \Delta t^2\left[\left(\frac{1}{2} - \beta\right)\ddot{\mathbf{x}}_k + \beta\,\ddot{\mathbf{x}}_{k+1}\right] \end{cases} \tag{8.298}$$

Obviously, with the introduction of parameter α, the stability and numerical damping properties of this method are governed by the three parameters α, β, and γ. For $\alpha = 0$, the method becomes the standard Newmark family of methods. Hence, this method can be implemented in the same way as with the standard Newmark method.

To study the numerical properties of this method, we consider the modal response of the system through performing modal decomposition. Ignoring physical damping and external loading, we can express the algorithm in the following recursive form:

$$\begin{Bmatrix} x_{k+1} \\ \Delta t\dot{x}_{k+1} \\ \Delta t^2\ddot{x}_{k+1} \end{Bmatrix} = [A] \begin{Bmatrix} x_k \\ \Delta t\dot{x}_k \\ \Delta t^2\ddot{x}_k \end{Bmatrix} \tag{8.299}$$

$$[A] = \frac{1}{D}\begin{bmatrix} \left(1+\alpha\beta\delta^2\right) & 1 & \left(\frac{1}{2}-\beta\right) \\ -\gamma\delta^2 & 1-(1+\alpha)(\gamma-\beta)\delta^2 & 1-\gamma-(1+\alpha)\left(\frac{1}{2}\gamma-\beta\right)\delta^2 \\ -\delta^2 & -(1+\alpha)\delta^2 & -(1+\alpha)\left(\frac{1}{2}-\beta\right)\delta^2 \end{bmatrix} \tag{8.300}$$

$$\begin{cases} D = 1+(1+\alpha)\beta\delta^2 \\ \delta = \omega\Delta t \end{cases} \tag{8.301}$$

The stability criterion is obtained by requiring that the spectral radius satisfy the condition $\rho(A) \le 1$. In this case, the characteristic equation is as follows:

$$\det(A-\lambda I) = \lambda^3 - A_1\lambda^2 + A_2\lambda + A_3 = 0 \tag{8.302}$$

$$\begin{cases} A_1 = \text{trace}(A) = 2 - \dfrac{\delta^2}{D}\left[(1+\alpha)\left(\gamma+\dfrac{1}{2}\right)-\alpha\beta\right] \\ A_2 = \sum \text{principal minors} = 1 - \dfrac{\delta^2}{D}\left[\gamma-\dfrac{1}{2}+2\alpha(\gamma-\beta)\right] \\ A_3 = \det(A) = \dfrac{\alpha\delta^2}{D}\left(\beta-\gamma+\dfrac{1}{2}\right) \end{cases} \tag{8.303}$$

By eliminating velocities and acceleration terms, we can obtain the difference equation in terms of displacements.

$$x_{k+1} - A_1x_k + A_2x_{k-1} + A_3x_{k-2} = 0 \tag{8.304}$$

Comparing this equation with the characteristic equation results in the following solution:

$$x_k = \sum_{j=1}^{3} c_j\lambda_j^k \tag{8.305}$$

The constants c_j are determined from initial conditions.

Numerical studies show that if the parameter α takes positive values, the dissipation introduced by the algorithm is not very effective. A more effective approach is to use negative values for α, and define a one parameter family of algorithms with $\beta = (1-\alpha)^2/4$ and $\gamma = 1/2 - \alpha$. By doing so, the invariants become as follows:

$$\begin{cases} A_1 = 2 - \dfrac{\delta^2}{D} + A_3 \\ A_2 = 1 + 2A_3 \\ A_3 = \dfrac{\alpha\delta^2}{4D}(1+\alpha)^2 \end{cases} \tag{8.306}$$

$$D = 1 + \frac{\delta^2}{4}(1-\alpha)(1-\alpha^2) \tag{8.307}$$

It can be shown that when $-1/3 \le \alpha \le 0$ we have $|\lambda| \le 1$ which means that the method is unconditionally stable. With the decreasing value of α, the magnitude of numerical damping increases. This method has the advantage that it preserves the unconditional stability of constant average acceleration method in Newmark family of methods and the numerical damping introduced can be controlled to the extent such that the lower modes are not affected significantly.

8.5.3 Collocation method

The basic idea of collocation for the solution of dynamic equations of motion was proposed by Wilson in the θ-method where the equation of motion is satisfied at the collocation point, $t_{k+\theta}$. In fact, the parameter θ in the Wilson's method can be considered as the collocation parameter. Here, we discuss a generalization of the Wilson's collocation idea originally developed by Ghaboussi and Wilson (1972) and later on reanalyzed by Hilber and Hughes (1978).

The generalized collocation method can be described in the way similar to Wilson's θ-method. First, the equation of motion is satisfied at the collocation point.

$$\mathbf{M}\ddot{\mathbf{x}}_{k+\theta} + \mathbf{C}\dot{\mathbf{x}}_{k+\theta} + \mathbf{K}\mathbf{x}_{k+\theta} = \mathbf{p}_{k+\theta} \tag{8.308}$$

The acceleration, velocity, displacement, and load vectors at the collocation point can be computed by following equations:

$$\begin{cases} \ddot{\mathbf{x}}_{k+\theta} = (1-\theta)\ddot{\mathbf{x}}_k + \theta\ddot{\mathbf{x}}_{k+1} \\ \dot{\mathbf{x}}_{k+\theta} = \mathbf{x}_k + \theta\Delta t\left[(1-\gamma)\ddot{\mathbf{x}}_k + \gamma\ddot{\mathbf{x}}_{k+\theta}\right] \\ \mathbf{x}_{k+\theta} = \mathbf{x}_k + \theta\Delta t\dot{\mathbf{x}}_k + (\theta\Delta t)^2\left[\left(\dfrac{1}{2} - \beta\right)\ddot{\mathbf{x}}_k + \beta\ddot{\mathbf{x}}_{k+\theta}\right] \\ \mathbf{p}_{k+\theta} = (1-\theta)\mathbf{p}_k + \theta\mathbf{p}_{k+1} \end{cases} \tag{8.309}$$

After substituting the above relations into the equation of motion, we obtain the static equivalent system of equations.

$$\bar{\mathbf{K}}\mathbf{x}_{k+\theta} = \bar{\mathbf{p}}_{k+\theta} \tag{8.310}$$

$$\bar{\mathbf{K}} = \frac{1}{\beta\theta^2\Delta t^2}\mathbf{M} + \frac{\gamma}{\beta\theta\Delta t}\mathbf{C} + \mathbf{K} \tag{8.311}$$

$$\begin{aligned}\bar{\mathbf{p}}_{k+\theta} = \mathbf{p}_{k+\theta} &+ \frac{1}{\beta\theta^2\Delta t^2}\mathbf{M}\left[\mathbf{x}_k + \theta\Delta t\dot{\mathbf{x}}_k + \theta^2\Delta t^2\left(\frac{1}{2} - \beta\right)\ddot{\mathbf{x}}_k\right] \\ &+ \frac{\gamma}{\beta\theta\Delta t}\mathbf{C}\left[\mathbf{x}_k + \theta\Delta t\left(1 - \frac{\beta}{\gamma}\right)\dot{\mathbf{x}}_k + \theta^2\Delta t^2\left(\frac{1}{2} - \frac{\beta}{\gamma}\right)\ddot{\mathbf{x}}_k\right]\end{aligned} \tag{8.312}$$

The accelerations, velocities, and displacements at t_{k+1} are found through the following relations that are derived from linear interpolation for acceleration and the Newmark formulas for the displacements and velocities:

$$\begin{cases}\ddot{\mathbf{x}}_{k+1} = \dfrac{1}{\beta\theta^3\Delta t^2}\left(\mathbf{x}_{k+\theta} - \mathbf{x}_k\right) - \dfrac{1}{\beta\theta^2\Delta t}\dot{\mathbf{x}}_k + \left(1 - \dfrac{1}{2\beta\theta}\right)\ddot{\mathbf{x}}_k \\[2ex] \dot{\mathbf{x}}_{k+1} = \dfrac{1}{\beta\theta^3\Delta t}\left(\mathbf{x}_{k+\theta} - \mathbf{x}_k\right) + \left(1 - \dfrac{\gamma}{\beta\theta^2}\right)\dot{\mathbf{x}}_k + \left(1 - \dfrac{\gamma}{2\beta\theta}\right)\Delta t\ddot{\mathbf{x}} \\[2ex] \mathbf{x}_{k+1} = \dfrac{1}{\theta^3}\mathbf{x}_{k+\theta} + \left(1 - \dfrac{1}{\theta^3}\right)\mathbf{x}_k + \left(1 - \dfrac{1}{\theta^2}\right)\Delta t\dot{\mathbf{x}}_k + \dfrac{1}{2}\left(1 - \dfrac{1}{\theta}\right)\Delta t^2\ddot{\mathbf{x}}_k\end{cases} \tag{8.313}$$

It is easy to observe that when $\theta = 1$, the collocation method reduces to the Newmark family of methods, and if $\theta > 1$, $\beta = 1/6$ and $\gamma = 1/2$, it becomes Wilson's θ-method. Therefore, new methods can be derived by selecting appropriate values for θ, β, and γ such that favorable numerical properties would be obtained. To do so, we need to study the stability characteristic of this method. It can be shown that the approximation or amplification matrix of the collocation method can be obtained as follows:

$$\mathbf{A} = \begin{bmatrix} 1 + \beta A_{31} & 1 + \beta A_{32} & \dfrac{1}{2} + \beta\left(A_{33} - 1\right) \\ \gamma A_{31} & 1 + \gamma A_{32} & 1 + \gamma\left(A_{33} - 1\right) \\ A_{31} & A_{32} & A_{33} \end{bmatrix} \tag{8.314}$$

$$\begin{cases} A_{31} = -\dfrac{\delta^2}{D} \\[2ex] A_{32} = -\dfrac{\theta\delta^2}{D} \\[2ex] A_{33} = 1 - \dfrac{2 + \theta^2\delta^2}{2D} \\[2ex] \delta = \omega\Delta t, \; D = \theta\left(1 + \beta\theta^2\delta^2\right) \end{cases} \tag{8.315}$$

The characteristic equation for $\mathbf{A}$ is as follows:

$$\lambda^3 - A_1\lambda^2 + A_2\lambda + A_3 = 0 \tag{8.316}$$

$$\begin{cases} A_1 = 2 - \dfrac{1}{D}\delta^2\left(\gamma + \theta - \dfrac{1}{2}\right) + A_3 \\ A_2 = 1 - \dfrac{1}{D}\delta^2\left(\gamma + \theta - \dfrac{3}{2}\right) + 2A_3 \\ A_3 = \dfrac{1}{D}(\theta - 1)\left\{1 + \delta^2\left[\beta\left(\theta^2 + \theta + 1\right) - \gamma - \dfrac{1}{2}(\theta - 1)\right]\right\} \end{cases} \tag{8.317}$$

The stability criterion is determined by satisfying the condition that the spectral radius satisfy the condition $\rho(A) \leq 1$. It can be shown that the conditions for unconditional stability of the collocation method are as follows:

$$\begin{cases} \gamma = \dfrac{1}{2}, \quad \theta \geq 1 \\ \dfrac{\theta}{2(\theta+1)} \geq \beta \geq \dfrac{2\theta^2 - 1}{4\left(2\theta^3 - 1\right)} \end{cases} \tag{8.318}$$

This method maintains second-order accuracy in the unconditionally stable conditions. A set of optimal collocation parameters are determined by ensuring complex conjugate principal roots of the characteristic polynomial as $\Delta t/T \to \infty$. This method has improved numerical property over the Wilson's θ-method. It is important to remember that although, the value of γ slightly more than 1/2 can be used beneficially to damp out the spurious oscillations associated with the higher frequencies of a discrete system, there is always the possibility of losing some of the important features of the response of the structural system.

8.5.4 Bossak's Alfa method

This method is also a relaxation-type method with similar idea behind the α-method (Wood et al., 1980). It still uses the Newmark method for temporal discretization. A relaxation parameter is introduced such that the following equilibrium equation is satisfied:

$$(1-\alpha_B)\mathbf{M}\ddot{\mathbf{x}}_{k+1} + \alpha_B\mathbf{M}\ddot{\mathbf{x}}_k + \mathbf{C}\dot{\mathbf{x}}_{k+1} + \mathbf{K}\mathbf{x}_{k+1} = \mathbf{p}_{k+1} \tag{8.319}$$

Obviously, by setting $\alpha_B = 0$, we obtain the Newmark method. The introduction of parameter α_B has a similar effect as the parameter α in the α-method, which induces numerical damping in the higher modes. The approximation or amplification matrix can be obtained as

$$[A] = \frac{1}{D}\begin{bmatrix} (1-\alpha_B) & (1-\alpha_B) & \left(\dfrac{1}{2} - \beta - \dfrac{\alpha_B}{2}\right) \\ -\gamma\delta^2 & 1 - \alpha_B + (\beta - \gamma)\delta^2 & 1 - \gamma - \alpha_B + \left(\beta - \dfrac{1}{2}\right)\delta^2 \\ -\delta^2 & -\delta^2 & -\alpha_B - \left(\dfrac{1}{2} - \beta\right)\delta^2 \end{bmatrix} \tag{8.320}$$

$$\delta = \omega\Delta t,\ D = 1 - \alpha_B + \beta\delta^2 \tag{8.321}$$

The stability condition of this method can be determined from the condition $\rho(A) \le 1$. The conditions for unconditional stability, positive damping ratio, and maintaining second-order accuracy of the method are as follows:

$$\alpha_B = \frac{1}{2} - \gamma \le 0 \ \text{ and } \ \beta > \frac{\gamma}{2} > \frac{1}{4} \tag{8.322}$$

It can be shown that the conditions for unconditional stability of the method in terms of one parameter α_B are as follows:

$$\begin{cases} \alpha_B \le 0 \\ \gamma = \dfrac{1}{2} - \alpha_B \\ \beta = \dfrac{1}{4}(1 - \alpha_B)^2 \end{cases} \tag{8.323}$$

Numerical simulations indicate that this method has similar numerical properties as of the α-method, especially when the values of α and α_B are small, such as in the range of $[-0.20, 0.0]$. We note that the two methods are the same for $\alpha_B = \alpha = 0$. Once the values of α and α_B are near the critical value -0.30 for α, then the α-method gives more accurate results. It is thus advisable that the value of α_B should also be in the range of $-1/3 \le \alpha_B \le 0$ in practical implementations.

8.5.5 SSpj method

This method was proposed (Zienkiewicz et al., 1984) for generating a series of single-step algorithms of order p in direct integration of diffusion equation and dynamic equation of motion. The basic idea of this algorithm starts with approximating the displacement **x** with a polynomial of degree p within a time step Δt.

$$\mathbf{x}(t) = \mathbf{x}_k + \dot{\mathbf{x}}_k t + \frac{1}{2}\ddot{\mathbf{x}}_k t^2 + \cdots + \boldsymbol{\alpha}_k^{(p)} t^p \frac{1}{p!} \tag{8.324}$$

The time is within the time step, $0 \le t \le \Delta t$ and $\boldsymbol{\alpha}$ is an unknown vector to be determined. A second-order approximation of displacements within a time step Δt can be schematically illustrated in Figure 8.17. The above equation can be expressed in the following compact form:

$$\mathbf{x}(t) = \sum_{q=0}^{p-1} \frac{1}{q!} t^q \mathbf{x}_k^{(q)} + \frac{1}{p!} t^p \boldsymbol{\alpha}_k^{(p)} \tag{8.325}$$

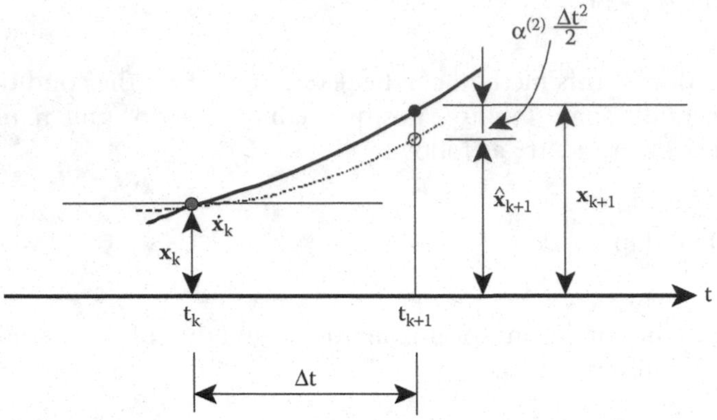

Figure 8.17 Assumption of a second-order approximation of displacements within a time step in the SSpj method.

The velocities and accelerations can be computed directly. For $t = \Delta t$, we can obtain the following relationships by carrying out direct derivatives:

$$\begin{cases} \mathbf{x}_{k+1} = \sum_{q=0}^{p-1} \frac{1}{q!}\Delta t^q \mathbf{x}_k^{(q)} + \frac{1}{p!}\Delta t^p \boldsymbol{\alpha}_k^{(p)} \\ \dot{\mathbf{x}}_{k+1} = \sum_{q=1}^{p-1} \frac{1}{(q-1)!}\Delta t^{q-1} \mathbf{x}_k^{(q)} + \frac{1}{(p-1)!}\Delta t^{p-1} \boldsymbol{\alpha}_k^{(p)} \end{cases} \tag{8.326}$$

We note that α can be considered as the average of the pth derivative of the function within the time interval, and the value of p can be considered as the order of approximation. This unknown vector can be determined from a weighted residual satisfaction of the dynamic equation of motion.

$$\frac{1}{\int_0^{\Delta t} \mathrm{W}dt}\int_0^{\Delta t} \mathrm{W}\left(\mathbf{M}\ddot{\mathbf{x}} + \mathbf{C}\dot{\mathbf{x}} + \mathbf{K}\mathbf{x} - \mathbf{p}\right)dt = 0 \tag{8.327}$$

This is the basic weighted residual approach used in finite element formulations. Now, we define the following:

$$\begin{gathered} \frac{1}{\int_0^{\Delta t} \mathrm{W}dt}\int_0^{\Delta t} \mathrm{W}t^q dt = \theta_q \Delta t^q, \text{ for } q = 1,\ldots, p \\ \theta_0 = 1,\ 0 \le \theta_q \le 1 \end{gathered} \tag{8.328}$$

After carrying out integrations in Equation 8.327, we have the following:

$$\begin{aligned} &\mathbf{M}\left(\bar{\mathbf{x}}_{k+1}^{(2)} + \frac{\theta_{p-2}\Delta t^{p-2}}{(p-2)!}\boldsymbol{\alpha}_k^{(p)}\right) + \mathbf{C}\left(\bar{\mathbf{x}}_{k+1}^{(1)} + \frac{\theta_{p-1}\Delta t^{p-1}}{(p-1)!}\boldsymbol{\alpha}_k^{(p)}\right) \\ &\qquad + \mathbf{K}\left(\bar{\mathbf{x}}_{k+1} + \frac{\theta_p \Delta t^p}{(p)!}\boldsymbol{\alpha}_k^{(p)}\right) = \bar{\mathbf{p}} \end{aligned} \tag{8.329}$$

$$\begin{cases} \bar{p} = \dfrac{1}{\int_0^{\Delta t} W dt} \int_0^{\Delta t} W p dt \\ \bar{\mathbf{x}}_{k+1} = \sum_{q=0}^{p-1} \dfrac{1}{q!} \theta_q \Delta t^q \mathbf{x}_k^{(q)} \\ \bar{\mathbf{x}}_{k+1}^{(1)} = \sum_{q=1}^{p-1} \dfrac{1}{(q-1)!} \theta_{q-1} \Delta t^{q-1} \mathbf{x}_k^{(q)} \\ \bar{\mathbf{x}}_{k+1}^{(2)} = \sum_{q=2}^{p-1} \dfrac{1}{(q-2)!} \theta_{q-2} \Delta t^{q-2} \mathbf{x}_k^{(q)} \end{cases} \tag{8.330}$$

The vector α can be determined by solving the following system of equations:

$$\begin{aligned} &\left(\frac{\Delta t^{p-2}}{(p-2)!} \theta_{p-2} \mathbf{M} + \frac{\Delta t^{p-1}}{(p-1)!} \theta_{p-1} \mathbf{C} + \frac{\Delta t^p}{(p)!} \theta_p \mathbf{K} \right) \alpha_k^{(p)} \\ &= \bar{p} - \mathbf{M}\bar{\mathbf{x}}_{k+1}^{(2)} - \mathbf{C}\bar{\mathbf{x}}_{k+1}^{(1)} - \mathbf{K}\bar{\mathbf{x}}_{k+1} \end{aligned} \tag{8.331}$$

It is obvious that because the displacement vector **x** within a time step Δt is approximated by a pth degree polynomial, then the error of approximation is on the order of $O(\Delta t^{p+1})$. This is a general form of one-step family of methods. By selecting different values for p, algorithms of different order of accuracy can be obtained. However, the lowest order of p for dynamic problems should be 2 and for diffusion problems should be 1. In the following, we discuss some of the algorithms from order 1 (linear algorithm) to order 3 (cubic algorithm).

8.5.5.1 Linear algorithm (SS11)

In this case, we are solving the diffusion equations. Hence, substituting $\mathbf{M} = 0$, and $p = 1$ into the above general SSpj algorithm, we have

$$\begin{cases} \bar{\mathbf{x}}_{k+1} = \mathbf{x}_k \\ \alpha_k^{(1)} = \left(\mathbf{C} + \theta_1 \Delta t \mathbf{K}\right)^{-1} \left(\bar{p} - \mathbf{K}\mathbf{x}_k\right) \\ \mathbf{x}_{k+1} = \mathbf{x}_k + \Delta t \alpha_k^{(1)} \end{cases} \tag{8.332}$$

This is the standard algorithm for diffusion equations. We have determined earlier that the unconditional stability of the algorithm requires that $1/2 \le \theta_1 \le 1$.

8.5.5.2 Quadratic algorithm (SS22)

This family of algorithms is obtained by setting $p = 2$. Subsequently, there are two parameters θ_1 and θ_2 in the algorithm.

$$\begin{cases} \bar{\mathbf{x}}_{k+1} = \mathbf{x}_k + \theta_1 \Delta t \dot{\mathbf{x}}_k \\ \bar{\mathbf{x}}_{k+1}^{(1)} = \dot{\mathbf{x}}_k \end{cases} \tag{8.333}$$

And the vector $\boldsymbol{\alpha}$ is determined by solving the following system of equations:

$$\left(\mathbf{M}+\theta_1\Delta t\mathbf{C}+\frac{1}{2}\theta_2\Delta t^2\mathbf{K}\right)\alpha_k^{(2)}=\bar{\mathbf{p}}-\mathbf{C}\bar{\mathbf{x}}_{k+1}^{(1)}-\mathbf{K}\bar{\mathbf{x}}_{k+1} \tag{8.334}$$

Then, the displacement and velocity at t_{k+1} are computed.

$$\begin{cases}\mathbf{x}_{k+1}=\mathbf{x}_k+\Delta t\dot{\mathbf{x}}_k+\dfrac{\Delta t^2}{2}\alpha_k^{(2)}\\ \dot{\mathbf{x}}_{k+1}=\dot{\mathbf{x}}_k+\Delta t\alpha_k^{(2)}\end{cases} \tag{8.335}$$

If we substitute $\theta_1 = \gamma$ and $\theta_2 = 2\beta$ into the above equations, we will arrive at the same formulas as in Newmark method. In that case, the stability properties of Newmark method can be directly applied to this algorithm. However, we note that the implementation of this algorithm is different from the Newmark method in that it requires only the starting values of displacement and velocity at the beginning of the time step.

8.5.5.3 Cubic algorithm (SS32)

This algorithm is obtained by setting $p = 3$, which results in three integration parameters θ_1, θ_2, and θ_3. The procedure can be described as follows.

$$\begin{cases}\bar{\mathbf{x}}_{k+1}=\mathbf{x}_k+\theta_1\Delta t\dot{\mathbf{x}}_k+\dfrac{1}{2}\theta_2\Delta t^2\ddot{\mathbf{x}}_k\\ \bar{\mathbf{x}}_{k+1}^{(1)}=\dot{\mathbf{x}}_k+\theta_1\Delta t\ddot{\mathbf{x}}_k\\ \bar{\mathbf{x}}_{k+1}^{(2)}=\ddot{\mathbf{x}}_k\end{cases} \tag{8.336}$$

The vector $\boldsymbol{\alpha}$ is determined by solving the following system of equations:

$$\left(\theta_1\Delta t\mathbf{M}+\frac{1}{2}\theta_2\Delta t^2\mathbf{C}+\frac{1}{6}\theta_3\Delta t^3\mathbf{K}\right)\alpha_k^{(3)}=\bar{\mathbf{p}}-\mathbf{M}\bar{\mathbf{x}}_{k+1}^{(2)}-\mathbf{C}\bar{\mathbf{x}}_{k+1}^{(1)}-\mathbf{K}\bar{\mathbf{x}}_{k+1} \tag{8.337}$$

The displacements, velocities, and accelerations at the end of the time step are computed as follows:

$$\begin{cases}\mathbf{x}_{k+1}=\mathbf{x}_k+\Delta t\dot{\mathbf{x}}_k+\dfrac{\Delta t^2}{2}\ddot{\mathbf{x}}_k+\dfrac{\Delta t^3}{6}\alpha_k^{(3)}\\ \dot{\mathbf{x}}_{k+1}=\dot{\mathbf{x}}_k+\Delta t\ddot{\mathbf{x}}_k+\dfrac{\Delta t^2}{2}\alpha_k^{(3)}\\ \ddot{\mathbf{x}}_{k+1}=\ddot{\mathbf{x}}_k+\Delta t\alpha_k^{(3)}\end{cases} \tag{8.338}$$

It can be shown that the cubic algorithms encompass a class of well-known methods, including the Houbolt method, the Wilson's θ-method, and the α-method. For example, the Houbolt method corresponds to $(\theta_1, \theta_2, \theta_3) = (2, 11/3, 6)$; the Wilson's θ-method corresponds to $(\theta_1, \theta_2, \theta_3) = (\theta, \theta^2, \theta^3)$; and, the α-method corresponds to $(\theta_1, \theta_2, \theta_3) = [1, 2/3 + 2\beta - 2\alpha^2, 6\beta(1 + \alpha)]$. The stability properties of these methods are preserved.

It should be noted that since the system of equations of motion is satisfied in an average sense within the time step Δt, then the load vector should be computed approximately as average within the time step.

8.5.6 Symmetric successive quadrature method

The successive symmetric quadrature for direct integration of dynamic system of equations of motion was proposed by Robinson (Robinson and Healy, 1984). Before we discuss the algorithm, let us recap on some basics in numerical integrations. In practical numerical integration, Gauss quadrature is most widely used especially in finite element analysis. In general, when it is difficult to carry out exact integration or when only a few points of the function f(x) within the interval [a, b] is known, which may be obtained from observations, a definite integral of f(x) over that interval can be computed using numerical integration or quadrature schemes. The basic idea is from the mean value theorem for integral.

$$I = \int_a^b f(x)dx = (b-a)f(\xi), \quad a \le \xi \le b \tag{8.339}$$

The value of $f(\xi)$ can be considered as the average functional value within the interval. If we approximate the function f(x) with a polynomial g_k of degree k which agrees with f(x) at (k + 1) discrete sampling points, then the above definite integration can be approximately computed.

$$g_k(x) = \sum_{j=0}^{k} f(x_j) l_j(x) \tag{8.340}$$

$$\begin{aligned} I &= \int_a^b f(x)dx \approx \int_a^b g_k(x)dx \\ &= \int_a^b \sum_{j=0}^{k} f(x_j) l_j(x)dx = \sum_{j=0}^{k} w_j f(x_j) \end{aligned} \tag{8.341}$$

$$w_j = \int_a^b l_j(x)dx \tag{8.342}$$

The function l_j is the jth Lagrange polynomial. We can always use a Lagrange polynomial (or shape functions) to represent a general polynomial within an interval, especially when the function values at discrete sampling points are known. For example, when k = 0, then we evaluate the function at only one sampling point within the interval [a, b]. If we choose the sampling point at the midpoint of the interval we obtain the *midpoint rule*.

$$\begin{cases} x_0 = \dfrac{b-a}{2} \\ l_0 = 1 \\ w_0 = \int_a^b l_0(x)dx = b-a \end{cases} \tag{8.343}$$

$$I = \int_a^b f(x)dx \approx (b-a)f\left(\frac{b-a}{2}\right) \tag{8.344}$$

The midpoint rule will give exact result if function f(x) is a straight line. For k = 1, we then evaluate the function f(x) at two sampling points within the interval. We obtain the *trapezoidal rule* if we choose the two boundary points.

$$\begin{cases} x_0 = a;\ x_1 = b \\ l_0 = 1 - \dfrac{x-a}{b-a};\ l_1 = \dfrac{x-a}{b-a} \\ w_0 = \displaystyle\int_a^b l_0(x)dx = b-a;\ w_1 = \int_a^b l_1(x)dx = b-a \end{cases} \tag{8.345}$$

$$I = \int_a^b f(x)dx \approx \frac{(b-a)}{2}\left[f(a) - f(b)\right] \tag{8.346}$$

This will give the exact result for f(x) being a polynomials of degree ≤ 1.

As the number of sampling points increases, the accuracy of the numerical integration improves. Depending on the way the sampling points are selected, we can obtain different quadrature methods. The Gauss quadrature is an optimal integration method in which the sampling points are determined so that the approximation accuracy is optimized. These sampling points are called Gauss points, which are located symmetrically with respect to the midpoint of the interval. Usually the definite integration with Gauss quadrature is for integration in the interval [−1, 1].

For example, the midpoint rule corresponds to one-point Gauss quadrature (k = 0), where $\xi_0 = 0$ and $w_0 = 2.0$. The sampling points and weights for the two-point Gauss quadrature are $\xi_{0,1} = \pm 1/\sqrt{3} = \pm 0.57735$ and $w_{0,1} = 1.0$. The two-point Gauss quadrature yields the exact solution for a polynomial of degree ≤ 3. In general, a polynomial of degree (2n − 1) can be integrated exactly by an n-point Gauss quadrature rule.

With this brief review on numerical integration methods, now we discuss symmetric successive quadrature for direct integration of equations of motion. This method was developed based on the constant average acceleration method ($\gamma = 1/2$, $\beta = 1/4$) in the Newmark family of methods, which is the only method that is unconditionally stable and conserves energy of the system as well. Although this method does not have amplitude decay, it still introduces numerical integration error in the form of period elongation. It is thus clear that if we can reduce the error in frequency distortion, we would have a more accurate direct integration method. We recall the discretization equations for the constant average acceleration method.

$$\begin{cases} \dot{\mathbf{x}}_{k+1} = \dot{\mathbf{x}}_k + \dfrac{\Delta t}{2}(\ddot{\mathbf{x}}_k + \ddot{\mathbf{x}}_{k+1}) \\ \mathbf{x}_{k+1} = \mathbf{x}_k + \Delta t\ \dot{\mathbf{x}}_k + \dfrac{\Delta t^2}{4}(\ddot{\mathbf{x}}_k + \ddot{\mathbf{x}}_{k+1}) \end{cases} \tag{8.347}$$

These equations can also be written in the following form:

$$\begin{cases} \dot{\mathbf{x}}_{k+1} = \dot{\mathbf{x}}_k + \dfrac{\Delta t}{2}(\ddot{\mathbf{x}}_k + \ddot{\mathbf{x}}_{k+1}) \\ \mathbf{x}_{k+1} = \mathbf{x}_k + \dfrac{\Delta t}{2}(\dot{\mathbf{x}}_k + \dot{\mathbf{x}}_{k+1}) \end{cases} \tag{8.348}$$

Comparing the above relations with the trapezoidal rule, we can observe that this special Newmark method appears to perform two successive quadratures on acceleration to obtain velocity and displacement within the time step Δt using the same trapezoidal rule. This observation gives rise to the conjecture that by using more accurate quadrature formulas successively, it may result in a more accurate integration method that preserves the properties of unconditional stability and energy conservation in the constant average acceleration method. It can be shown that the stability and accuracy property of the method can be preserved provided that the form of the quadrature is *symmetric*. However, in this case, Gauss quadrature may not be a suitable choice because the Gauss sampling points exclude the boundary points.

Naturally, to obtain a more accurate quadrature formula, a simple and straightforward approach would be to introduce a finite number of intermediate sampling points within the time step Δt. For example, with one intermediate point within the time interval, the quadrature formulas can exactly integrate a parabola, which is more accurate than the trapezoidal rule, which is exact for a linear function. Of course, with the introduction of intermediate points, additional sets of unknowns are also introduced. However, by satisfying the equation of motion at the intermediate points, we have as many additional equations as the unknowns. Thus, at least in theory, the solution process can be carried out. During this solution process, the same quadrature rule is also used successively as in the constant average acceleration method. In general, with (m − 1) intermediate sampling points, m = 1, 2, ..., the quadrature formulas can be expressed as follows ($h = \Delta t$):

$$I = \int_{t_{n0}}^{t_{nk}} f(t)dt \approx h\sum_{j=0}^{k} f_j\alpha_{jk};\ k = 1, 2, \ldots, m \tag{8.349}$$

$f_j = f(x_{nj})$ and α_{jk} are dimensionless weights. Obviously, $\alpha_{j0} = 0$.

For example, for m = 1, and k = 1, there is no intermediate point within the time interval.

$$\begin{cases} f_0 = f(x_{n0}) \\ f_1 = f(x_{n1}) \end{cases} \tag{8.350}$$

$$f(t) = \frac{h-t}{h}f_0 + \frac{t}{h}f_1 \tag{8.351}$$

$$\begin{aligned} \int_{t_{n0}}^{t_{n1}} f(t)dt &\approx \int_{o}^{h}\left(\frac{h-1}{h}f_0 + \frac{t}{h}f_1\right)dt \\ &= h\left(\frac{1}{2}f_0 + \frac{1}{2}f_1\right) \end{aligned} \tag{8.352}$$

From the above relation, we get $\alpha_{01} = \alpha_{11} = 1/2$, which results in the trapezoidal rule. The trapezoidal rule is the lowest order method in the family of symmetric successive quadrature methods.

For m = 2, and k = 1, 2, this is one intermediate point within the time interval $h = \Delta t$. The conservation of numerical properties of the average acceleration method requires that the intermediate point must be symmetrically located. This means we can only use the midpoint of the interval as the intermediate point. In this case, we can express the function f(t) within the interval $h = \Delta t$ in terms of functional values at the three nodal points as

$$\begin{cases} f_0 = f(x_{n0}) \\ f_1 = f(x_{n1}) \\ f_2 = f(x_{n2}) \end{cases} \tag{8.353}$$

$$f(t) = \frac{(2t-h)(t-h)}{h^2} f_0 + \frac{4t(h-t)}{h^2} f_1 + \frac{t(2t-h)}{h^2} f_2 \tag{8.354}$$

$$\begin{aligned} \int_{t_{n0}}^{t_{n1}} f(t)dt &\approx \int_{0}^{h/2} f(t)dt \\ &= h\left(\frac{5}{24} f_0 + \frac{8}{24} f_1 - \frac{1}{24} f_2\right) \end{aligned} \tag{8.355}$$

$$\begin{aligned} \int_{t_{n0}}^{t_{n2}} f(t)dt &\approx \int_{0}^{h} f(t)dt \\ &= h\left(\frac{1}{6} f_0 + \frac{4}{6} f_1 - \frac{1}{6} f_2\right) \end{aligned} \tag{8.356}$$

It can be shown that with this quadrature formulas, the numerical integration error in period elongation becomes as follows:

$$\begin{aligned} &\frac{T}{T_{ps}} = 1 + \frac{\delta^4}{720} + O(\delta^6) \\ &\delta = \omega \Delta t \end{aligned} \tag{8.357}$$

We recall the following period elongation error in the Newmark constant average acceleration method:

$$\frac{T}{T_{ps}} = 1 + \frac{\delta^2}{12} + O(\delta^4) \tag{8.358}$$

Obviously, the new quadrature scheme is much more accurate, which is even better than the best scheme (in terms of period elongation error) in the Newmark family methods with $\beta = 1/12$. It can be shown that for more than one intermediate points within the time interval, the Lobatto points give the optimum location. With two intermediate points, $m = 3$, the corresponding quadrature formulas using Lobatto points has the following numerical integration error in period elongation:

$$\frac{T}{T_{ps}} = 1 + \frac{\delta^6}{100{,}800} + O(\delta^{12}) \tag{8.359}$$

This shows that these higher order quadrature formulas can be highly accurate. However, for implicit method these formulas require the satisfaction of equation of motion at intermediate points within the time step and they require more computational efforts.

In application to structural dynamic problems, a successive quadrature scheme with one (symmetric) intermediate point can provide sufficient accuracy. In this case, the symmetric successive quadrature formulas for computing the velocities and displacements are as follows:

$$\begin{cases} \dot{x}_{k+1/2} = \dot{x}_k + \dfrac{\Delta t}{24}(5\ddot{x}_k + 8\ddot{x}_{k+1/2} - \ddot{x}_{k+1}) \\ \dot{x}_{k+1} = \dot{x}_k + \dfrac{\Delta t}{6}(\ddot{x}_k + 4\ddot{x}_{k+1/2} - \ddot{x}_{k+1}) \end{cases} \tag{8.360}$$

$$\begin{cases} x_{k+1/2} = x_k + \dfrac{\Delta t}{24}(5\dot{x}_k + 8\dot{x}_{k+1/2} - \dot{x}_{k+1}) \\ x_{k+1} = x_k + \dfrac{\Delta t}{6}(\dot{x}_k + 4\dot{x}_{k+1/2} - \dot{x}_{k+1}) \end{cases} \tag{8.361}$$

This method can be implemented in a predictor–corrector format. It should be implemented using the original iterative framework proposed by Newmark. By so doing, there are several iterations within each time step. The solution process within each time step can proceed as follows:

1. Assume initial trial values for $\ddot{x}_{k+1/2}$ and $\ddot{x}_{k+1}$ and then compute velocities and displacements at $t_{k+1/2}$ and t_{k+1} using the above quadrature formulas.
2. Satisfy the equation of motion at $t_{k+1/2}$ and t_{k+1}, to compute new accelerations.
3. Repeat the calculation using successive quadrature formulas.

This iterative process continues until convergence is reached. Numerical simulations indicate that this procedure takes about four times more computational efforts than required with a standard Newmark method. However, the most important advantage of this method is that it can accommodate fairly large time step specifically for diffusion type problems where the response of the system computed from numerical direct integration is not limited by the highest frequency of the system, and it has high accuracy in terms of period elongation.

8.6 CONCLUDING REMARKS

This chapter discusses some of the most commonly used numerical methods for the solution of dynamic equations of motions of a structural system. For linear problems, the principle of superposition always applies leading to the mode superposition method in dynamics of structural systems. For practical problems in structural dynamics such as earthquake engineering, after carrying out spatial discretization of the structural system with finite element method, very often the dynamic responses are dominated by the few lowest modes. The contribution from the highest modes can often be neglected. However, with wave propagation-type problems, all the modes contribute to the response of the system. This shows that judgment must be exercised on the use of different numerical solution methods depending on the dynamic characteristic of the problem and the type of loading encountered.

A heuristic rule is that for structural dynamic problems whose responses are determined by the first few lowest modes, an implicit unconditionally stable method would be more advantageous than an explicit conditionally stable algorithm. Although for stability consideration alone, there is no restriction on the time step with an unconditionally stable algorithms, attention should be paid to the accuracy of the algorithm, which can be measured by errors in terms of amplitude decay and period elongation. The Newmark β-method with $\gamma = 1/2$ and $\beta = 1/4$ is the only unconditionally stable algorithm that does not have amplitude decay, but it introduces numerical error in the form of period elongation. It can be argued that for a system with dominant lower mode responses, the introduction of certain amount of numerical dissipation (exhibited as amplitude decay) would be of advantage to damp out the spurious higher modes arisen from spatial discretization using finite element or finite difference methods. For example, the Wilson's θ-method and the α-method can effectively damp out some highest mode responses. However, as pointed out by Dobbs (1974), this does not mean that the introduction of numerical damping to the structural system is always advantageous. If an algorithm introduces excessive numerical damping to the system, in order to maintain accuracy, a small step size compared with the shortest period of the system would be required. In addition, when the step size is comparable to the shortest period of the system, the Wilson's θ-method may result in significant errors exposed as spikes in the first few cycles of computation, even on a simple simulation problem, whereas this *overshoot* behavior is not present in the Newmark constant average acceleration method.

For wave propagation type of problems, it is advisable to use explicit conditionally stable algorithm such as the second central difference method. For this kind of problem, the critical time step is the time for signals to propagate across one element, which is usually the smallest element. If the time step used in computation is larger than this critical time step, it would result in divergence in the solution process. This critical time step exists regardless of the integration algorithms used. If an implicit unconditionally stable method is used with a time step larger than the critical time step, we do not expect any problem with the stability of the algorithm, but the results will not be accurate. For very large problems in contact mechanics, such as the discrete element method (Ghaboussi and Barbosa, 1990) and discrete finite element method (Ghaboussi, 1988; Barbosa and Ghaboussi, 1990) in analysis of granular materials, because of the very large amount of computation involved in determining the contact mechanisms, it is not feasible to use implicit methods in carrying out the dynamic response analysis. For these problems conditionally stable explicit integration methods are used.

Chapter 9

Generalized difference method

9.1 INTRODUCTION

In Chapter 8 we have discussed a number of well-known direct integration methods. During the past several decades, a number of direct integration methods have been proposed, and many of them have been widely used in the field of structural dynamics. We observe from the history of development of the direct integration methods that the formulation process of these methods may be basically divided into two main approaches: *finite difference-type formulation* and *finite element-type formulation*. Attempts have been made to search for the common basis in the derivation of some well-known computational methods and some general methods have emerged as a result of these studies (Zienkiewicz, 1977; Zienkiewicz et al., 1984; Ghaboussi, 1987; Wu and Ghaboussi,1992). However, in-depth research on the intrinsic relationships among different approaches or methods has not been thoroughly and systematically exploited yet, especially in the detailed relation between the finite element approach and the finite difference approach with reference to the basis of approximations.

In this chapter, we will discuss *the generalized difference method* in the domain of n-equidistant time stations to analyze the formulation process of numerical integration methods for structural dynamics problem as well as for diffusion equation. The stability properties of the generalized schemes, and the relationship between the generalized difference approach and the finite element weighted residual approach in terms of the formulation process will be discussed.

9.2 WEIGHTED RESIDUAL APPROACH

In Chapter 8, we have discussed direct integration methods for the solution of a system of dynamic equations of motion.

$$\mathbf{M}\ddot{\mathbf{x}}(t)+\mathbf{C}\dot{\mathbf{x}}(t)+\mathbf{K}\mathbf{x}(t)=\mathbf{p}(t) \tag{9.1}$$

Direct integration methods are time stepping algorithms that result in displacements, velocities, and accelerations at discrete time stations that are a time step Δt apart. Most practical algorithms such as the Newmark family of methods and Wilson's θ-method are one-step methods which relates the displacements, velocities, and accelerations within a time step at successive time stations. For example, the Newmark β-method uses the following temporal discretization equations within a time step Δt from t_k to t_{k+1}:

$$\begin{cases}\dot{\mathbf{x}}_{k+1}=\dot{\mathbf{x}}_k+\Delta t\left[(1-\gamma)\ddot{\mathbf{x}}_k+\gamma\ddot{\mathbf{x}}_{k+1}\right]\\ \mathbf{x}_{k+1}=\mathbf{x}_k+\Delta t\,\dot{\mathbf{x}}_k+\Delta t^2\left[\left(\frac{1}{2}-\beta\right)\ddot{\mathbf{x}}_k+\beta\ddot{\mathbf{x}}_{k+1}\right]\end{cases} \tag{9.2}$$

The dynamic equation of motion is satisfied at time station t_{k+1}. These equations are rearranged such that the displacements, velocities, and accelerations at the end of the time step are computed from their known values at the beginning of the time step. It is in this time stepping form that these methods are usually used in practice.

The Newmark method can also be written in the form of recurrence formula which relates the displacements at three successive time stations. This can be accomplished by using the Newmark temporal discretization formulas in Equation 9.2 at the time steps from t_{k-1} to t_k and from t_k to t_{k+1}. In addition, the dynamic equation of motion is satisfied at the three time stations. By so doing, we can eliminate the six variables (vectors of variables) associated with velocities and accelerations from the seven equations and obtain a recurrence formula in the following general form:

$$\mathbf{G}_1\mathbf{x}_{k-1} + \mathbf{G}_2\mathbf{x}_k + \mathbf{G}_3\mathbf{x}_{k+1} = \bar{\mathbf{p}} \tag{9.3}$$

The matrices $\mathbf{G}_1$, $\mathbf{G}_2$, and $\mathbf{G}_3$ are functions of $\mathbf{M}$, $\mathbf{C}$, $\mathbf{K}$, Δt, and the integration constants, γ and β; and $\bar{\mathbf{p}}$ is a suitable discretization of the load vector $\mathbf{p}(t)$ (Zienkiewicz, 1977).

$$\begin{aligned}
&\left[\mathbf{M} + \gamma\Delta t\mathbf{C} + \beta\Delta t^2\mathbf{K}\right]\mathbf{x}_{k+1} \\
&+\left[-2\mathbf{M} + (1-2\gamma)\Delta t\mathbf{C} + \left(\frac{1}{2} + \gamma - 2\beta\right)\Delta t^2\mathbf{K}\right]\mathbf{x}_k \\
&+\left[\mathbf{M} - (1-\gamma)\Delta t\mathbf{C} + \left(\frac{1}{2} - \gamma + \beta\right)\Delta t^2\mathbf{K}\right]\mathbf{x}_{k-1} \\
&= \beta\Delta t^2\mathbf{p}_{k+1} + \left(\frac{1}{2} + \gamma - 2\beta\right)\Delta t^2\mathbf{p}_k + \left(\frac{1}{2} - \gamma + \beta\right)\Delta t^2\mathbf{p}_{k-1}
\end{aligned} \tag{9.4}$$

It can be seen that this formula gives an equivalent representation of the Newmark β-method. We will show later that these recurrence formulae are of importance in revealing some of the fundamental properties of direct integration methods.

Zienkiewicz (1977) has shown that this formula can be derived from a finite element weighted residual process applied to the dynamic equation of motion by considering the domain of the problem as consisting of three nodes (time stations) in time t_{k-1}, t_k, and t_{k+1} as shown in Figure 9.1. Therefore, the displacements, velocities, and accelerations within the time domain ($2\Delta t$) can be interpolated using the standard Lagrangian shape functions.

$$\begin{cases}
\mathbf{x} = \sum_{j=0}^{2} N_{k-1+j}\,\mathbf{x}_{k-1+j} \\
\dot{\mathbf{x}} = \sum_{j=0}^{2} \dot{N}_{k-1+j}\,\mathbf{x}_{k-1+j} \\
\ddot{\mathbf{x}} = \sum_{j=0}^{2} \ddot{N}_{k-1+j}\,\mathbf{x}_{k-1+j}
\end{cases} \tag{9.5}$$

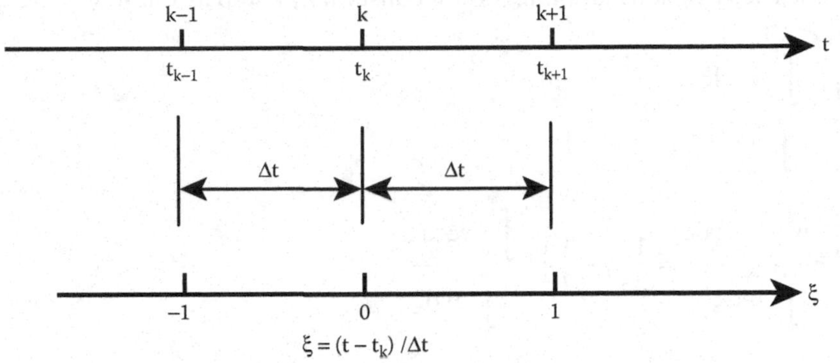

Figure 9.1 **Three time stations used in the finite element weighted residual process.**

$$\begin{cases} \mathbf{N}_{k-1} = -\dfrac{1}{2}\xi(1-\xi) \\ \mathbf{N}_{k} = (1-\xi)(1+\xi) \\ \mathbf{N}_{k+1} = \dfrac{1}{2}\xi(1+\xi) \end{cases} \tag{9.6}$$

$$\xi = \frac{t - t_k}{\Delta t}; \quad -1 \le \xi \le 1$$

The weighted residual equation is the following integral over the domain of $(2\Delta t)$ with the weighting function w:

$$\int_0^{2\Delta t} w\left[\mathbf{M}\ddot{\mathbf{x}}(t) + \mathbf{C}\dot{\mathbf{x}}(t) + \mathbf{K}\mathbf{x}(t) - \mathbf{p}(t)\right]dt = 0 \tag{9.7}$$

$$\int_0^{2\Delta t} w\left[\mathbf{M}\sum_j \ddot{\mathbf{N}}_j\mathbf{x}_j + \mathbf{C}\sum_j \dot{\mathbf{N}}_j\mathbf{x}_j + \mathbf{K}\sum_j \mathbf{N}_j\mathbf{x}_j - \sum_j \mathbf{N}_j\mathbf{p}_j\right]dt = 0 \tag{9.8}$$

for $j = k-1,\ k,\ k+1$

$$\begin{aligned} \int_0^{2\Delta t} w[&(\mathbf{M}\ddot{\mathbf{N}}_{k+1} + \mathbf{C}\dot{\mathbf{N}}_{k+1} + \mathbf{K}\mathbf{N}_{k+1})\mathbf{x}_{k+1} \\ &+ (\mathbf{M}\ddot{\mathbf{N}}_{k} + \mathbf{C}\dot{\mathbf{N}}_{k} + \mathbf{K}\mathbf{N}_{k})\mathbf{x}_{k} \\ &+ (\mathbf{M}\ddot{\mathbf{N}}_{k-1} + \mathbf{C}\dot{\mathbf{N}}_{k-1} + \mathbf{K}\mathbf{N}_{k-1})\mathbf{x}_{k-1} \\ &- (\mathbf{N}_{k+1}\mathbf{p}_{k+1} + \mathbf{N}_{k}\mathbf{p}_{k} + \mathbf{N}_{k-1}\mathbf{p}_{k-1})]dt = 0 \end{aligned} \tag{9.9}$$

We can define the two Newmark integration constants γ and β as follows:

$$\begin{cases} \gamma = \dfrac{1}{2} + \dfrac{\int_{-1}^{1} w\xi d\xi}{\int_{-1}^{1} w d\xi} \\ \beta = \dfrac{\int_{-1}^{1} w\xi(1+\xi) d\xi}{2\int_{-1}^{1} w d\xi} = \dfrac{1}{2}\left(\gamma - \dfrac{1}{2}\right) + \dfrac{\int_{-1}^{1} w\xi^2 d\xi}{2\int_{-1}^{1} w d\xi} \end{cases} \tag{9.10}$$

With these definitions of γ and β, we can validate that Equation 9.9 becomes exactly the same as Equation 9.4. This shows that the Newmark family of methods can be directly derived from a finite element weighted residual approach. Obviously, in this case, the values of γ and β depend on the choice of the weighting function, w. It is observed that $\gamma = 1/2$ corresponds to symmetric weighting functions within the time domain of $(2\Delta t)$.

The finite element weighted residual approach is in fact a general method for deriving numerical integration methods. For example, many of the other well-known methods, such as the Houbolt method and Galerkin method, can be considered as special cases of higher order formulas using a larger number of time stations. For example, if we use a set of cubic shape functions defined over four consecutive time stations $k-2$, $k-1$, k, and $k+1$, we have the Lagrangian shape functions in natural coordinate.

$$\begin{cases} \mathbf{N}_{k-2} = -\dfrac{1}{6}(\xi-1)(\xi-2)(\xi-3) \\ \mathbf{N}_{k-1} = \dfrac{1}{2}\xi(\xi-2)(\xi-3) \\ \mathbf{N}_{k} = -\dfrac{1}{2}\xi(\xi-1)(\xi-3) \\ \mathbf{N}_{k+1} = \dfrac{1}{6}\xi(\xi-1)(\xi-2) \end{cases} \tag{9.11}$$

$$\xi = \frac{t - t_{k-2}}{\Delta t}; \quad 0 \le \xi \le 3$$

The recursive equation for numerical integration methods corresponding to this higher order formula can be directly derived from the weighted residual approach. In this case, there are three integration parameters which are defined as follows:

$$\begin{cases} \alpha = \left(\int_0^3 w\xi^3 d\xi\right) / \left(\int_0^3 w d\xi\right) \\ \bar{\beta} = \left(\int_0^3 w\xi^2 d\xi\right) / \left(\int_0^3 w d\xi\right) \\ \bar{\gamma} = \left(\int_0^3 w\xi d\xi\right) / \left(2\int_0^3 w d\xi\right) \end{cases} \tag{9.12}$$

The recursive formula is in the following form:

$$\begin{aligned}
&\left[(\bar{\gamma}-1)\mathbf{M}+\left(\frac{1}{2}\bar{\beta}-\bar{\gamma}+\frac{1}{3}\right)\Delta t\mathbf{C}+\left(\frac{1}{6}\alpha-\frac{1}{2}\bar{\beta}+\frac{1}{3}\bar{\gamma}\right)\Delta t^2\mathbf{K}\right]\mathbf{x}_{k+1}\\
&\quad-\left[(3\bar{\gamma}-4)\mathbf{M}+\left(\frac{3}{2}\bar{\beta}-4\bar{\gamma}+\frac{3}{2}\right)\Delta t\mathbf{C}+\left(\frac{1}{2}\alpha-2\bar{\beta}+\frac{3}{2}\bar{\gamma}\right)\Delta t^2\mathbf{K}\right]\mathbf{x}_k\\
&\quad+\left[(3\bar{\gamma}-5)\mathbf{M}+\left(\frac{3}{2}\bar{\beta}-5\bar{\gamma}+3\right)\Delta t\mathbf{C}+\left(\frac{1}{2}\alpha-\frac{5}{2}\bar{\beta}+3\bar{\gamma}\right)\Delta t^2\mathbf{K}\right]\mathbf{x}_{k-1}\\
&\quad-\left[(\bar{\gamma}-2)\mathbf{M}+\left(\frac{1}{2}\bar{\beta}-2\bar{\gamma}+\frac{11}{6}\right)\Delta t\mathbf{C}+\left(\frac{1}{6}\alpha-\bar{\beta}+\frac{11}{6}\bar{\gamma}-1\right)\Delta t^2\mathbf{K}\right]\mathbf{x}_{k-2}\\
&=\left(\frac{1}{6}\alpha-\frac{1}{2}\bar{\beta}+\frac{1}{3}\bar{\gamma}\right)\Delta t^2\mathbf{p}_{k+1}-\left(\frac{1}{2}\alpha-2\bar{\beta}+\frac{3}{2}\bar{\gamma}\right)\Delta t^2\mathbf{p}_k\\
&\quad+\left(\frac{1}{2}\alpha-\frac{5}{2}\bar{\beta}+3\bar{\gamma}\right)\Delta t^2\mathbf{p}_{k-1}-\left(\frac{1}{6}\alpha-\bar{\beta}+\frac{11}{6}\bar{\gamma}-1\right)\Delta t^2\mathbf{p}_{k-2}
\end{aligned} \tag{9.13}$$

It can be seen that this formula contains infinite number of numerical integration methods depending on the values of the three integration parameters, or the choice of the weighting functions. For example, the Houbolt method corresponds to a simple Dirac delta weighting function (a pulse function) at time station or node (k + 1), which in turn results in the following values of parameters:

$$\alpha = 27,\ \bar{\beta} = 9,\ \bar{\gamma} = 3 \tag{9.14}$$

The Wilson's θ-method is a special case with the following parameters:

$$\begin{cases}\alpha = 2+4\theta+3\theta^2+\theta^3\\ \bar{\beta} = \dfrac{4}{3}+2\theta+\theta^2\\ \bar{\gamma} = 1+\theta\end{cases} \tag{9.15}$$

and the Newmark family of method is obtained when $\theta = 1.0$. In this case, the four level (four time stations) scheme is reduced to the three level (three time stations) scheme derived earlier.

9.3 NONUNIQUENESS IN APPROXIMATION OF DISPLACEMENTS

A close examination of the basic assumptions for the Zienkiewicz's three level formula and the original Newmark family of methods indicates that the approximation of displacements is nonunique. We observe that the displacements within the three time stations are approximated with second-order polynomials (quadratic functions) in the Zienkiewicz's three level formula. However, in the Newmark family of methods, for instance, with the linear acceleration method, the corresponding displacements are approximated with third-order polynomials.

Nevertheless, the three level formula derived from Zienkiewicz's finite element weighted residual method can have an equivalent form with Newmark family of methods on the time domain of three time stations. This clearly indicates that the approximation of displacements is not unique, which in turn indicates that the approximation of displacements in direct integration methods is immaterial. This observation provides the motivation for developing the generalized difference method, which will be discussed in the following sections.

9.4 GENERALIZED DIFFERENCE OPERATORS

Similar to approaches used in Zienkiewicz's finite element weighted residual method, we will discuss the recurrence relation form of direct integration methods. The generalized difference method we proposed is a consistent method for deriving these recurrence relations. Unlike the well-known direct integration methods such as the Newmark family of methods, the generalized difference method does not make any assumptions on the variation of displacements, velocities, and accelerations within a time step. It uses only the discrete values of the displacements at the time stations within a time domain and the generalized differences to derive the recurrence relation for direct integration methods.

Using direct integration methods, at best, we can compute an approximation to the solution of the dynamic equation of motion. However, if we neglect the round-off error, the direct integration methods can be considered to be giving the exact solution of the discrete equivalent of the dynamic equation of motion. The general form of the discrete dynamic equation of motion of a structural system is as follows:

$$\mathbf{MD}^2(\tau) + \mathbf{CD}^1(\tau) + \mathbf{KD}^0(\tau) = \mathbf{D}^p(\tau) \tag{9.16}$$

The generalized differences $\mathbf{D}^j(\tau)$, $j = 0, 1, 2$ are the discrete equivalents of the displacement and its derivatives, velocities, and accelerations. These generalized difference operators are defined over a field of n time stations, as shown in Figure 9.2. The discrete approximation to displacements at the n time stations is $\{\mathbf{d}_{k+1}, \mathbf{d}_k, \ldots, \mathbf{d}_{k-n+2}\}$. The generalized differences are continuous functions of a normalized time, τ, defined within the field of n time stations.

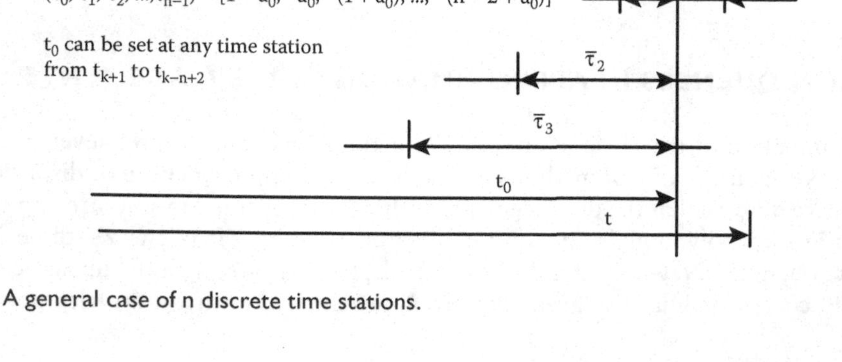

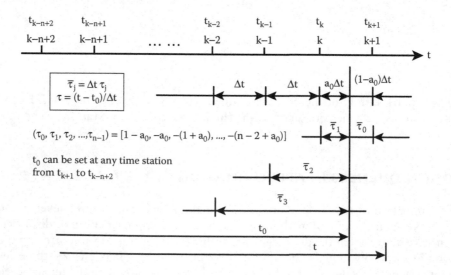

Figure 9.2 A general case of n discrete time stations.

$$\tau = \frac{1}{\Delta t}(t - t_0) \tag{9.17}$$

The origin of τ within the field of n time stations is fixed by the position of t_0.

We consider n discrete time stations. The generalized differences are defined over this field of n discrete time stations and we recognize that during direct integration of the equation of motions this field moves forward in time. In order to free the formulation from the absolute position of the field of n time stations, we introduce the normalized time τ as defined above, which is relative to an absolute time t_0 which defines the origin of τ. For $\tau = 0$, we define the basic generalized differences 0 by the following expression:

$$\mathbf{D}^j(0)\Delta t^j = \sum_{i=0}^{n-1} a_{ij}\mathbf{d}_{k+1-i};\ j=0, 1,\ldots, n-1 \tag{9.18}$$

In this expression, $\{\mathbf{d}_{k+1}, \mathbf{d}_k, \ldots, \mathbf{d}_{k-n+2}\}$ are discrete approximations to displacements at n time stations, $\{\mathbf{x}(t_{k+1}), \mathbf{x}(t_k), \ldots, \mathbf{x}(t_{k-n+2})\}$; $\mathbf{D}^j(0)$ is the generalized difference approximation to the jth time derivative of $\mathbf{x}(t)$, and a_{ij} are coefficients obtained from the generalized difference method. The above equation can be written in the following matrix form:

$$\mathbf{D}_0 = \mathbf{T}\mathbf{A}^T\mathbf{d} \tag{9.19}$$

$$\left\{\begin{aligned} &\mathbf{D}_0 = \left[\mathbf{D}^0(0), \mathbf{D}^1(0),\ldots, \mathbf{D}^{n-1}(0)\right]^T \\ &\mathbf{T} = \text{diag}\left[1,\ \Delta t^{-1},\ \ldots,\ \Delta t^{-(n-1)}\right] \\ &\mathbf{A} = \left[a_{ij}\right]_{n\times n} \\ &\mathbf{d} = \left[\mathbf{d}_{k+1}, \mathbf{d}_k, \ldots, \mathbf{d}_{k-n+2}\right]^T \end{aligned}\right. \tag{9.20}$$

The coefficients of the generalized differences, a_{ij} i, j = 0, 1, 2, ..., n − 1, are determined by satisfying the *consistency conditions* of discretization at the n time stations, which can be expressed as follows:

$$\left\{\begin{aligned} &\mathbf{d}_{k+1-i} = \mathbf{f}_m(\tau_i); && i = 0, 1, \ldots, n-1 \\ &\mathbf{f}_m(\tau) = \sum_{j=0}^{m}\frac{1}{j!}\mathbf{b}_j(t - t_0)^j; && m = 0, 1, \ldots, n-1 \end{aligned}\right. \tag{9.21}$$

As shown in Figure 9.2, the normalized time at discrete time stations are τ_i. The vectors $\mathbf{b}_j$ are constant.

The consistency conditions mean that the generalized difference operators must be consistent with the conditions of *constant* displacement, *constant* velocity, *constant* acceleration, and so on. Satisfying the consistency conditions in the domain of n time stations leads to the following matrix equation for the determination of the coefficient matrix $\mathbf{A}_n$ which is an n × n matrix:

$$\mathbf{V}_n\mathbf{A}_n = \mathbf{E}_n \tag{9.22}$$

V_n is an n × n Vandermonde matrix and E_n is an n × n diagonal matrix. For $\tau = 0$, these two matrices are of the following forms:

$$V_n = \begin{bmatrix} 1 & 1 & \cdots & 1 \\ \tau_0 & \tau_1 & \cdots & \tau_{n-1} \\ \vdots & \vdots & \vdots & \vdots \\ \tau_0^{n-1} & \tau_1^{n-1} & \cdots & \tau_{n-1}^{n-1} \end{bmatrix} \tag{9.23}$$

$$E_n = \begin{bmatrix} 1 & & & & \\ & 1 & & & \\ & & 2! & & \\ & & & \ddots & \\ & & & & (n-1)! \end{bmatrix} \tag{9.24}$$

It can be seen that the V matrix is a standard Vandermonde matrix. Obviously, the Vandermonde matrix depends on our choice of the value of t_0. We note that to determine matrix **A** we need to invert the Vandermonde matrix or solve a system of equation with it. There exists a method for inverting the Vandermonde matrix (Macon and Spitzbart, 1957).

Once the coefficient matrix **A** is determined by solving Equation 9.22, the generalized difference operators are then defined. In other words, the generalized differences for the displacement and its derivatives of order up to n − 1 can be determined, after the consistency conditions for temporal discretization is satisfied. It may appear unusual to develop a generalized difference operator for the displacement itself. However, as will be seen later, it is a central part of a consistent way of using generalized differences. The complete set of generalized difference operators for a field of n discrete time stations can be written in closed form, which will be discussed later in this section.

Now, we will illustrate the determination of the coefficient matrix of the generalized differences from consistency conditions through a simple example. We consider a field of three discrete time stations, t_{k+1}, t_k, t_{k-1} and n = 3, as shown in Figure 9.3. Assuming that we are considering a one-dimensional problem, we can consider the displacement and generalized operators as scalar quantities. From Equation 9.18, the generalized differences are as follows:

$$\begin{cases} D^0 = a_{00}d_{k+1} + a_{01}d_k + a_{02}d_{k-1} \\ \Delta t D^1 = a_{10}d_{k+1} + a_{11}d_k + a_{12}d_{k-1} \\ \Delta t^2 D^2 = a_{20}d_{k+1} + a_{21}d_k + a_{22}d_{k-1} \end{cases} \tag{9.25}$$

In this case, there are three consistency conditions that need to be satisfied. First, the condition of constant displacement is expressed by the following relations:

$$\begin{cases} d_{k+1} = b_0;\ d_k = b_0;\ d_{k-1} = b_0 \\ D^0 = b_0;\ D^1 = 0;\ D^2 = 0 \end{cases} \tag{9.26}$$

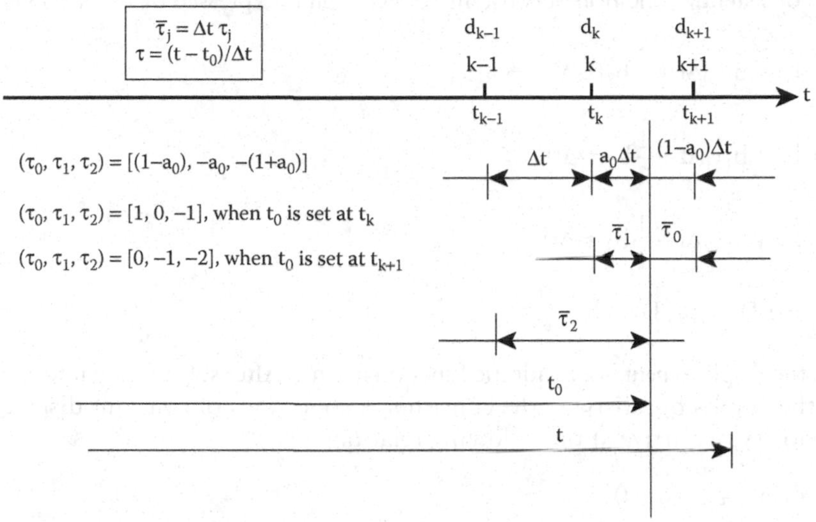

Figure 9.3 Example of a three time station case for generalized differences.

The constant value of displacement is b_0. This is obvious because with constant displacement, the corresponding velocity and acceleration are zero. Substituting these equations in Equation 9.25 results in the following relations:

$$\begin{cases} a_{00} + a_{01} + a_{02} = 1 \\ a_{10} + a_{11} + a_{12} = 0 \\ a_{20} + a_{21} + a_{22} = 0 \end{cases} \tag{9.27}$$

Similarly, the consistency condition of constant velocity b_1 is expressed by the following relations:

$$\begin{cases} d_{k+1} = b_0 + b_1\tau_0\Delta t \\ d_k = b_0 + b_1\tau_1\Delta t \\ d_{k-1} = b_0 + b_1\tau_2\Delta t \\ D^0 = b_0;\ D^1 = b_1;\ D^2 = 0 \end{cases} \tag{9.28}$$

After substituting into Equation 9.25 and using the results for satisfying the condition of constant displacement, we arrive at the following relations:

$$\begin{cases} \tau_0 a_{00} + \tau_1 a_{01} + \tau_2 a_{02} = 0 \\ \tau_0 a_{10} + \tau_1 a_{11} + \tau_2 a_{12} = 1 \\ \tau_0 a_{20} + \tau_1 a_{21} + \tau_2 a_{22} = 0 \end{cases} \tag{9.29}$$

Finally, the consistency condition of constant acceleration is expressed by the following relations:

$$\begin{cases} d_{k+1} = b_0 + b_1\tau_0\Delta t + \frac{1}{2}b_2(\tau_0\Delta t)^2 \\ d_k = b_0 + b_1\tau_1\Delta t + \frac{1}{2}b_2(\tau_1\Delta t)^2 \\ d_{k-1} = b_0 + b_1\tau_2\Delta t + \frac{1}{2}b_2(\tau_2\Delta t)^2 \\ D^0 = b_0;\ D^1 = b_1;\ D^2 = b_2 \end{cases} \tag{9.30}$$

In this case, the displacement is a quadratic function in time. After substituting into Equation 9.25 and using the results of satisfying the consistency conditions of constant displacement and constant velocity, we arrive at the following relations:

$$\begin{cases} \tau_0^2 a_{00} + \tau_1^2 a_{01} + \tau_2^2 a_{02} = 0 \\ \tau_0^2 a_{10} + \tau_1^2 a_{11} + \tau_2^2 a_{12} = 0 \\ \tau_0^2 a_{20} + \tau_1^2 a_{21} + \tau_2^2 a_{22} = 2 \end{cases} \tag{9.31}$$

The consequences of satisfying all the consistency conditions can be summarized in the following matrix equation, which is the result of combing Equations 9.27, 9.29, and 9.31:

$$\begin{bmatrix} 1 & 1 & 1 \\ \tau_0 & \tau_1 & \tau_2 \\ \tau_0^2 & \tau_1^2 & \tau_2^2 \end{bmatrix} \begin{bmatrix} a_{00} & a_{01} & a_{02} \\ a_{10} & a_{11} & a_{12} \\ a_{20} & a_{21} & a_{22} \end{bmatrix} = \begin{bmatrix} 1 & & \\ & 1 & \\ & & 2 \end{bmatrix} \tag{9.32}$$

$$\mathbf{V}_3\mathbf{A}_3 = \mathbf{E}_3 \tag{9.33}$$

The Vandermonde matrix is clearly dependent on the origin of the normalized discrete time τ, determined by t_0. As seen in Figure 9.3, it is t_0 that determines the numerical value of the set $[\tau_0, \tau_1, \tau_2]$. As a specific example, we will consider the case when $t_0 = t_k$, which results in the set $[\tau_0, \tau_1, \tau_2] = [1, 0, -1]$. Substitution of this set into the Vandermonde matrix and the solution of the corresponding system of equations results in the following matrix of coefficients:

$$\mathbf{A}_3 = \begin{bmatrix} 0 & 1/2 & 1 \\ 1 & 0 & -2 \\ 0 & -1/2 & 1 \end{bmatrix} \tag{9.34}$$

The columns of this matrix are recognized as the coefficients of the central difference operators corresponding to the following generalized difference operators:

$$\begin{cases} D^0(0) = d_k \\ D^1(0) = \frac{1}{2\Delta t}(d_{k+1} - d_{k-1}) \\ D^2(0) = \frac{1}{\Delta t^2}(d_{k+1} - 2d_k + d_{k-1}) \end{cases} \tag{9.35}$$

It is obvious that these generalized difference operators change if we shift the origin of τ. Consider shifting the origin to coincide with the station at t_{k+1}, with the resulting set $[\tau_0, \tau_1, \tau_2] = [0, -1, -2]$. After substituting in the Vandermonde matrix in Equation 9.32 and solving the resulting system of equations, we obtain the following coefficient matrix **A** and the corresponding generalized difference operators:

$$\mathbf{A}_3 = \begin{bmatrix} 1 & 3/2 & 1 \\ 0 & -2 & -2 \\ 0 & 1/2 & 1 \end{bmatrix} \tag{9.36}$$

$$\begin{cases} D^0(0) = d_{k+1} \\ D^1(0) = \dfrac{1}{\Delta t}\left(\dfrac{3}{2}d_{k+1} - 2d_k + \dfrac{1}{2}d_{k-1}\right) \\ D^2(0) = \dfrac{1}{\Delta t^2}(d_{k+1} - 2d_k + d_{k-1}) \end{cases} \tag{9.37}$$

This set of generalized difference operators is recognized as backward difference operators. Similarly, for $t_0 = t_{k-1}$, we have $[\tau_0, \tau_1, \tau_2] = [2, 1, 0]$. This leads to the following coefficient matrix **A** and the corresponding generalized difference operators:

$$\mathbf{A}_3 = \begin{bmatrix} 0 & -1/2 & 1 \\ 0 & 2 & -2 \\ 1 & -3/2 & 1 \end{bmatrix} \tag{9.38}$$

$$\begin{cases} D^0(0) = d_{k-1} \\ D^1(0) = \dfrac{1}{\Delta t}\left(-\dfrac{1}{2}d_{k+1} + 2d_k - \dfrac{3}{2}d_{k-1}\right) \\ D^2(0) = \dfrac{1}{\Delta t^2}(d_{k+1} - 2d_k + d_{k-1}) \end{cases} \tag{9.39}$$

This set of generalized difference operators is the forward difference operators.

This simple example on a domain of three time stations shows that by shifting the origin of the time axis at different time stations, we obtain some standard difference operators. This does not mean that the origin of τ should always be at a nodal point (time station).

As a matter of fact, the origin of τ given by t_0 can be at any point within the field of n time stations, or even outside the field. This suggests that the set of generalized differences are continuous functions of τ. For a field of n discrete time stations and n discrete values of a function at these time stations, we can construct a continuous function and its derivatives of up to order $n - 1$. This set of n continuous functions is the complete set of generalized difference functions, expanded in terms of the generalized differences at t_0. Using Taylor's expansion in the time domain, we can obtain the complete set of generalized difference functions.

$$D^j(\bar{\tau}) = D^j(0) + \sum_{k=j+1}^{n-1} \frac{1}{(k-j)!} \bar{\tau}^{(k-j)} D^k(0); \text{ for } j = 0, 1, \ldots, n-1 \tag{9.40}$$

$$\bar{\tau} = \tau \Delta t$$

We note that here we used a discrete equivalent of the Taylor series without making any assumptions about the variation of displacement. The difference between Taylor series and its discrete counterpart is that the discrete Taylor series is exact, whereas the Taylor series is an infinite series or a finite series with a remainder term. The introduction of $\bar{\tau}$ is to make sure that the discrete dynamic equation of motion is satisfied at any point in the time domain.

As an example, we consider the case of three time stations for which we have the following generalized difference functions:

$$\begin{cases} D^0(\bar{\tau}) = D^0(0) + \bar{\tau} D^1(0) + \dfrac{\bar{\tau}^2}{2} D^2(0) \\ D^1(\bar{\tau}) = D^1(0) + \bar{\tau} D^2(0) \\ D^2(\bar{\tau}) = D^2(0) \end{cases} \tag{9.41}$$

In general, Equation 9.40 can be expressed in the following matrix form:

$$D(\bar{\tau}) = L^T(\bar{\tau}) D(0) \tag{9.42}$$

$$D(\bar{\tau}) = \left[D^0(\bar{\tau}),\ D^1(\bar{\tau}), \ldots,\ D^{n-1}(\bar{\tau}) \right] \tag{9.43}$$

$$L^T(\bar{\tau}) = \begin{bmatrix} 1 & & & & \\ \bar{\tau} & 1 & & & \\ \vdots & \vdots & \ddots & & \\ \bar{\tau}^{n-2}/(n-2)! & \ldots & \ldots & 1 & \\ \bar{\tau}^{n-1}/(n-1)! & \ldots & \ldots & \bar{\tau} & 1 \end{bmatrix} \tag{9.44}$$

This relation can also be written in a more general form as follows:

$$D(t + \bar{\tau}) = L^T(\bar{\tau}) D(t) \tag{9.45}$$

Obviously, the above equation is the discrete analog of the Taylor series. Although Taylor series operate on infinite dimensional space of continuous functions and their derivatives, this equation operates on a finite dimensional space of generalized difference functions. This discrete equivalent is exact and does not have a remainder term.

In order to prove the validity of Equation 9.45 we start by expressing the generalized differences at t and t + τ in terms of their coefficient matrices, determined from the corresponding Vandermonde matrices.

$$\begin{cases} D(t) = TA^T(t)d \\ V(t)A(t) = E \\ D(t+\tau) = TA^T(t+\tau)d \\ V(t+\tau)A(t+\tau) = E \end{cases} \tag{9.46}$$

In the above equations, the Vandermonde matrix V(t) is defined by the set $[\tau_0, \tau_1, \ldots, \tau_{n-1}]$ whereas the Vandermonde matrix $V(t+\tau)$ is defined by the set $[(\tau_0 - \tau), (\tau_1 - \tau), \ldots, (\tau_{n-1} - \tau)]$. After combining these equations and substituting into Equation 9.45 we obtain the following relation:

$$TEV^{-T}(t+\tau)d = L^T(\tau)TEV^{-T}(t)d \tag{9.47}$$

This equation can be simplified by observing that it must remain valid for any arbitrary vector **d**. As a consequence, we obtain the following relation between the Vandermonde matrices:

$$V(t+\tau) = \bar{L}^{-1}(\tau)V(t) \tag{9.48}$$

$$\bar{L}(\tau) = ETL(\bar{\tau})T^{-1}E^{-1} \tag{9.49}$$

It can be verified that $\bar{L}(\tau)$ is a unit lower triangular matrix and that it is a function τ. The terms of this matrix are as follows:

$$L_{ij}(\tau) = \begin{cases} 0 & i < j \\ 1 & i = j \\ \tau^{(i-1)} & j = 1 \\ \dfrac{(i-1)!}{(j-1)!(i-j)!}\tau^{(i-j)} & i > j > 1 \end{cases} \tag{9.50}$$

The relation in Equation 9.48 can be verified by showing that the terms of columns of both sides of the equation are equal. The terms of jth column of the matrix which results from multiplication of the two matrices on the right-hand side are as follows:

$$\bar{V} = \bar{L}^{-1}(\tau)V(t) \tag{9.51}$$

$$\begin{aligned} \bar{V}_{1j} &= 1 \\ \bar{V}_{2j} &= \tau_j - \tau \\ &\vdots \\ \bar{V}_{ij} &= \tau_j^{(i-1)} - \tau^{(i-1)} - \sum_{k=2}^{i-1} \frac{(i-1)!}{(k-1)!(i-k)!}\tau^{(i-1)}\bar{V}_{kj} \\ &= (\tau_j - \tau)^{(i-1)} \end{aligned} \tag{9.52}$$

The above relations are recognized as the terms of the jth column of the Vandermonde matrix $V(t + \tau)$. Thus, the relation in Equation 9.48 is verified.

9.5 DIRECT INTEGRATION METHODS

It was stated earlier that we can think of direct integration methods as either finding an approximate solution for the matrix equation of motion or determining the exact solution (neglecting the round-off error) of the discrete equivalent of the matrix equation of motion. We can now further develop this discrete equation by substituting for the generalized difference functions of Equation 9.16, the expressions from Equation 9.40. The result is the following recurrence relation, which relates the displacements at the n time stations:

$$\mathbf{G}_1(\tau)\mathbf{d}_{k+1} + \mathbf{G}_2(\tau)\mathbf{d}_k + \cdots + \mathbf{G}_n(\tau)\mathbf{d}_{k-n+2} = \mathbf{d}^p(\tau) \tag{9.53}$$

In the above equation, the coefficient matrices $\mathbf{G}_i$ are polynomials in τ and the coefficients of these polynomials are functions of **M**, **C**, and **K**.

Before proceeding further we will consider the example of three time stations. With the use of generalized differences, the discrete equation of motion for this case is as follows:

$$\mathbf{M}\mathbf{D}^2(0) + \mathbf{C}\left[\mathbf{D}^1(0) + \overline{\tau}\mathbf{D}^2(0)\right] + \mathbf{K}\left[\mathbf{D}^0(0) + \overline{\tau}\mathbf{D}^1(0) + \frac{\overline{\tau}^2}{2}\mathbf{D}^2(0)\right] = \mathbf{d}^p(\tau) \tag{9.54}$$

The generalized differences in this equation are at the origin of τ and therefore depend on our choice of value of t_0. As a specific example, we will choose $t_0 = t_k$, and therefore the generalized differences, $\mathbf{D}^0(0)$, $\mathbf{D}^1(0)$, and $\mathbf{D}^2(0)$ are the second central difference operators. Substituting the second central difference operators from Equation 9.35 into Equation 9.54, we obtain the following relation:

$$\begin{aligned}
&\frac{1}{\Delta t^2}\mathbf{M}\left[d_{k+1} - 2d_k + d_{k-1}\right] \\
&+ \frac{1}{\Delta t}\mathbf{C}\left[\left(\frac{1}{2}d_{k+1} - \frac{1}{2}d_{k-1}\right) + \tau\left(d_{k+1} - 2d_k + d_{k-1}\right)\right] \\
&+ \mathbf{K}\left[d_k + \tau\left(\frac{1}{2}d_{k+1} - \frac{1}{2}d_{k-1}\right) + \frac{\tau^2}{2}\left(d_{k+1} - 2d_k + d_{k-1}\right)\right] = \mathbf{d}^p(\tau)
\end{aligned} \tag{9.55}$$

The role of generalized difference functions in this equation is obvious. Now we rearrange the terms of this equation and rewrite it in a form similar to Equation 9.53.

$$\begin{aligned}
&\left[\left(\frac{1}{\Delta t^2}\mathbf{M} + \frac{1}{2\Delta t}\mathbf{C}\right) + \left(\frac{1}{\Delta t}\mathbf{C} + \frac{1}{2}\mathbf{K}\right)\tau + \left(\frac{1}{2}\mathbf{K}\right)\tau^2\right]d_{k+1} \\
&+ \left[\left(-\frac{2}{\Delta t^2}\mathbf{M} + \mathbf{K}\right) + \left(-\frac{2}{\Delta t}\mathbf{C}\right)\tau + (-\mathbf{K})\tau^2\right]d_k \\
&+ \left[\left(\frac{1}{\Delta t^2}\mathbf{M} - \frac{1}{2\Delta t}\mathbf{C}\right) + \left(\frac{1}{\Delta t}\mathbf{C} - \frac{1}{2}\mathbf{K}\right)\tau + \left(\frac{1}{2}\mathbf{K}\right)\tau^2\right]d_{k-1} = \mathbf{d}^p(\tau)
\end{aligned} \tag{9.56}$$

The quantities in the square brackets are $G_1(\tau)$, $G_2(\tau)$, and $G_3(\tau)$ in Equation 9.53.

Different direct integration methods can now be obtained by satisfying the recurrence equation at any point within or outside the field of three time stations. Three well-known methods are obtained by satisfying the recurrence equation at the three time stations. We obtain the second central difference method by satisfying Equation 9.55 at t_k ($\tau = 0$). Backward difference and forward difference methods are obtained by satisfying Equation 9.55 at t_{k+1} and t_{k-1} ($\tau = 1$ and $\tau = -1$), respectively. Any other value of τ between the time stations will give an equally valid direct integration method. However, these methods have not been studied thoroughly.

In passing, it should be pointed out that the single degree of freedom equivalent of Equation 9.55 is of special interest in the study of the stability and accuracy of the direct integration methods. The discrete single degree of freedom equation of motion, in absence of loads, is written in the following form:

$$D^2(0) + 2\xi\omega\left[D^1(0) + \overline{\tau}D^2(0)\right] + \omega^2\left[D^0(0) + \overline{\tau}D^1(0) + \frac{\overline{\tau}^2}{2}D^2(0)\right] = 0 \tag{9.57}$$

In this equation, ω is the natural frequency and ξ is the damping ratio of the system. After introducing the generalized differences for $t_0 = t_k$ we obtain the following:

$$G_1(\tau)d_{k+1} + G_2(\tau)d_k + G_3(\tau)d_{k-1} = 0 \tag{9.58}$$

$$\begin{cases} G_1(\tau) = 1 + \xi\omega + \left(\frac{1}{2}\omega^2 + 2\xi\omega\right)\Delta t\tau + \frac{1}{2}\omega^2\Delta t^2\tau^2 \\ G_2(\tau) = -2 + \omega^2 - 4\xi\omega\Delta t\tau - \omega^2\Delta t^2\tau^2 \\ G_3(\tau) = 1 - \xi\omega + \left(-\frac{1}{2}\omega^2 + 2\xi\omega\right)\Delta t\tau + \frac{1}{2}\omega^2\Delta t^2\tau^2 \end{cases} \tag{9.59}$$

We also note that the characteristic equation is as follows:

$$G_1(\tau)\lambda^2 + G_2(\tau)\lambda + G_3(\tau) = 0 \tag{9.60}$$

The roots of the characteristic equation determine the basic properties of the integration method, including the stability and accuracy. However, this aspect of direct integration methods has been extensively covered in Chapter 8.

Next, we will consider the case of four time stations, t_{k+1}, t_k, t_{k-1} and t_{k-2}, as shown in Figure 9.4. In this case we use discrete operators, $\mathbf{D}^0$, $\mathbf{D}^1$, $\mathbf{D}^2$, and $\mathbf{D}^3$ and from Equation 9.40, we have the following expressions for the generalized differences:

$$\begin{cases} \mathbf{D}^0(\overline{\tau}) = \mathbf{D}^0(0) + \overline{\tau}\mathbf{D}^1(0) + \frac{\overline{\tau}^2}{2}\mathbf{D}^2(0) + \frac{\overline{\tau}^3}{6}\mathbf{D}^3(0) \\ \mathbf{D}^1(\overline{\tau}) = \mathbf{D}^1(0) + \overline{\tau}\mathbf{D}^2(0) + \frac{\overline{\tau}^2}{2}\mathbf{D}^3(0) \\ \mathbf{D}^2(\overline{\tau}) = \mathbf{D}^2(0) + \overline{\tau}\mathbf{D}^3(0) \end{cases} \tag{9.61}$$

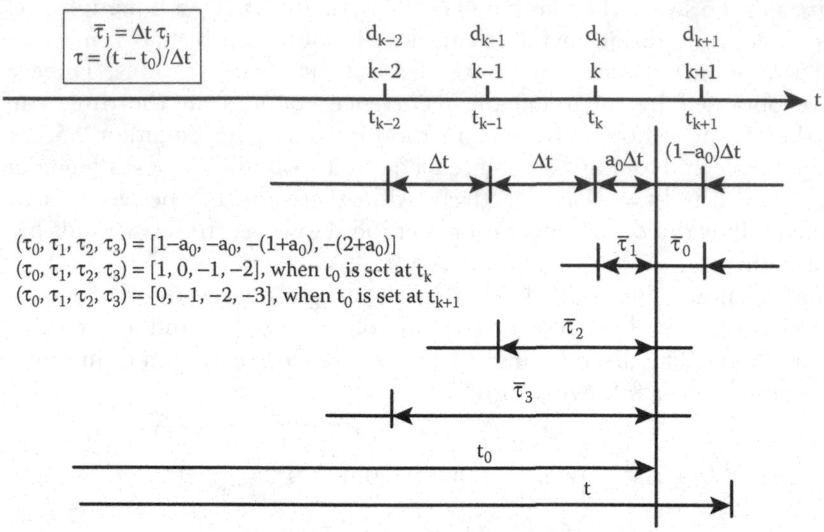

Figure 9.4 Example of a four time station case for generalized differences.

The discrete equation of motion is obtained by introducing the following generalized difference functions:

$$\begin{aligned} &\mathbf{M}\left[\mathbf{D}^2(0) + \bar{\tau}\mathbf{D}^3(0)\right] + \mathbf{C}\left[\mathbf{D}^1(0) + \bar{\tau}\mathbf{D}^2(0) + \frac{\bar{\tau}^2}{2}\mathbf{D}^3(0)\right] \\ &+ \mathbf{K}\left[\mathbf{D}^0(0) + \bar{\tau}\mathbf{D}^1(0) + \frac{\bar{\tau}^2}{2}\mathbf{D}^2(0) + \frac{\bar{\tau}^3}{6}\mathbf{D}^3(0)\right] = \mathbf{d}^p(\tau) \end{aligned} \tag{9.62}$$

The coefficients of the generalized differences at t_0 are determined by inverting the appropriate Vandermonde matrix. If we choose t_0 at the middle of the field of four time stations, that is, $t_0 = (t_k + t_{k-1})/2$ then the Vandermonde matrix is defined by the set $[\tau_0, \tau_1, \tau_2, \tau_3] = [1.5, 0.5, -0.5, -1.5]$. The coefficient matrix for the generalized differences is the solution of the following matrix equation:

$$\mathbf{V}_4\mathbf{A}_4 = \mathbf{E}_4 \tag{9.63}$$

$$\mathbf{V}_4 = \begin{bmatrix} 1 & 1 & 1 & 1 \\ \frac{3}{2} & \frac{1}{2} & -\frac{1}{2} & -\frac{3}{2} \\ \frac{9}{4} & \frac{1}{4} & \frac{1}{4} & \frac{9}{4} \\ \frac{27}{8} & \frac{1}{8} & -\frac{1}{8} & -\frac{27}{8} \end{bmatrix}; \quad \mathbf{E}_4 = \text{diag}\begin{bmatrix} 1 \\ 1 \\ 2 \\ 6 \end{bmatrix} \tag{9.64}$$

$$
A_4 = \begin{bmatrix} -\frac{1}{16} & -\frac{1}{24} & \frac{1}{2} & 1 \\ \frac{9}{16} & \frac{9}{8} & -\frac{1}{2} & -3 \\ \frac{9}{16} & -\frac{9}{8} & -\frac{1}{2} & 3 \\ -\frac{1}{16} & \frac{1}{24} & \frac{1}{2} & -1 \end{bmatrix} \tag{9.65}
$$

The columns of this matrix are the coefficients of the generalized differences for this particular value of t_0.

$$
\begin{cases} D^0(0) = -\frac{1}{16}d_{k+1} + \frac{9}{16}d_k + \frac{9}{16}d_{k-1} - \frac{1}{16}d_{k-2} \\ \Delta t D^1(0) = -\frac{1}{24}d_{k+1} + \frac{9}{8}d_k - \frac{9}{8}d_{k-1} + \frac{1}{24}d_{k-2} \\ \Delta t^2 D^2(0) = \frac{1}{2}d_{k+1} - \frac{1}{2}d_k - \frac{1}{2}d_{k-1} + \frac{1}{2}d_{k-2} \\ \Delta t^3 D^3(0) = d_{k+1} - 3d_k + 3d_{k-1} - d_{k-2} \end{cases} \tag{9.66}
$$

The discrete equation of motion for the case of four time stations is obtained by introducing the generalized differences into Equation 9.62. The result will be a recurrence equation similar to Equation 9.53 for n = 4. This equation can be satisfied anywhere within or outside the field of four time stations by substitution of the appropriate value of τ and various direct integration methods will result. Of all these methods only one is well known and that is Houbolt method which is obtained by satisfying Equation 9.62 at t_{k+1} ($\tau = 1.5$).

Up to this point only methods which can be directly obtained from difference operators have been discussed. However, some of the most well-known implicit direct integration methods do not fall into this category. Among these implicit methods we can name the Newmark family of methods; Wilson's θ-method; and Hilber, Hughes, and Taylor's α method. Next, we will show how these methods fit within the framework presented so far, in terms of the generalized differences. First, we will consider the Newmark family of methods which are based on three time stations. The discrete equation of motion for three time stations is given in Equation 9.54. In this equation, the normalized time appears in $\bar{\tau}$ and $\bar{\tau}^2$. We propose to treat these as independent parameters and replace them with $\bar{\gamma}\Delta t$ and $\bar{\beta}\Delta t^2$, resulting in the following equation:

$$
\begin{aligned} &\mathbf{M}\mathbf{D}^2(0) + \mathbf{C}\left[\mathbf{D}^1(0) + \bar{\gamma}\Delta t \mathbf{D}^2(0)\right] \\ &+ \mathbf{K}\left[\mathbf{D}^0(0) + \bar{\gamma}\Delta t \mathbf{D}^1(0) + \frac{1}{2}\bar{\beta}\Delta t^2 \mathbf{D}^2(0)\right] = \mathbf{d}^p(\tau) \end{aligned} \tag{9.67}
$$

These parameters are related to the parameters of Newmark family of methods, γ and β.

$$\begin{cases} \bar{\gamma} = \gamma - \dfrac{1}{2} \\ \bar{\beta} = 2\beta - \gamma + \dfrac{1}{2} \end{cases} \tag{9.68}$$

It can be easily verified that after introducing Equation 9.68 into Equation 9.67, we obtain a three time station equation which is identical to the three time station recurrence equation given by Zienkiewicz (1977).

It is interesting to further examine the generalized difference operators involved in Newmark method. First, we will define the new generalized differences for Newmark family of methods.

$$\begin{cases} \mathbf{D}^0 = \mathbf{D}^0(0) + \bar{\gamma}\Delta t\mathbf{D}^1(0) + \dfrac{1}{2}\bar{\beta}\Delta t^2\mathbf{D}^2(0) \\ \mathbf{D}^1 = \mathbf{D}^1(0) + \bar{\gamma}\Delta t\mathbf{D}^2(0) \\ \mathbf{D}^2 = \mathbf{D}^2(0) \end{cases} \tag{9.69}$$

As a specific example, we choose the constant average acceleration method. This method corresponds to values of parameters, $\gamma = 1/2$ and $\beta = 1/4$, which corresponds to $\bar{\gamma} = 0$ and $\beta = 1/2$ For these values of the parameters, we get the following difference operators:

$$\begin{cases} \mathbf{D}^0 = \dfrac{1}{4}\mathbf{d}_{k+1} + \dfrac{1}{2}\mathbf{d}_k + \dfrac{1}{4}\mathbf{d}_{k-1} \\ \Delta t\mathbf{D}^1 = \dfrac{1}{2}\mathbf{d}_{k+1} - \dfrac{1}{2}\mathbf{d}_{k-1} \\ \Delta t^2\mathbf{D}^2 = \mathbf{d}_{k+1} - 2\mathbf{d}_k - \mathbf{d}_{k-1} \end{cases} \tag{9.70}$$

It is indeed surprising to note that this very special method which is unconditionally stable and does not have any numerical damping (amplitude decay) uses standard generalized difference operators for velocity and acceleration terms, but uses a nonstandard difference operator for displacements.

In a similar way, a four time station expression can be constructed by replacing the terms $\bar{\tau}$, $\bar{\tau}^2$, and $\bar{\tau}^3$ in Equation 9.62 with $\alpha_1\Delta t$, $\beta_1\Delta t^2$, and $\gamma_1\Delta t^3$, respectively.

$$\begin{aligned} &\mathbf{M}\left[\mathbf{D}^2(0) + \alpha_1\Delta t\mathbf{D}^3(0)\right] + \mathbf{C}\left[\mathbf{D}^1(0) + \alpha_1\Delta t\mathbf{D}^2(0) + \beta_1\Delta t^2\mathbf{D}^3(0)\right] \\ &+ \mathbf{K}\left[\mathbf{D}^0(0) + \alpha_1\Delta t\mathbf{D}^1(0) + \beta_1\Delta t^2\mathbf{D}^2(0) + \gamma_1\Delta t^3\mathbf{D}^3(0)\right] = \mathbf{d}^p(\tau) \end{aligned} \tag{9.71}$$

This is similar to Zienkiewicz's four time station recurrence relation that can be obtained from Equation 9.71, if we use the generalized differences defined for $t_0 = t_{k-2}$. Zienkiewicz has shown that Wilson's θ-method can be obtained by proper choice of α_1, β_1, and γ_1 in terms of parameter θ. It can also be shown similarly that a set of $[\alpha_1, \beta_1, \gamma_1]$ exists such that Equation 9.71 yields the Hilber, Hughes, and Taylor's α-method.

9.6 CONCLUDING REMARKS

In this chapter we presented a new way of looking at direct integration of dynamic equations of motion. In derivation of most well-known direct integration methods, often the starting point is making an assumption about the variation of some variable, such as accelerations, within the time step. In actual computation using these methods we are determining the values of the variables at distinct time station Δt apart. With the generalized difference method, we are dealing with distinct time stations from the beginning and making no assumptions about the variation within the time step. We end up developing a discrete equivalent of the continuous dynamic equation of motion. The exact solution of this discrete equation of motion (neglecting the round-off error) is the approximate solution of the actual equation of motion. In arriving at the discrete equation of motion, we have developed discrete equivalents of displacements and its time derivatives, as well as the discrete equivalent of the Taylor series, which has finite number of terms and no remainder. This shows that we can operate at the discrete equivalent of the continuous system and only deal with the values of the variables at the discrete time stations. We have shown that all the existing direct integration method, and many other new methods, can be developed within the discrete domain using the generalize difference method.

References

Barbosa, R.; Ghaboussi, J. Discrete finite element method for multiple deformable bodies, *Journal of Finite Elements in Analysis and Design*, 7, 145–158 (1990).

Bathe, K. J.; Wilson, E. L. Stability and accuracy analysis of direct integration methods, *Earthquake Engineering and Structural Dynamics*, 1, 283–291 (1973).

Bergan, P. G.; Horrigmoe, G.; Brakeland, B.; Soreide, T. H. Solution techniques for non-linear finite element problems, *International Journal for Numerical Methods in Engineering*, 12, 1677–1696 (1978).

Crank, J.; Nicolson, P. A practical method for numerical evaluation of solutions of partial differential equations of the heat conduction type, *Proceedings of the Cambridge Philosophical Society*, 43(1), 50–67 (1947).

Dobbs, M. W. Comments on stability and accuracy analysis of direct integration methods, *Earthquake Engineering & Structural Dynamics*, 2, 295–299 (1974).

Fletcher, R.; Reeves, C. M. Function minimization by conjugate gradient, *The Computer Journal*, 7, 149–153 (1964).

Fox, L.; Goodwin, E. T. Some new methods for numerical integration of ordinary differential equations, *Mathematical Proceedings of the Cambridge Philosophical Society*, 45(3), 373–388 (1949).

Gantmacher, F. R. *The Theory of Matrices*, Chelsea Publishing Co., New York (1959).

Ghaboussi, J. Generalized differences in direct integration methods for transient analysis, *Proceedings, International Conference on Numerical Method in Engineering: Theory and Applications*, Swansea, England (NUMETA-87) (1987).

Ghaboussi, J. Fully deformable discrete element analysis using a finite element approach, *International Journal of Computers and Geotechnics*, 5(3), 175–195 (1988).

Ghaboussi, J.; Barbosa, R. Three-dimensional discrete element method for granular materials, *International Journal for Numerical and Analytical Methods in Geomechanics*, 14, 451–472 (1990).

Ghaboussi, J.; Wilson, E. L. Variational formulation of dynamics of fluid-saturated porous elastic solids. *Journal of Engineering Mechanics*, ASCE, 4, 947–963 (1972).

Golub, G. H.; Van Loan, C. F. An analysis of the total least square problems, *SIAM Journal on Numerical Analysis*, 17(6), 883–893 (1980).

Guyan, R. J. Reduction of stiffness and mass matrix, *AIAA Journal*, 3(2), 380–382 (1965).

Hilber, H. M.; Hughes, T. J. R. Collocation, dissipation and overshoot for time integration schemes in structural dynamics, *Earthquake Engineering & Structural Dynamics*, 6, 99–117 (1978).

Hilber, H. M.; Hughes, T. J. R.; Taylor, R. L. Improved numerical dissipation for time integration algorithms in structural dynamics, *Earthquake Engineering & Structural Dynamics*, 5, 283–292 (1977).

Hoffman, A. J.; Wielandt, H. W. The variation of the spectrum of a normal matrix, *Duke Mathematical Journal*, 20, 57–39 (1953).

Houbolt, J. C. A recurrence matrix solution for the dynamic response of elastic aircraft, *Journal of the Aeronautical Sciences*, 17, 540–550 (1950).

Householder, A. S. Unitary triangularization of a non-symmetric matrix, *Journal of ACM*, 5(4), 339–342 (1958).

Hughes, T. J. R.; Levit, I.; Winget, J. M. An element-by-element solution algorithm for problems of structural and solid mechanics, *Computer Methods in Applied Mechanics*, 36, 231–254 (1983).

Lanczos, C. An iteration method for the solution of the eigenvalue problem of linear differential and integral operators, *Journal of Research of the National Bureau of Standards*, 45, 225–282 (1950).

Lax, P. D.; Richtmyer, R. D. Survey of the stability of linear finite difference equations, *Communications on Pure and Applied Mathematics*, 9, 267–293 (1956).

Macon, N.; Spitzbart, A. Inverse of Vandermonde matrices, *The American Mathematical Monthly*, 64, 79–82 (1957).

Meijerink, J. A.; van der Vorst, H. A. An iterative solution method for linear systems of which the coefficient matrix is a symmetric M-matrix, *Mathematics of Computation* (American Mathematical Society), 31(137), 148–162 (1977).

Newmark, N. M. A method of computation for structural dynamics, *Journal of the Engineering Mechanics*, ASCE, 3, 67–94 (1959).

Oettli, W.; Prager, W. Compatibility of approximate solution of linear equations with given error bounds for coefficients and right-hand sides, *Numerische Mathematik*, 6, 405–409 (1964).

Paige, C. C. The computation of eigenvalues and eigenvectors of very large sparse matrices, PhD thesis, University of London (1971).

Parlett, B. N. *The Symmetric Eigenvalue Problem*, Prentice-Hall, Englewood Cliffs, NJ (1980).

Parlett, B. N.; Scott, D. S. The Lanczos algorithm with selective orthogonalization, *Mathematics of Computation*, 33(145), 217–238 (1979).

Polak, E.; Ribiere, G. Note sur la convergence de methods de direction conjuguces, *Revue Française Information Recherche Opérationnelle*, 16, 35–43 (1969).

Riks, E. An incremental approach to the solution of buckling and snapping problems, *The International Journal of Solids and Structures*, 15, 524–551 (1979).

Robinson, A. R.; Healey, T. J. Symmetric successive quadrature: A new approach to the integration of ordinary differential equations, *Engineering Mechanics in Civil Engineering, Proceeding of the 5th Engineering Mechanics Division Specialty Conference*, ASCE, Laramie, WY, 176–179 (1984).

Rutishauser, H. Solution of eigenvalue problems with the LR-transformation, *National Bureau of Standards Applied Mathematics Series*, 49, 47–81 (1958).

Sharifi, P.; Popov, E. P. Nonlinear buckling analysis of sandwich arches, *Journal of the Engineering Mechanics Division*, ASCE, 91, 1397–1312 (1971).

Sharifi, P.; Popov, E. P. Nonlinear finite element analysis of sandwich shells of revolution, *AIAI Journal*, 11(5), 715–722 (1973).

Wilkinson, J. H. *The Algebraic Eigenvalue Problem*, Oxford University Press, Oxford, UK (1965).

Wilson, E. L.; Farhoomand, I.; Bathe, K. J. Nonlinear dynamic analysis of complex structures, *Earthquake Engineering & Structural Dynamics*, 1(2), 241–252 (1972).

Wood, W. L.; Bossak, M.; Zienkiewicz, O. C. An alpha modification of Newmark's method, *International Journal for Numerical Methods in Engineering*, 15, 1562–1566 (1980).

Wu, X.; Ghaboussi, J. A unified approach to temporal discretization in transient analysis: A generalized difference formulation, *Proceedings, International Conference on Computational Methods in Engineering*, Singapore (1992).

Zienkiewicz, O. C. A new look at the Newmark, Houbolt and other time stepping formulas, a weighted residual approach, *Earthquake Engineering & Structural Dynamics*, 5, 413–418 (1977).

Zienkiewicz, O. C.; Wood, W. L.; Hine, N. W.; Taylor, R. L. A unified set of single step algorithms. Part 1: General formulation and applications, *International Journal for Numerical Methods in Engineering*, 20, 1529–1552 (1984).

Index

Note: Page numbers followed by f refer to figures, respectively.

Printed in the United States
by Baker & Taylor Publisher Services